AF470578

UNSEEN COMPANIONS
Big Views of Tiny Creatures

First published 2007

LAST REFUGE Ltd.

Photography © David & Madeleine Spears/www.lastrefuge.co.uk

Additional Photography: Paul Cook, © Clouds Hill Imaging

Text © Adrian Warren

Photography colourists: Dae Sasitorn & Will Brett

This book was designed and produced by

Last Refuge Ltd.

Batch Farm, Panborough, near Wells, Somerset, BA5 1PN, UK

Telephone: 01934 712556, e-mail: info@lastrefuge.co.uk, www.lastrefuge.co.uk

Designer: Dae Sasitorn

Front cover: House Fly, *Musca domestica*
Back cover: Handprint with associated bacteria; Garden ant, *Lasius niger*; Psocid, *Lepinotus sp.*; Honeybee, *Apis mellifera*;
Lesser stag beetle, *Dorcus parallelopipedus*; Human crab louse, *Phthirus pubis*; Jumping spider, *Portia sp.*;
Backswimmer, *Notonecta glauca;* American cockroach, *Periplaneta americana;* Hydra, *Hydra sp.*;
Dust mite, *Dermatophagoides pteronyssinus*; Red admiral butterfly's scales,*Vanessa atalanta*

Dust mite, *Dermatophagoides pteronyssinus* (right)
Fungus gnat, *Bradysia paupera* (overleaf, left)
Earwig, *Forficula auricularia* (overleaf, right)
A cultured fungal colony (contents page)

All photographs in this book are available for prints, publication and all other usage from www.lastrefuge.co.uk

ISBN: 0-9544350-4-4

Printed and bound by 1010 Printing International, China

UNSEEN COMPANIONS
Big Views of Tiny Creatures

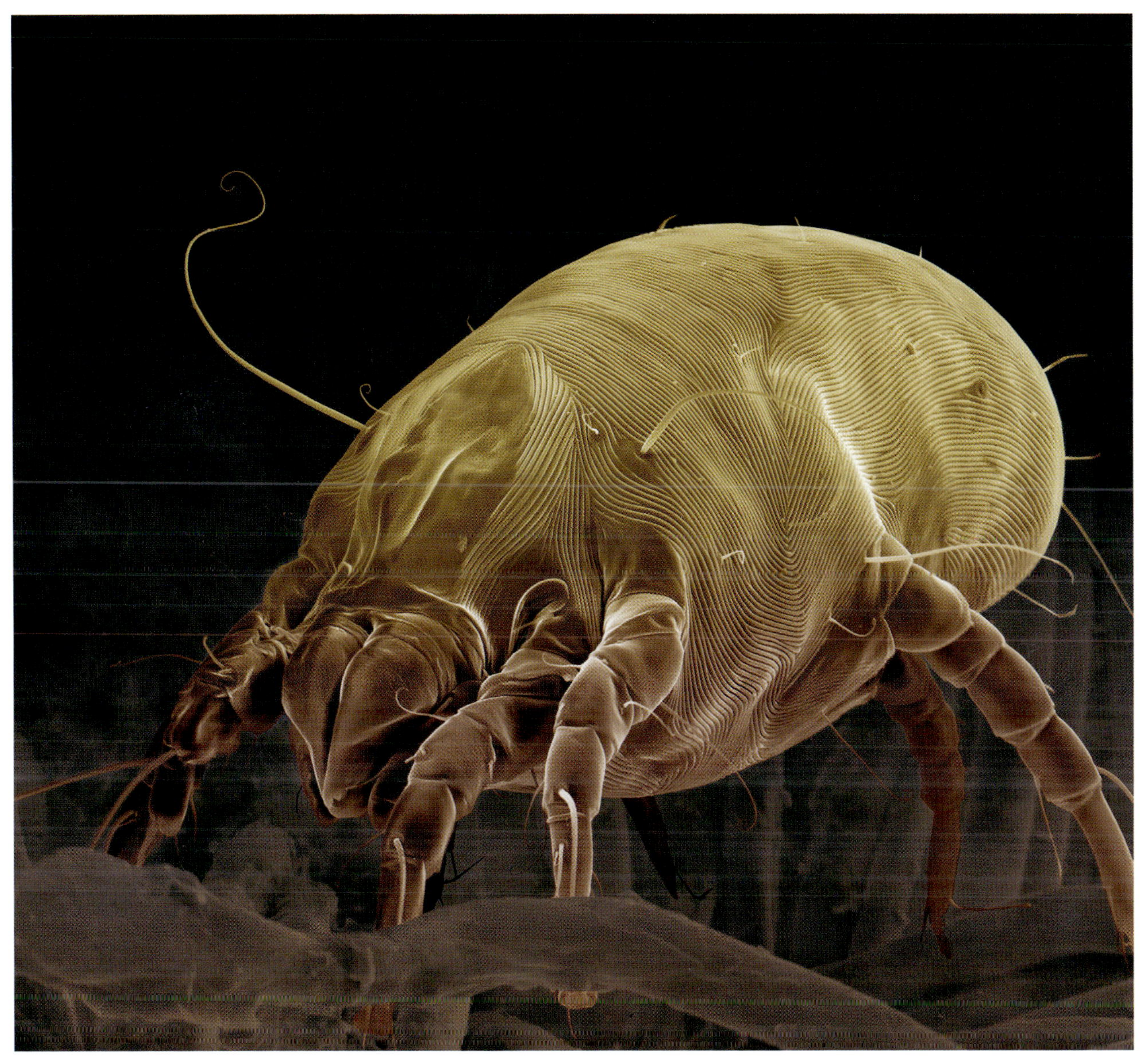

Photography by David & Madeleine Spears

Text by Adrian Warren

LAST REFUGE

To Tim and Judy

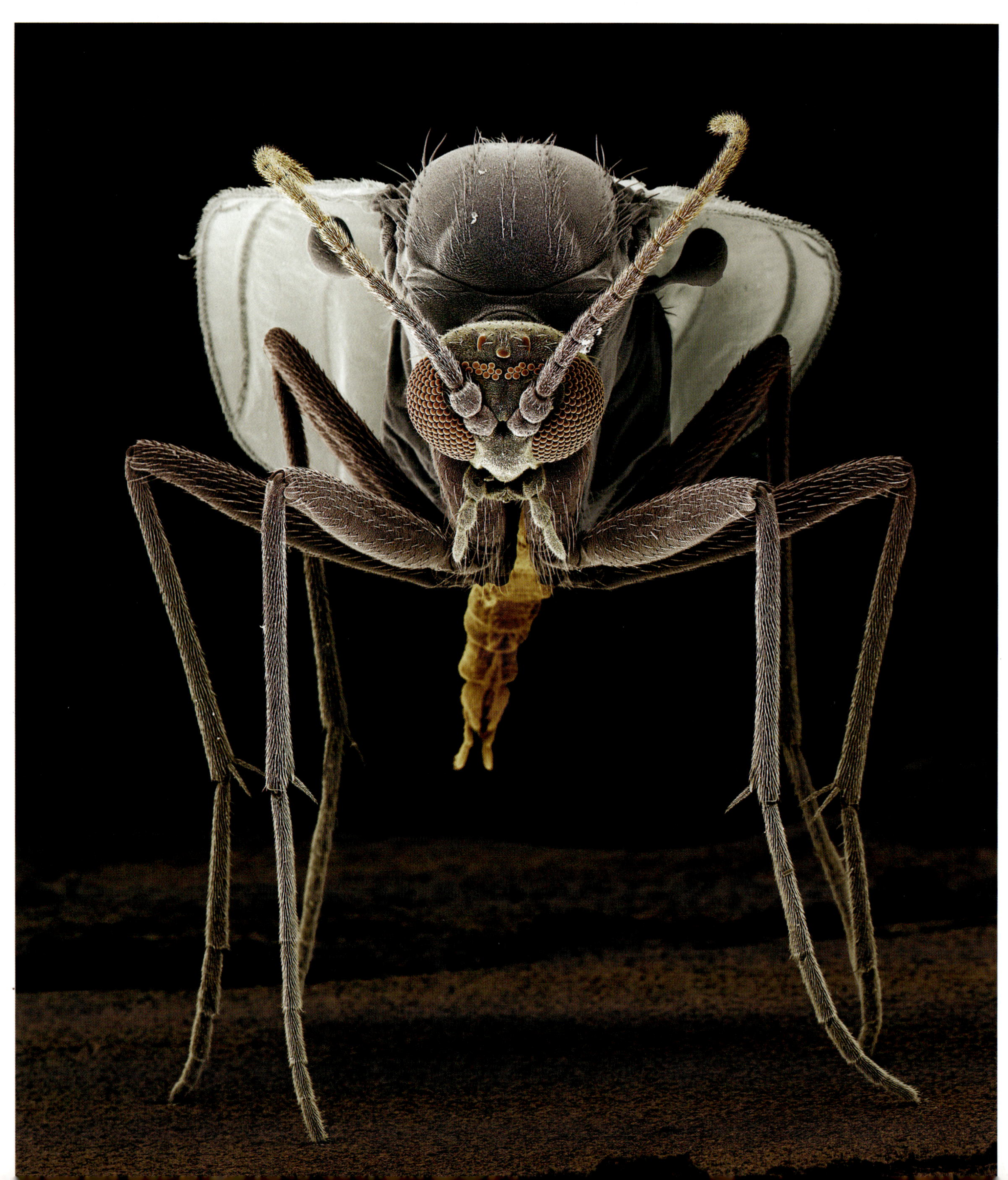

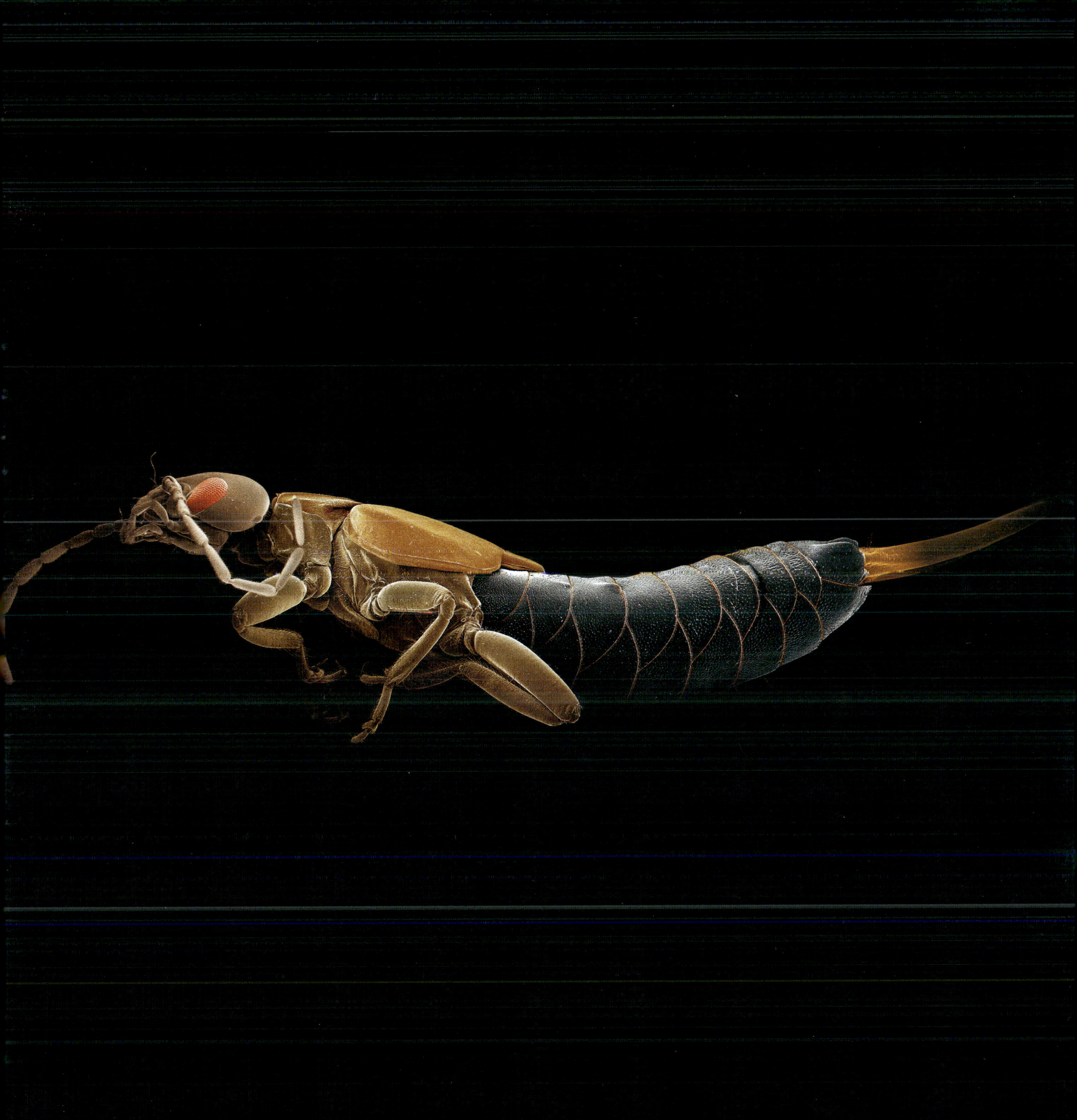

Contents

FOREWORD

For as long as we both care to remember, we have been enchanted by the tiny creatures that share our world. Such wonders of evolution, each of them so different, lured us as children armed with net and jar to every accessible pond, stream or hedgerow. These early adventures introduced us to the miniature unseen universe of tiny plants and animals. With eager anticipation, we would examine drops of water under a lens or microscope to watch the twirling shapes that even the most imaginative artist would be challenged to create. As a boy, David lived in Cheshire and Lancashire in northwest England, spending his summers on farms of friends while collecting all kinds of animals. It was unwise to open any matchbox in the house in case some hungry beetle that David had caught launched itself in a bid for freedom. Madeleine spent her childhood in Bassaleg and St Mellons in South Wales, where the woods, fields and ponds provided her with an endless source of small creatures. This fascination with wildlife led us both to London where we met while studying on the same course for our degrees in zoology. The photography of minute organisms became a passion. By the time David started his degree, he had mastered the art of fitting a camera onto a microscope and began producing images of amoebae, euglenids, diatoms and many other tiny organisms that filled him with a sense of wonder.

We should perhaps first explain a little of the process of capturing and enhancing the images contained in this book. As youngsters, we used light microscopes like the ones found in schools with optical lenses made of glass, which focused visible light rays onto a specimen mounted on a glass slide. This was a good introduction to the more sophisticated light microscopes we now work with, and helped us to learn about the properties and limitations of glass lenses, and the maximum magnifications we could achieve.

For many of the photographs in this book, however, we used a Scanning Electron Microscope (SEM). Instead of light rays, SEMs focus a beam of electrons onto the specimen that has been coated with a very thin layer of a metal such as gold or palladium. The electrons are reflected from the surface of the specimen, are collected, and form a black and white image through a photo-multiplier, which drives electronic amplifiers. Details of the preparation and imaging processes are explained in the technical notes towards the end of this book. The advantage of scanning electron microscopy is that much greater magnifications are possible than with light microscopy, and the images have greater depth of field and definition, appearing to be almost three-dimensional. However, since the original SEM photographs are black and white, a considerable amount of post-production work had to be accomplished to prepare the images for this book. With the aid of computers, Dae Sasitorn has expertly reintroduced colour and lighting to subtly simulate the natural appearance of the organisms.

This book is not intended as a comprehensive work, identification guide or biology textbook. Its aim is to show the strange beauty and intricacy of a few organisms with which, given the nature of our intimate relationships with them, we should certainly be better acquainted.

David & Madeleine Spears
Somerset, England
January 2007

Red Admiral Butterfly's Wing Scales, *Vanessa atalanta*

Magnification: x645   Scanning Electron Micrograph

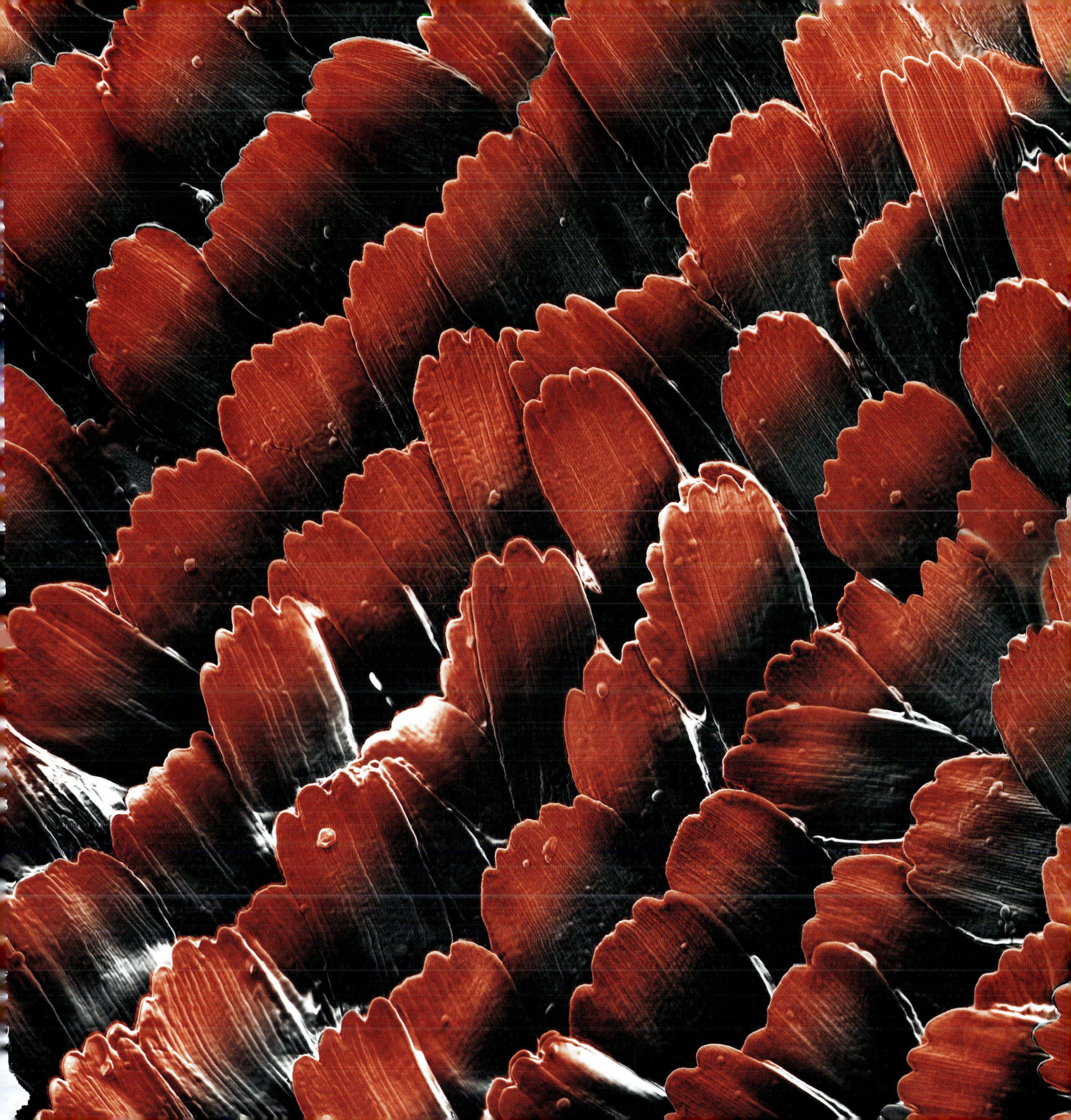

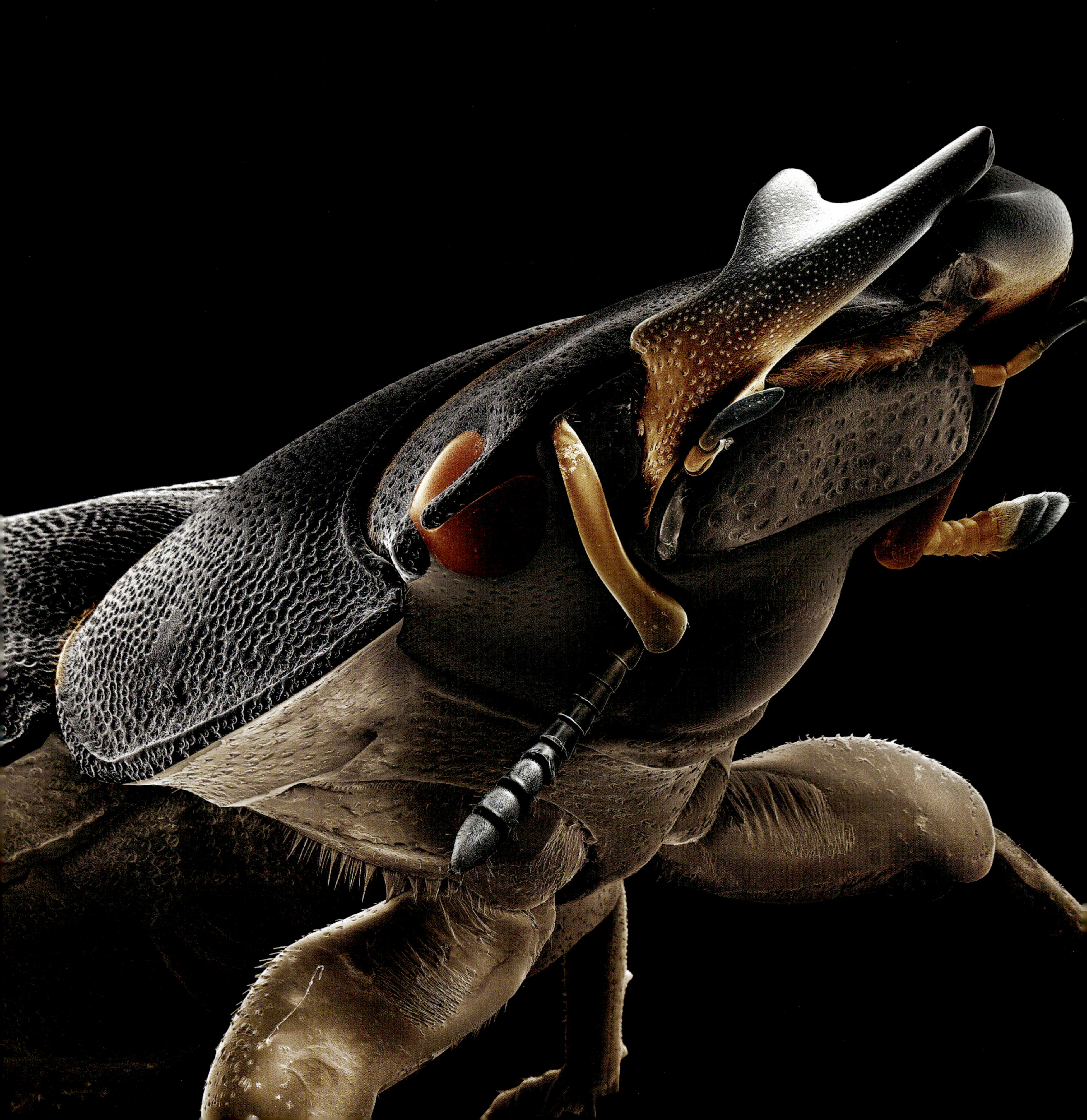

INTRODUCTION

There are few places on Earth where the environment is so extreme that the enterprising powers of nature have failed to find a niche. From the abyssal depths of the oceans to high mountain peaks, from frozen polar wastes to sizzling temperatures in hot springs, and from deep underground to high in the sky, organisms somehow manage to find a way to survive. Even the most apparently barren place is usually teeming with life, much of it microscopic. Tiny creatures are everywhere, often in their thousands, together with their eggs or spores. Many are invisible to the naked eye, and indeed some are so small that even conventional light microscopes are inadequate. However, by using high-powered scanning electron microscopes to increase the magnification and resolution, we can venture into a different world where the bodies of minute animals and plants, and the surfaces on which they live, are revealed in unimagined detail and texture. As we examine them, we can begin to understand how these organisms relate to each other and their environment, and the way in which they have evolved to survive in the most unexpected places.

Our planet depends on all the organisms on its surface, humans included, to live and work together in balance to maintain a healthy environment that can sustain life. While most living organisms are adapted to the environments that nature provides, we as humans have the ability to alter our surroundings radically to suit our needs. In doing so, we have created many new environmental niches for wild creatures to exploit. We do not have to look far. Our gardens are like teeming jungles, our homes are battlegrounds, and even our bodies provide accommodation for a wealth of opportunists. While some have taken advantage of our settled lifestyles, others have been our companions for millennia; their ancestors were probably carried by our ancestors as humans emerged from Africa to spread to every corner of the world. Since then, the human population has grown to over six billion, and we have not only changed the face of the world we live in but have started to reach out to other planets. Yet, our little companions are still with us, evolving and growing from strength to strength, adapting to every new habitat or niche that we create.

We may stare in wonder at the creatures presented in this book, but when we discover our uninvited guests in real life our attitude may not always be so welcoming. While it is true that some can be a nuisance, even dangerous, most do us no harm and may even be beneficial to our wellbeing. Nevertheless, we try to isolate our world from theirs by sealing ourselves in our houses and spraying insecticides, fungicides and herbicides in an effort to sanitise our surroundings. Since any chemical assault is harmful to ourselves as well as to helpful creatures, an understanding of their lifestyles might enable us to find less destructive methods to discourage those that are pests, and perhaps encourage the more friendly species.

The separate sections of this book relate to the habitats of these organisms and the special adaptations that allow them to survive and reproduce. Creatures that share our homes are often overlooked because of their shy, retiring natures. They scuttle off into sink plugholes, or into dark corners and crevices when we open little-used cupboards, or are nocturnal. Flies and bluebottles buzz around rooms, usually looking for a way out after scavenging on food or domestic waste. Dust mites teem in their thousands on mattresses and pillows, feeding on flakes of dead skin. The surface of our skin is frequented by a variety of creatures, among them bloodsuckers such as mosquitoes, fleas, lice and bedbugs, while inside our bodies there are microscopic organisms and, occasionally, larger parasites. All of them are superbly adapted to their way of life, many using stealth or small size, while others have evolved special ways of invading or attaching themselves to our bodies.

Our gardens, too, are havens for thousands of different species, most of them unseen: just a handful of garden soil contains many millions of microscopic bacteria, nematodes, algae and other life forms. These, the smallest of our garden companions, help to rot down decaying vegetation, dead animals and droppings into simple compounds, the nutrients that enable the healthy growth of vegetables and decorative plants, which in turn support the larger creatures, such as butterflies, bees, wasps, ants, and aphids. Frogs and goldfish might be the most noticeable inhabitants in our ponds, but they are outnumbered and outclassed by the variety of plants and tiny invertebrates that share their world. Most of them possess amazing adaptations and survival strategies, and all of them contribute in important ways to the wellbeing of a pond.

The following pages illustrate just a few of the creatures that share our immediate world. Some of them may be familiar, but the techniques used for the final imagery in this book show them in ways rarely seen before. The pictures may make us shudder or itch but, at the same time, we should marvel at the intricate structure and texture of their bodies, and the evolutionary design of their eyes, mouthparts and appendages, all of which give clues that help our understanding of their ways of life.

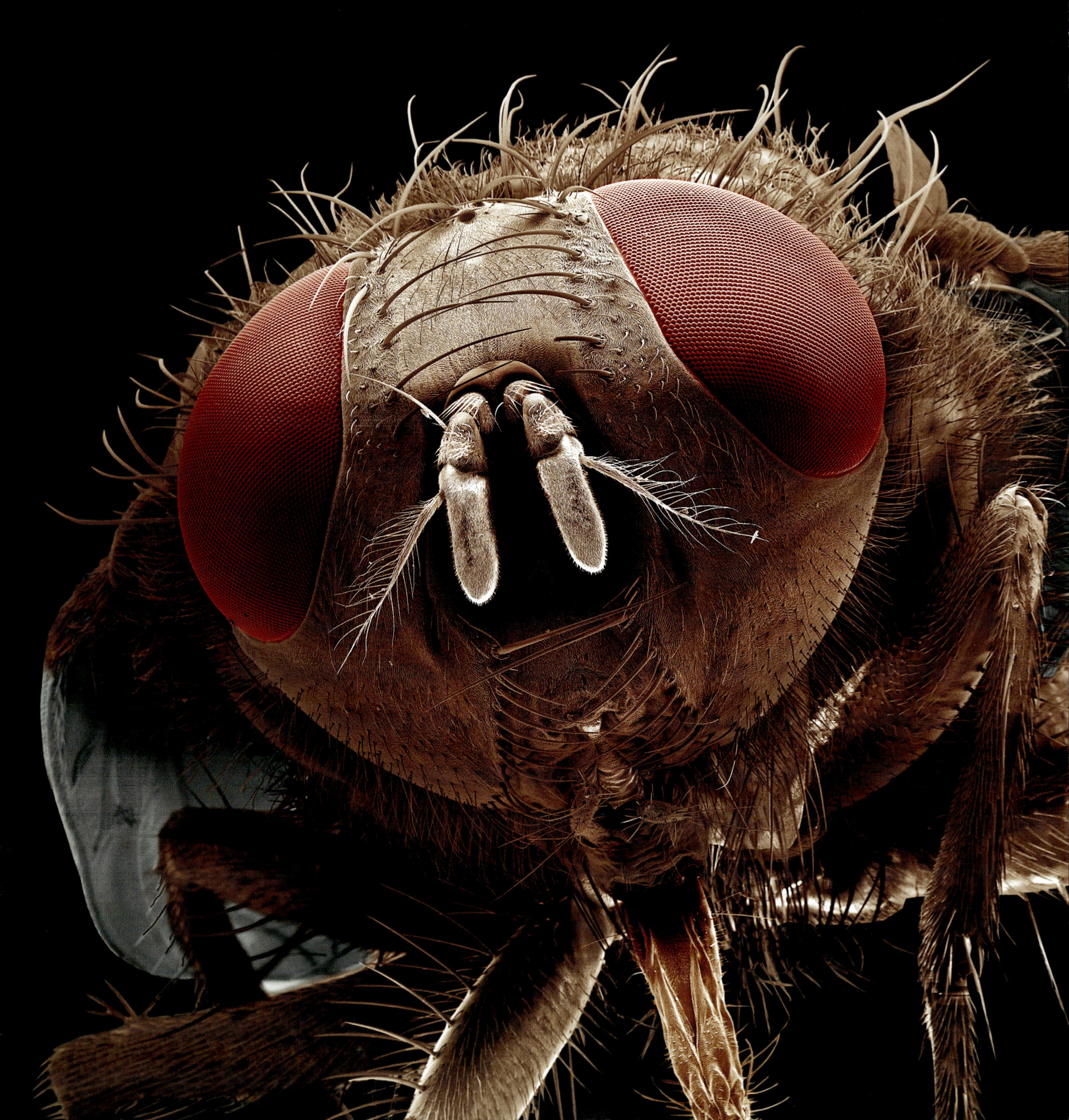

INVASION OF THE HOME

Ever since early humans learned how to construct dwellings using wood, mud and thatch to provide shelter and safety from the world outside, we have shared our homes with a host of small opportunistic creatures, which arrive seeking food, a place to hide or a hunting ground. A few are haphazard visitors, flying, crawling or wafting in on air currents through open windows or doorways, many of them eventually becoming trapped or exhausted on a window sill somewhere. Some enter as hitchhikers on shoes, clothing, domestic animals, firewood, groceries, houseplants, or other articles that we carry into our homes. Long ago, our ancestors quickly learned to deter uninvited guests by allowing the pungent smoke from open fires to impregnate their belongings and building materials. They also learned that it was impossible to be rid of them all; in fact, even today, some tribes in the tropics regularly burn and rebuild their houses every few weeks in an attempt to control infestations. In medieval Britain, householders, especially in cities, waged a constant war against pests. Open sewers and streets littered with household waste and animal dung were ideal breeding grounds for vermin and the diseases that they carried. Infestations of rats, lice, fleas and bacteria reached epidemic proportions, with very little that could be done to keep them under control.

Today we use modern, pest-resistant building materials, together with windows and doors that seal tight when closed, making it as difficult as possible for creepy crawlies to enter. In spite of this and the chemicals and cleaning fluids that help us to keep our surroundings relatively clean and pest free, there are many small creatures that still find a way in to live with us for once inside, our homes provide a wealth of crevices, corners and surfaces that are attractive living accommodation. Since we also control the climate in our houses with central heating, we unwittingly extend the summer for some creatures that might otherwise become dormant or die during the winter months. The warmth that we find so welcome, however, dries the air, a death sentence for some organisms that require a certain level of humidity to survive. But any established home, at any time of year, plays host to thousands upon thousands of uninvited guests, which feed, breed, die and decompose - usually invisibly - right in front of us, and given the right conditions, minor invasions can quickly become major infestations.

Many of our houseguests do not cause us any harm, but a few create a health hazard or cause structural damage. In our old farmhouse in Somerset in south western England, we find many small harmless creatures that have wandered in merely to seek shelter or a place to hibernate through the cold months of winter. It is an old house, built from stone, and the centuries-old mortar within the thick walls has crumbled in places, creating tunnels and passageways for tiny creatures to crawl in from outside. On the floor, we find woodlice, earwigs and garden beetles; in the winter, in crevices near windows and doors, we find the 7-spot ladybird, *Coccinella septempunctata*, the 22-spot ladybird, *Psyllobora vigintiduopunctata*, as well as butterflies, normally the small tortoiseshell, *Aglais urticae.* All year round there are bats in the roof, and plenty of spiders. Apart from droppings, which might present a health hazard if not cleaned up, none of these creatures present any serious problems, and in fact we welcome them. Ladybirds are friends of the gardener since they kill and eat aphids, and we encourage butterflies on our land by making sure that there are plenty of suitable food plants, such as nettles, available to them. The bats in the roof, like the house martins we have nesting under the eaves, help to control midges and flies, as do the spiders, which also eat other household nuisances.

<table>
<tr><td>

House Fly, *Musca domestica* (left)

Associated with poor sanitation and disease, the housefly carries pathogens from place to place on its hairy body and feet. Huge eyes give the housefly all-round vision and lightning-fast reactions, helping it to evade capture, and sucking mouthparts ingest liquid food; saliva is secreted on the surface of solid food to liquefy it before sucking it up the proboscis.

Magnification: x65   *Scanning Electron Micrograph*

</td><td>

Lesser Stag Beetle,
Dorcus parallelopipedus (previous page)

A small version of the stag beetle, its larger and rarer cousin, adult lesser stag beetles and their larvae, frequent soft, rotting tree trunks and can sometimes be found in compost heaps.

Magnification: x30   *Scanning Electron Micrograph*

</td></tr>
</table>

Some of the most obvious household companions are spiders. It is a shame that so many people have an aversion to them since, here in Britain, none can do us any real harm. Even those large attention-grabbing spiders with very long legs that run at high speed across the carpet or become stranded in the bath, as the cool night air of September and October draws them into the warmth of the home, are not aggressive. These familiar, hairy spiders are called *Tegenaria*; they typically build large distinctive webs comprising a woven platform of silk, with a funnel-shaped retreat. It is a 'trip web', not sticky like those of many spiders, but works by transmitting vibrations when an insect prey walks across it, sending a signal to the spider waiting in ambush.

In Britain, there are five species of *Tegenaria* spiders that visit our homes. The largest of these, if legs are included, is the giant house spider, *T. gigantea*, where the female can measure, toe-tip to toe-tip, up to 9 cm across, with a body length of nearly 2 cm (the male is slightly smaller). However, the cardinal spider, *T. parietina*, which is so-called because it terrified Cardinal Thomas Wolsey at Hampton Court in the 16th Century, has a body length that is even bigger. The cardinal spider, and *T. atrica*, another similar species, both occur in the south of England. Further north, the largest house spiders are likely to be another species, *T. saeva*. The smallest of them all, the common house spider, *T. domestica*, which has a body length of around 1 cm, can be found in our homes all year round. All these species are specialists in survival, being able to withstand long periods without water or food. Their bite can be a nasty nip, not surprising if you look at their huge fangs, but they will only attack if handled carelessly. Our son, who is not even two years old, managed to pick one up without harming it, by holding one of its legs gently between his finger and thumb, and the spider did not even try to bite. In the autumn, we frequently lift them out of the bath with our hands and carefully move them to another site without any problems. Interestingly, the *Tegenaria* group includes the hobo spider, *T. agrestis*, which has a brown hairy body up to 1.5 cm long, and is notorious in North America for being quite aggressive and is venomous. In southern Europe, this same species does not seem to be as unfriendly. Where it occurs, it is found in gardens, in woodpiles and loose stonework, and local people know it is a good idea to check firewood for unwelcome hitchhikers in case they carry them into the house. In southern England, the giant house spider lays its eggs in April in a spherical sac, which it covers with a thin protective layer of soil, wood fragments, the remains of prey, or whatever debris there is nearby. This is then covered with a layer of silk, and kept out of harm's reach within the web for the three weeks or so that it takes the young spiders to hatch. When mature, males seek females and may even live together with them for several weeks after mating. Males die in the autumn, and are eaten by the female, providing protein and carbohydrate to help her through the winter months. She may survive for several years.

Another familiar spider to be found in our homes is the 'daddy-long-legs' or cellar spider, *Pholcus phalangioides*, a fragile-looking creature with long, spindly legs. It resembles the winged 'daddy-long-legs', or crane fly, for example *Tipula paludosa*, and the long-legged harvestman, *Phalangium opilio*, but *Pholcus* is a true spider, usually found under a shelf, in a corner, or near the ceiling, dangling upside down in an untidy web that appears to be a random maze of threads. The web is cleverly designed however; a three-dimensional trap that enables the spider to capture prey much larger and stronger than itself. When a potential victim comes into contact with it, *Pholcus* vibrates rapidly, effectively tangling the victim in the sticky web that surrounds it, then rushes to the spot, immobilising it and wrapping it up alive to be eaten later. This same technique is used to outwit predators, the rapid vibration rendering the spider almost invisible. Remarkably, this delicate spider will even devour large *Tegenaria* spiders, given the chance. Both *Pholcus* and *Tegeneria* should be welcome in the house, for they consume many nuisance creatures, and will even capture and eat wood-boring beetles, which are some of the most serious of household pests.

Common House Spider, *Tegenaria domestica* (right and overleaf)

The familiar large spider that runs across the carpet or becomes stranded in the bath is packed with sensory detectors. Even the hairs on its legs (see right) gather information, some detecting touch, while others are sensitive to vibrations or air currents. The powerful chelicerae, or mouthparts (see overleaf, in the centre of each photograph left and right), have base segments that support large fangs that are tucked away underneath. On either side are the palps, which are covered in sensory hairs that are chemosensitive for smelling or tasting prey. The palps of the male (see overleaf, right) are very large and used to transfer his sperm to the female's sexual organs.

Magnification: x20   *Scanning Electron Micrograph* (right)
Magnification: x35   *Scanning Electron Micrograph* (overleaf, left)
Magnification: x50   *Scanning Electron Micrograph* (overleaf, right)

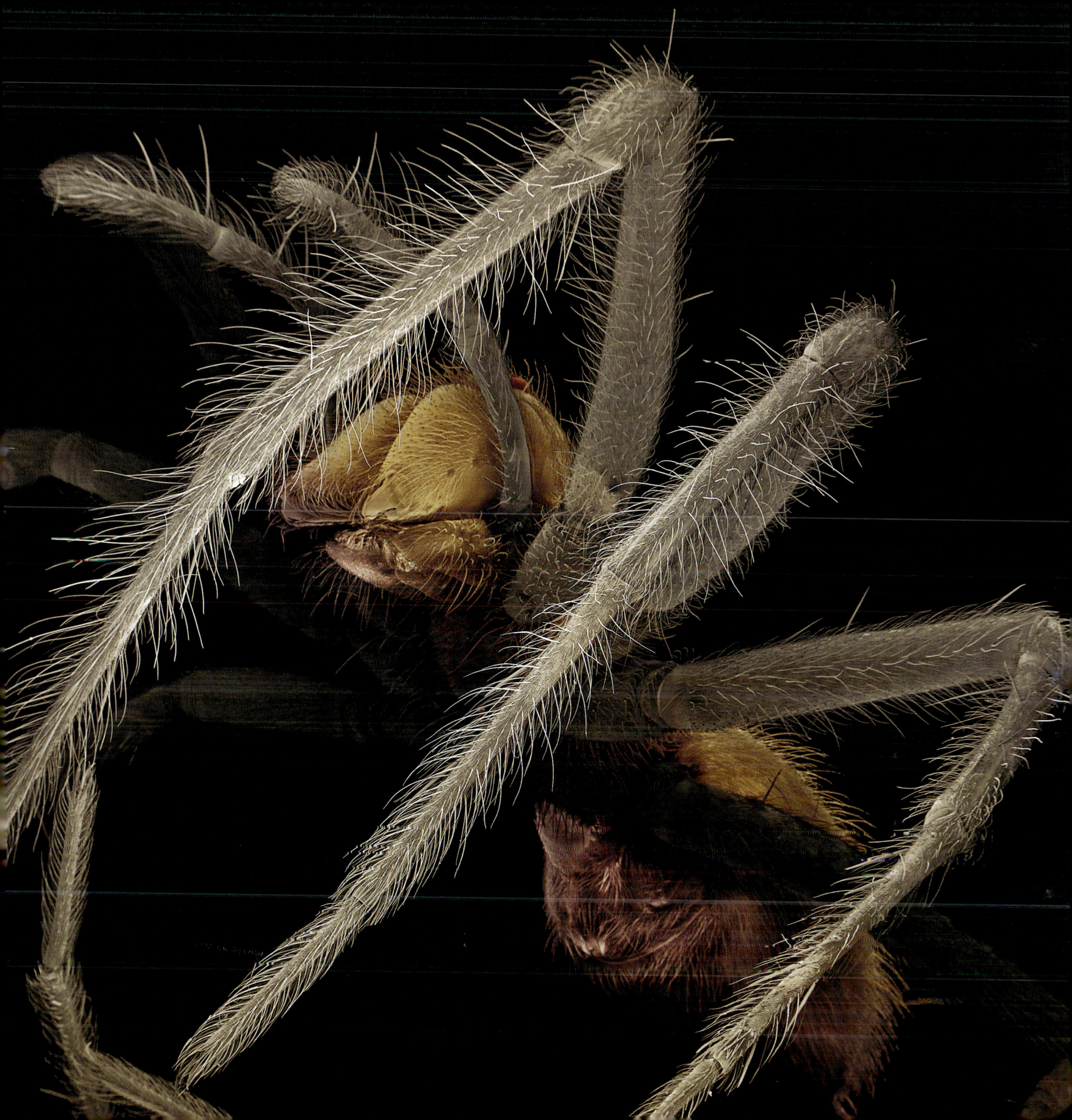

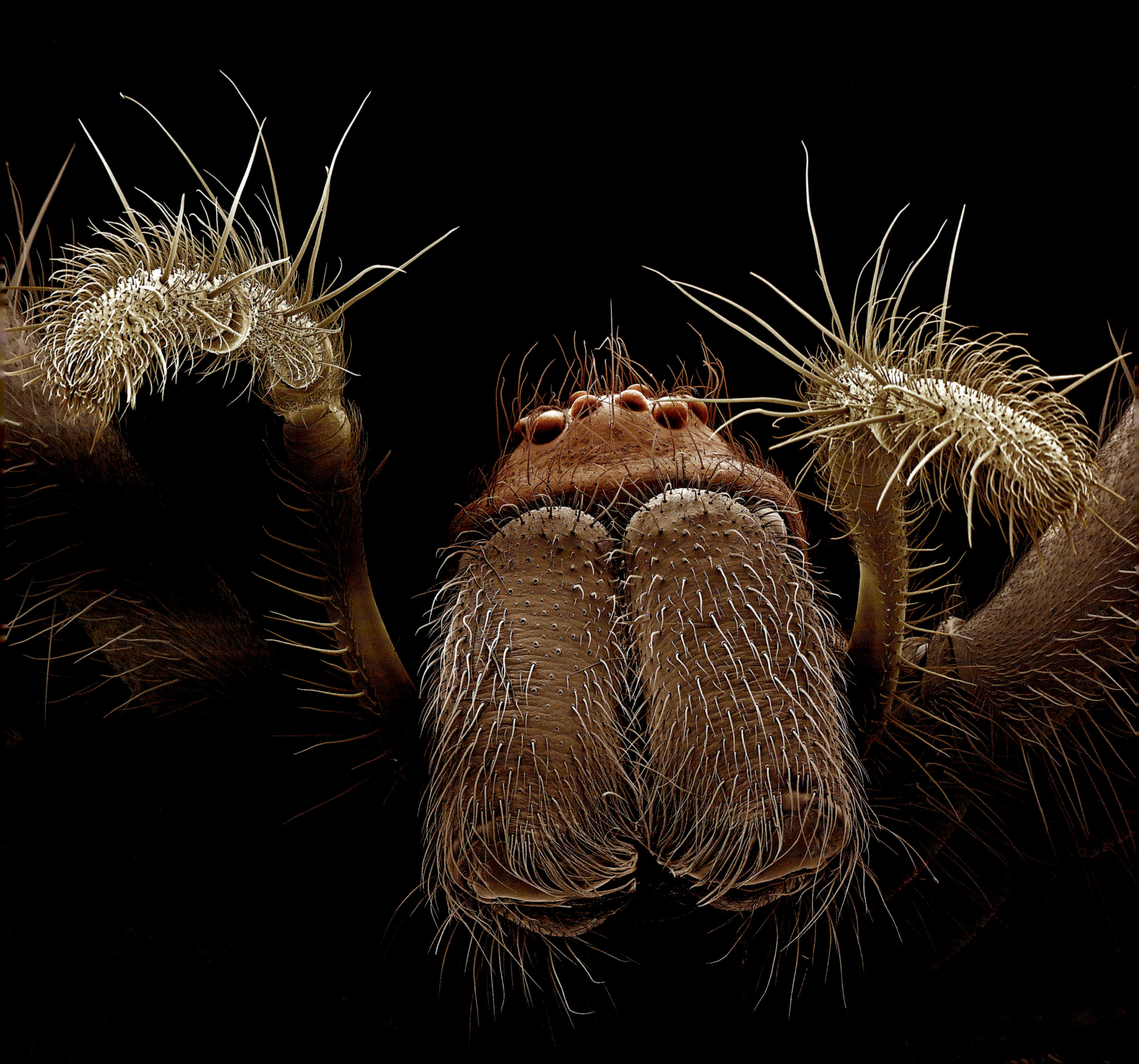

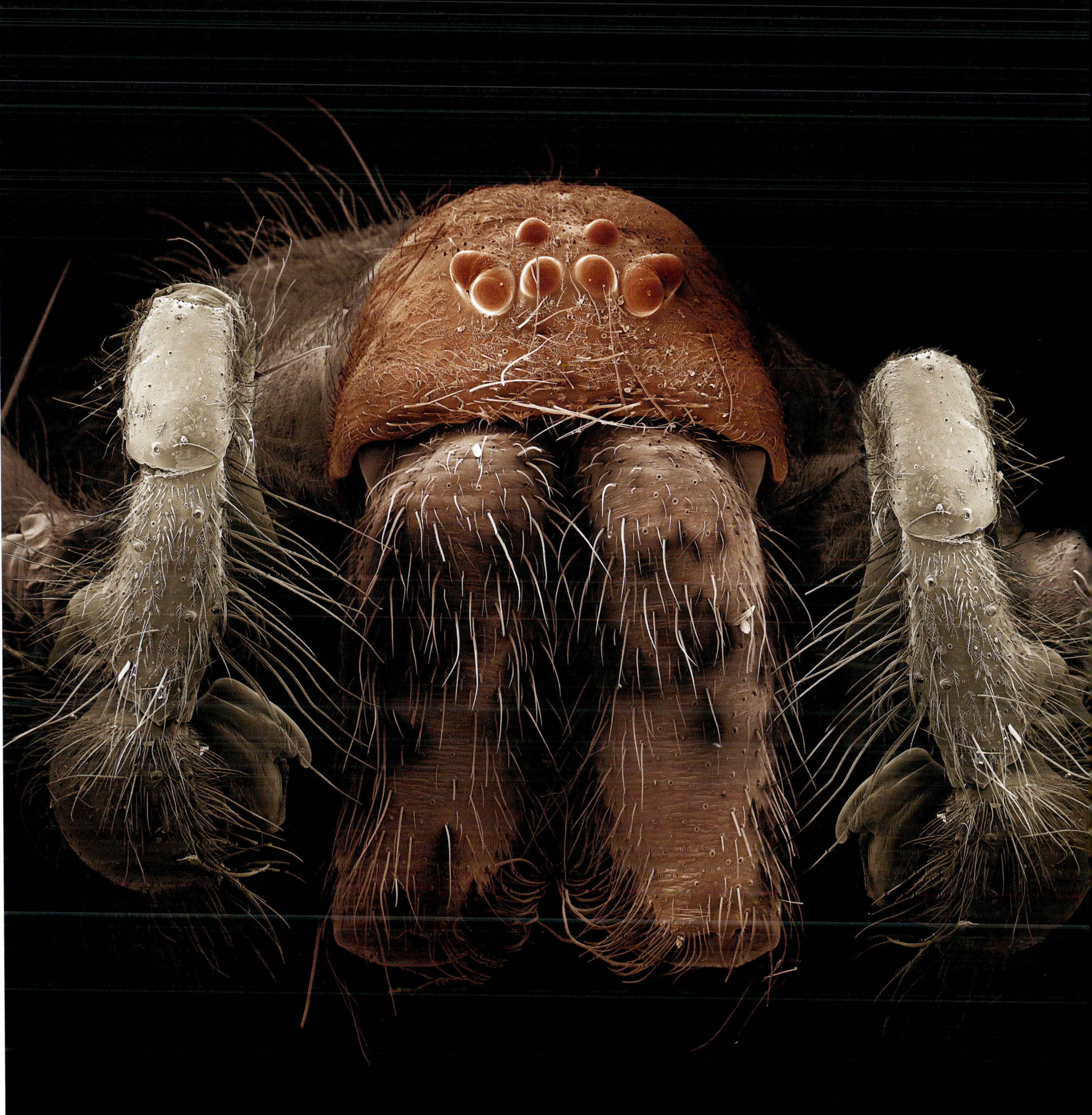

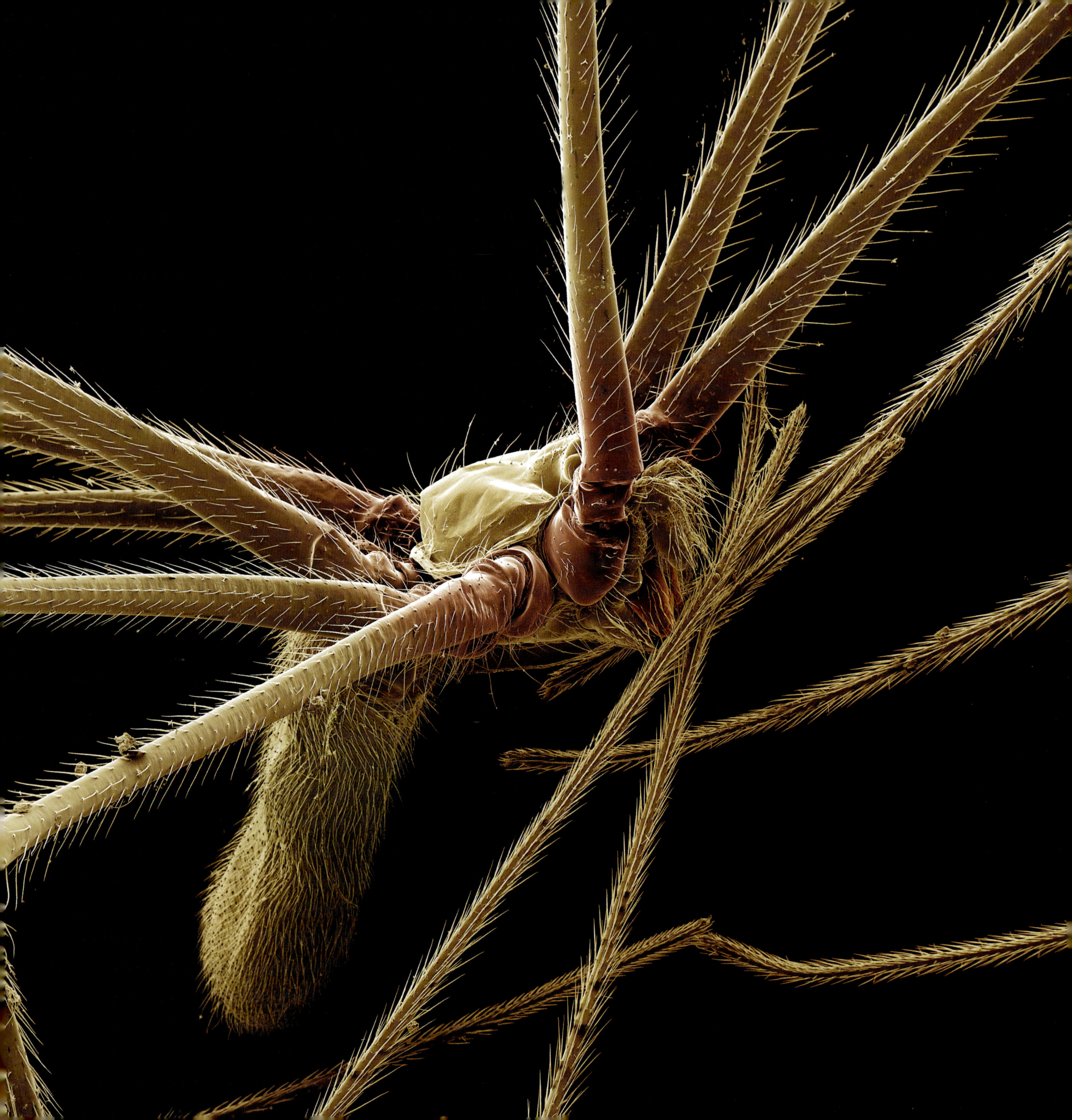

Cellar or 'Daddy-long-legs' Spider,

Pholcus phalangiodes

This large, fragile-looking spider frequents undisturbed corners in the house. It is often seen hanging upside down within its three-dimensional web and can trap prey even larger than itself. The female (see left) has very small palps, in comparison to those of the male, which are huge and bulbous (above right). When mature, female spiders are larger than males. A foot of *Pholcus* (below right) has comb-like 'claws' used to grasp the threads of its web.

Magnification: x50
Scanning Electron Micrograph (left)
Magnification: x50
Scanning Electron Micrograph (above right)
Magnification: x1200
Scanning Electron Micrograph (below right)

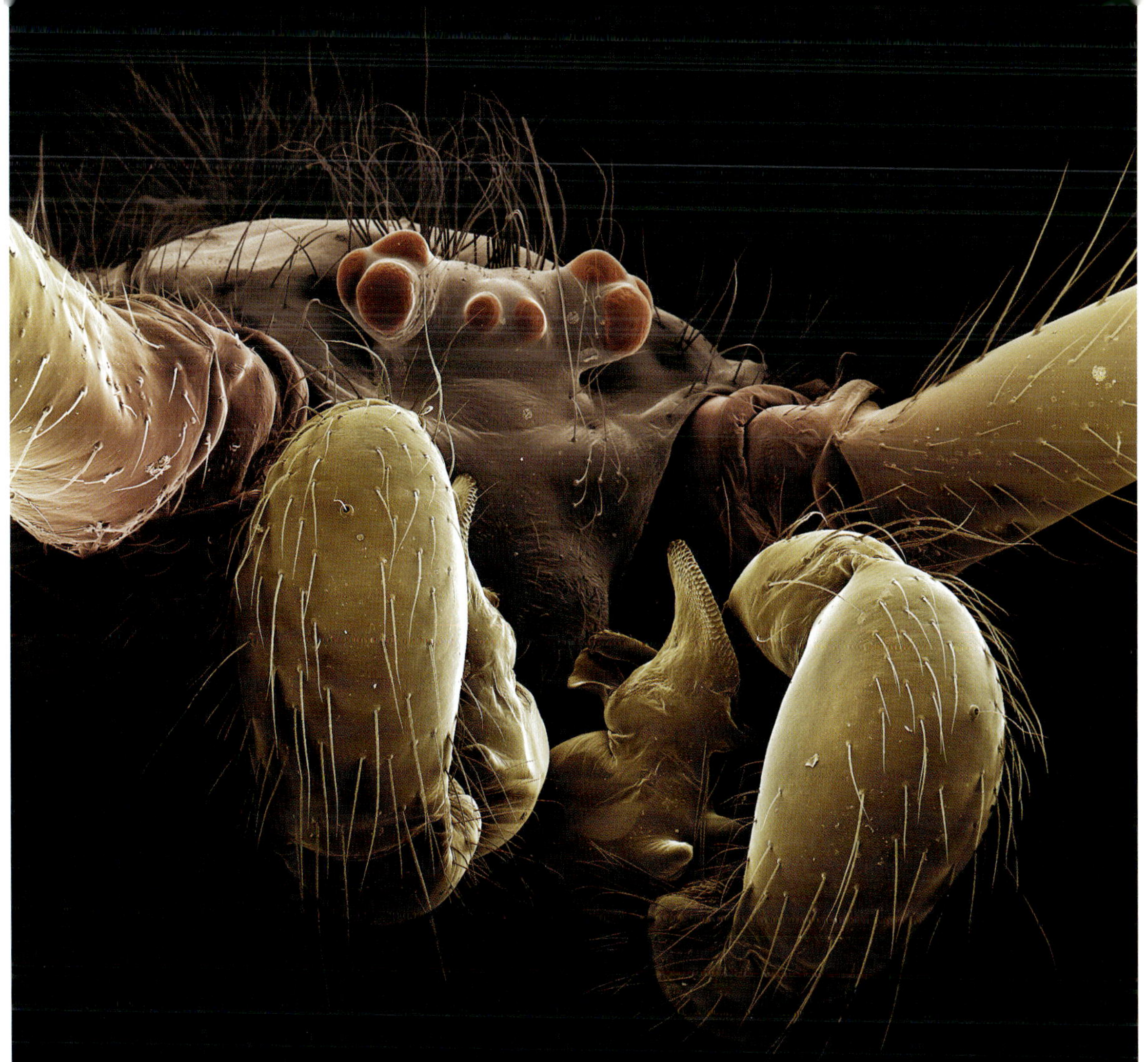

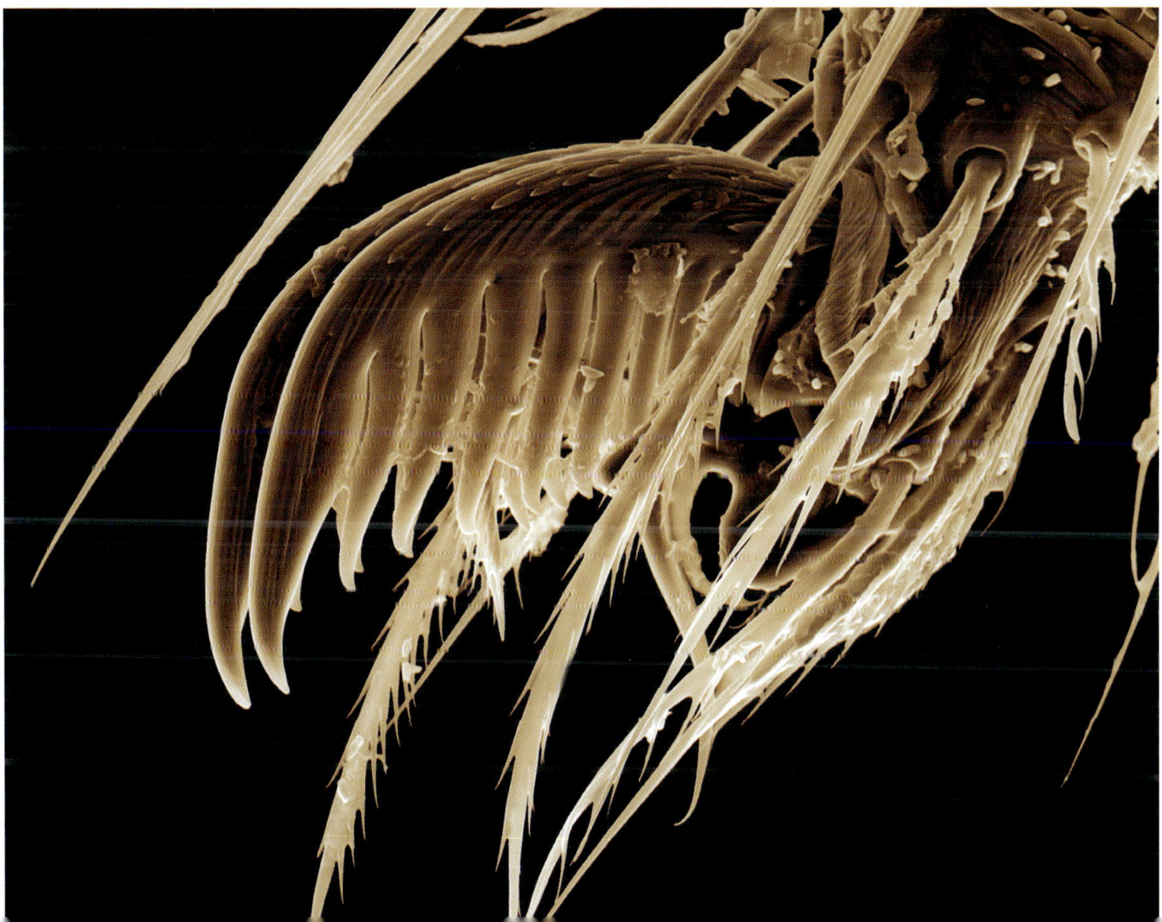

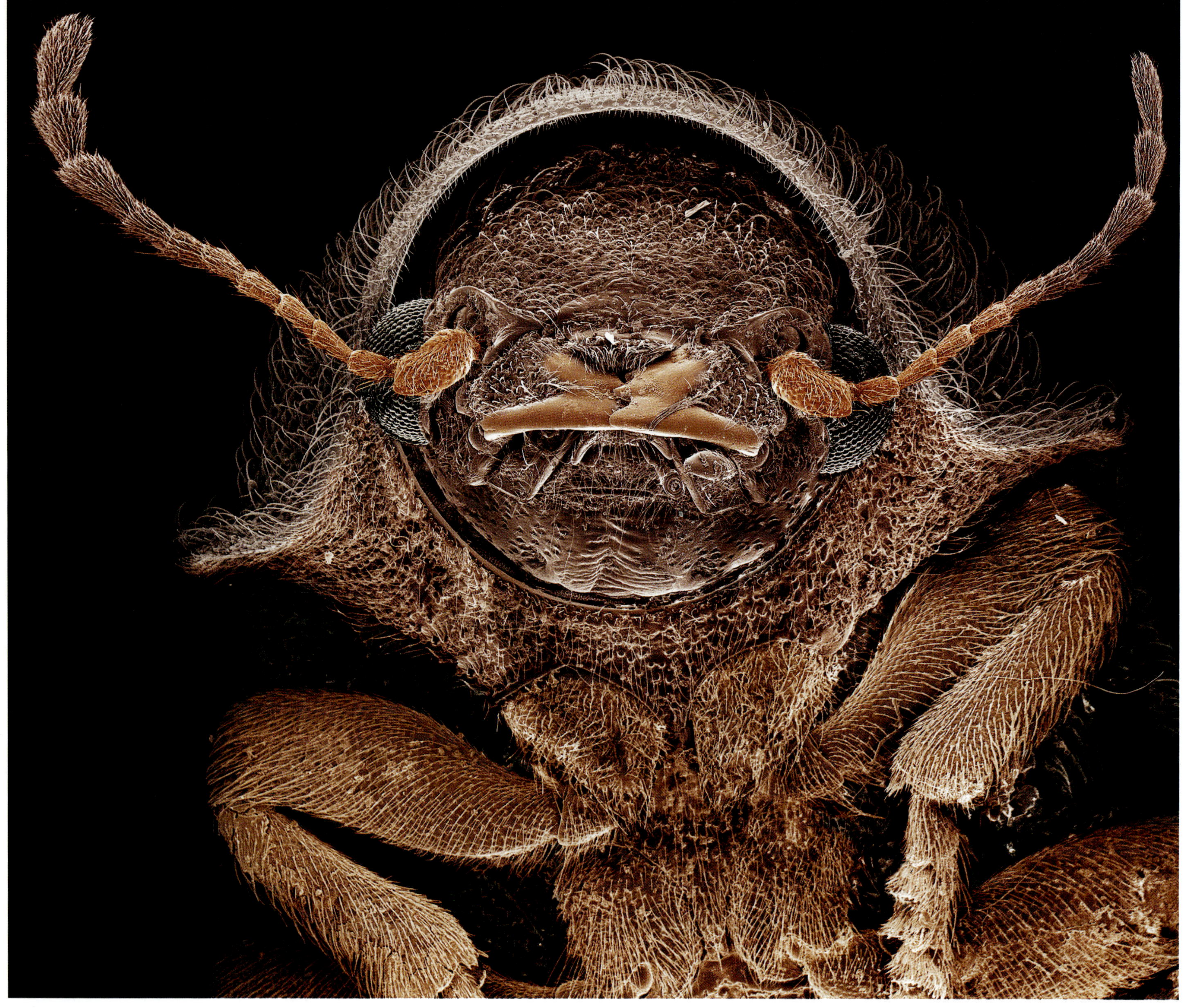

Deathwatch Beetle, *Xestobium villosum* (above)

Like the smaller furniture beetle, this species normally lives in woodland but enters buildings where its larvae prefer to infest hardwood, especially oak. The characteristic hooded thorax of the adult beetle protects its head. During courtship, males and females rhythmically tap their heads on wood, the strange sound at dead of night the basis for many ghost tales.

Magnification: x85   *Scanning Electron Micrograph*

Woodworm or Furniture Beetle, *Anobium punctatum* (right)

This small, heavily-armoured beetle lives in woodland where its larvae feed on dead wood but sometimes it burrows into furniture or the structural timber of houses providing conditions are a little damp. Severe infestations can cause catastrophic damage.

Magnification: x130   *Scanning Electron Micrograph*

WOODBORERS

Wood-boring beetles are among the most potent menaces in our homes; mostly never seen, their presence is obvious and sometimes devastating. Once they achieve a foothold, they can run amok, destroying precious furniture, and even threatening the structural integrity of a house. There are many animals that destroy wood but beetles are the most significant woodborers. Given the value that we place on wood in everyday life, the causes of decay are of major economic importance. Worldwide, there are many thousands of species of beetle that bore tunnels into living trees as well as dead or decaying wood. By using natural materials in the construction of our homes, we provide a potential food source for the creatures that normally attack it in nature. In Britain, the woodworm or furniture beetle, *Anobium punctatum,* and the deathwatch beetle, *Xestobium rufovillosum,* are closely related insects with similar habits. Both live in woodlands and gardens where their larvae feed on the dead parts of standing trees. Sometimes the beetles will fly or be transported into buildings where they may infest structural timbers, joinery or furniture if conditions are suitable. Both insects will attack sapwood, the outer part of the tree trunk, because that never has any durability, but they will only damage the heartwood of the timbers commonly used in our buildings if the wood chemistry has been modified by fungus. Deathwatch beetles are usually associated with oak in buildings, whilst furniture beetles will attack both oak and pine. Neither insect will thrive unless the wood is a little damp and good building maintenance is the best way to prevent and control attack.

The most common species is the tiny woodworm or furniture beetle, *Anobium punctatum*, about 3 mm long, whose larvae inflict most of the damage, feeding by tunnelling their way into timber, digesting cellulose and quickly reducing so much of it to dust that it becomes physically unsafe. The presence of the woodborers may not be noticed until the adult beetles emerge, leaving behind circular exit holes

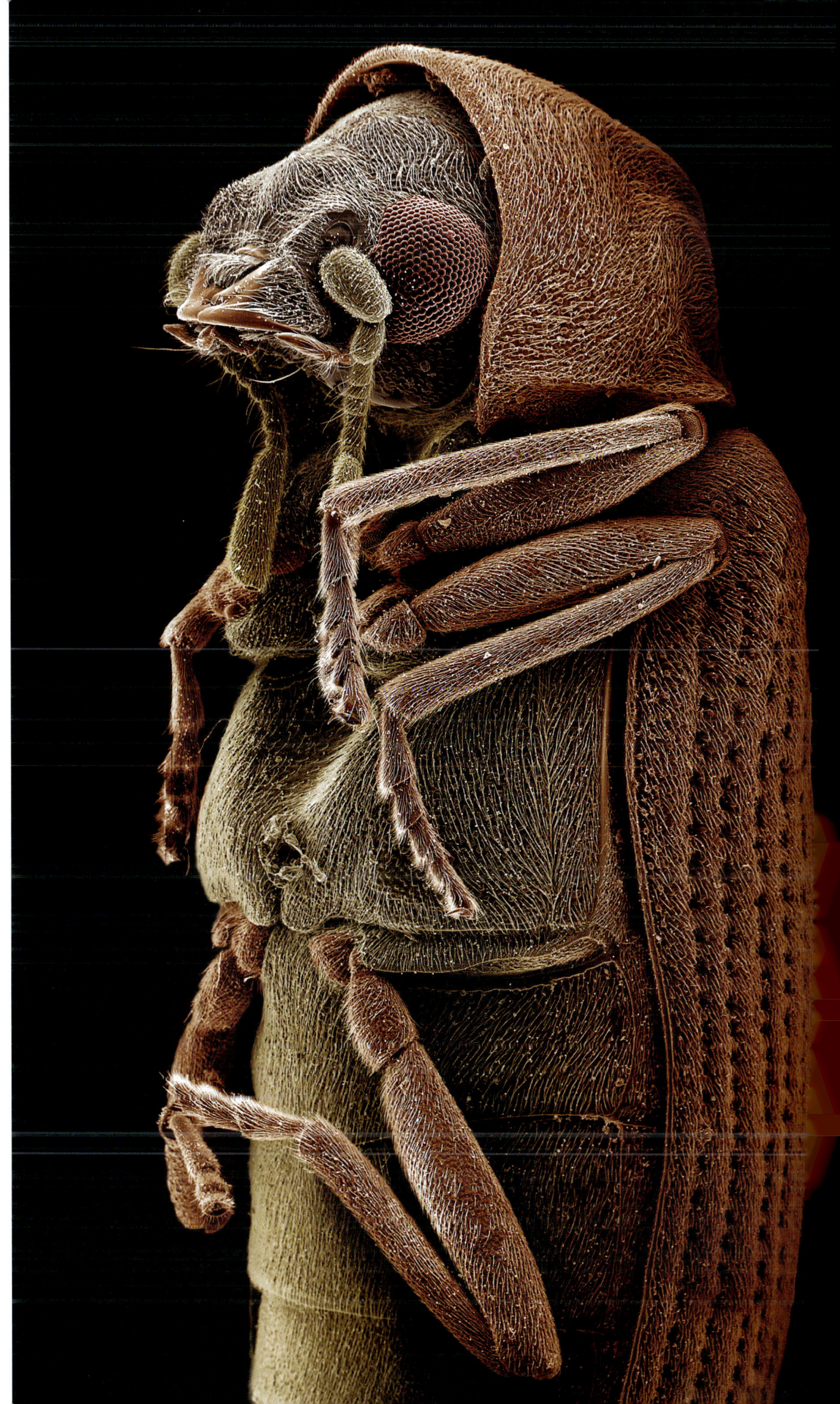

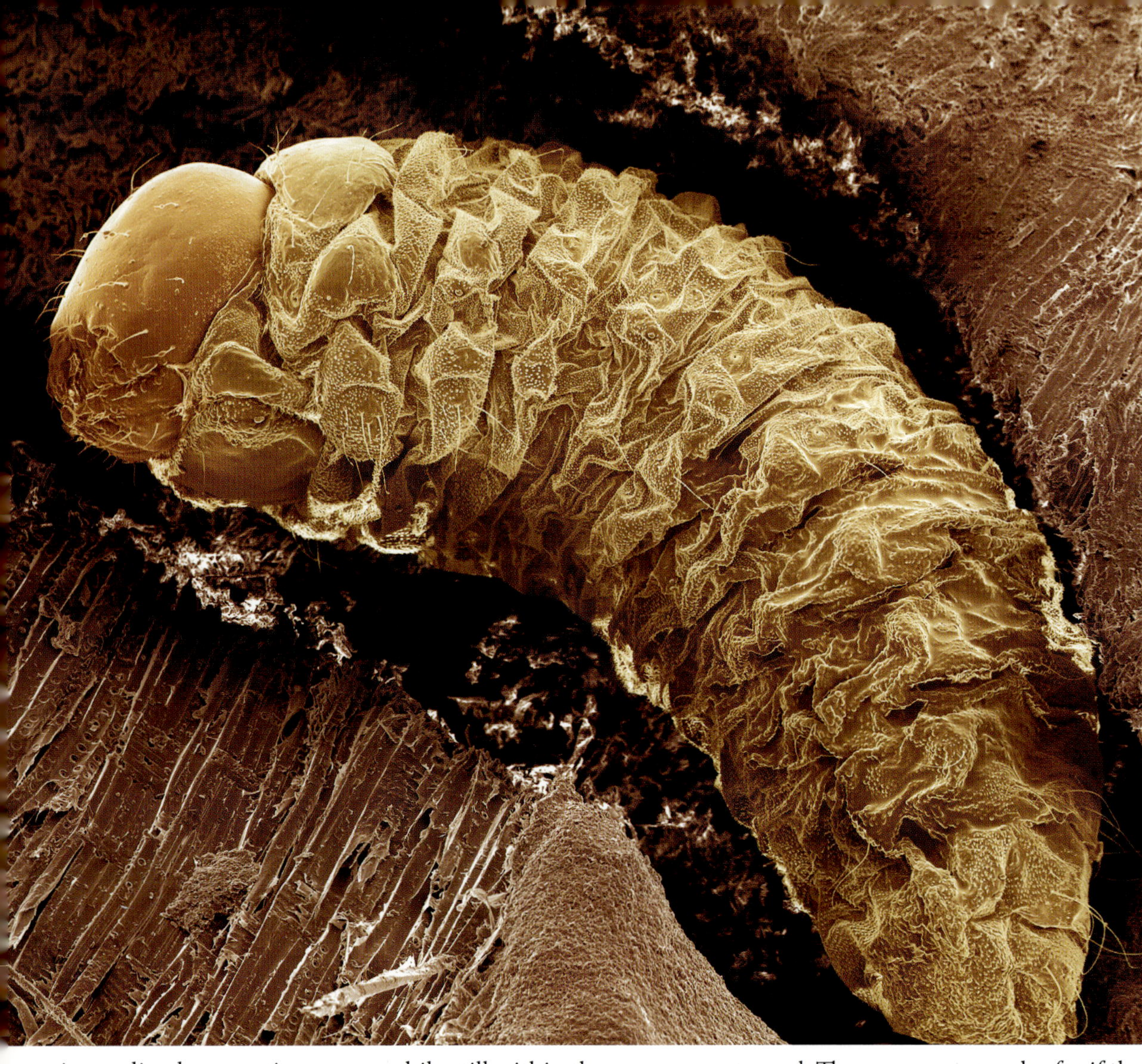

about 1-2 mm across, usually associated with small piles of pale wood dust. Woodworm infestations became a serious problem in Britain during the first half of the 20th Century, a period when two World Wars and economic restrictions distracted the attention from such matters. Building maintenance was neglected, and shortage of fuel resulted in homes becoming cold and damp. If the moisture content is above about 15%, woodworm thrive, as well as other woodborers. Today, with more efficient insulation methods and central heating in homes, living conditions are much drier and therefore infestations of woodworm have significantly decreased.

Under the right conditions, a female woodworm beetle can lay some twenty eggs a day, depositing them on the surface of wood, preferably where it is uneven. The hatching larva burrows into the wood through the base of the egg, picking up yeasts deposited by the female on the egg surface that, once established in the larva's digestive tract, helps to provide essential nutrients. The larva tunnels, feeds and grows for three years or so before excavating a small chamber near the surface, where it will pupate. The pupal stage lasts for about eight weeks before the adult beetle emerges, usually between May and August, to mate almost immediately, sometimes even while still within the emergence tunnel. They may not wander far if there is plenty of wood in suitable condition, but woodworm beetles are active fliers and can therefore disperse if necessary to find new feeding and breeding grounds.

The deathwatch beetle, *Xestobium rufovillosum*, is much larger, an emerging adult leaving an exit hole at least 3 mm in diameter. It usually attacks oak, among other hardwoods, but has also been known to feed upon decaying softwood timbers, the larvae tunnelling for up to twelve years before pupating. The adult female beetle produces around 40 eggs. The famous tapping, caused by the beetle striking its head on wood, is a mating communication call made during the flight season, between March and June. Males initiate a sequence of taps, to which females reply. More audible in the quiet of night, the sound of the deathwatch beetle was for a long time associated with ghosts.

Woodboring weevils, for example *Euophryum confine* and *E. rufum*, *Pentarthrum huttoni*, and *Caulotrupodes aeneopice*, are similar in size to the woodworm beetle, but with the characteristic long snout of all weevils. Although they only attack timber that is already damp and somewhat decayed by wood-rotting fungi, these species are prolific. Both the larvae and adults bore tunnels, and can cause serious deterioration of timber. The two *Euophryum* species were introduced into Britain from New Zealand during the 1930s, becoming widespread and extremely common in London. The house longhorn beetle, *Hylotrupes bajulus*; powder-post beetles, *Lyctus spp.*; and wharf borer, *Nacerdes melanura*, all occur in Britain but are not so common. Closely related to some of our British wood-boring beetles, the tropical *Bostrychoplites cornutus* is about 15 mm long, and has large, and distinctive thoracic horns. It is found in parts of Africa and Arabia, but is often imported to Britain in timber and specimens occasionally emerge from wooden articles purchased abroad as souvenirs.

All woodborers are prone to predation. The larvae of a small clerid beetle, *Korynetes caeruleus*, pursue woodborer larvae through their wood tunnels and devour them. Two tiny species of wasp also prey on woodborers: a braconid wasp, *Spathius exarator*, and the normally wingless pteromalid wasp, *Theocolax formiciformis*. *Spathius* females probe the tunnels of infected wood with a long egg-tube, or ovipositor, and lay eggs on the larvae. When the eggs hatch, the wasp

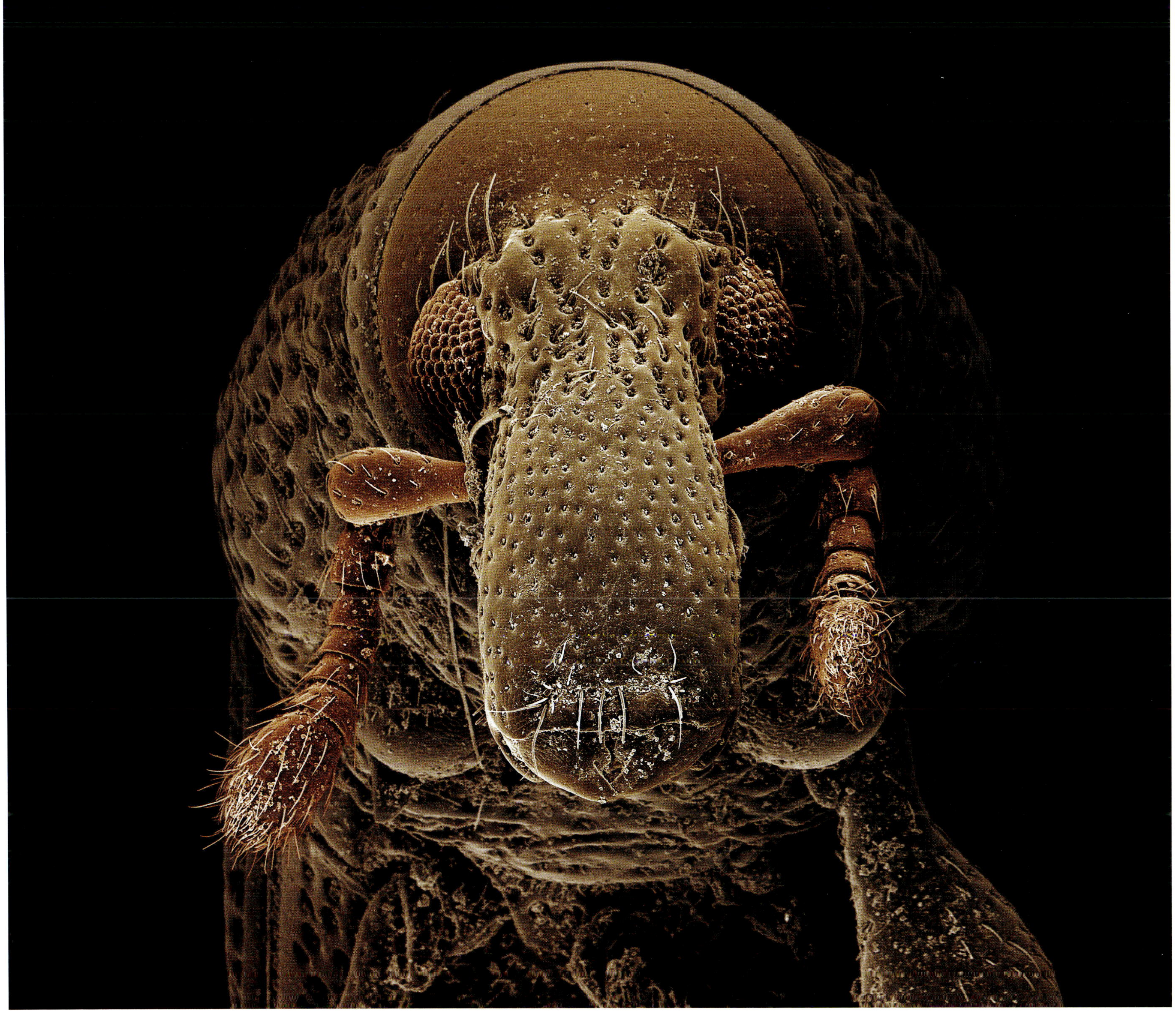

Woodworm or Furniture Beetle, *Anobium punctatum* (left)

The larva inside the wood tunnel it has bored.

Magnification: x90   *Scanning Electron Micrograph*

Woodboring Weevil, *Euophryum confine* (above)

Recognised by its strange-looking elongated head, characteristic of weevils, this woodborer only attacks wood if it has already decayed through the action of fungi.

Magnification: x240   *Scanning Electron Micrograph*

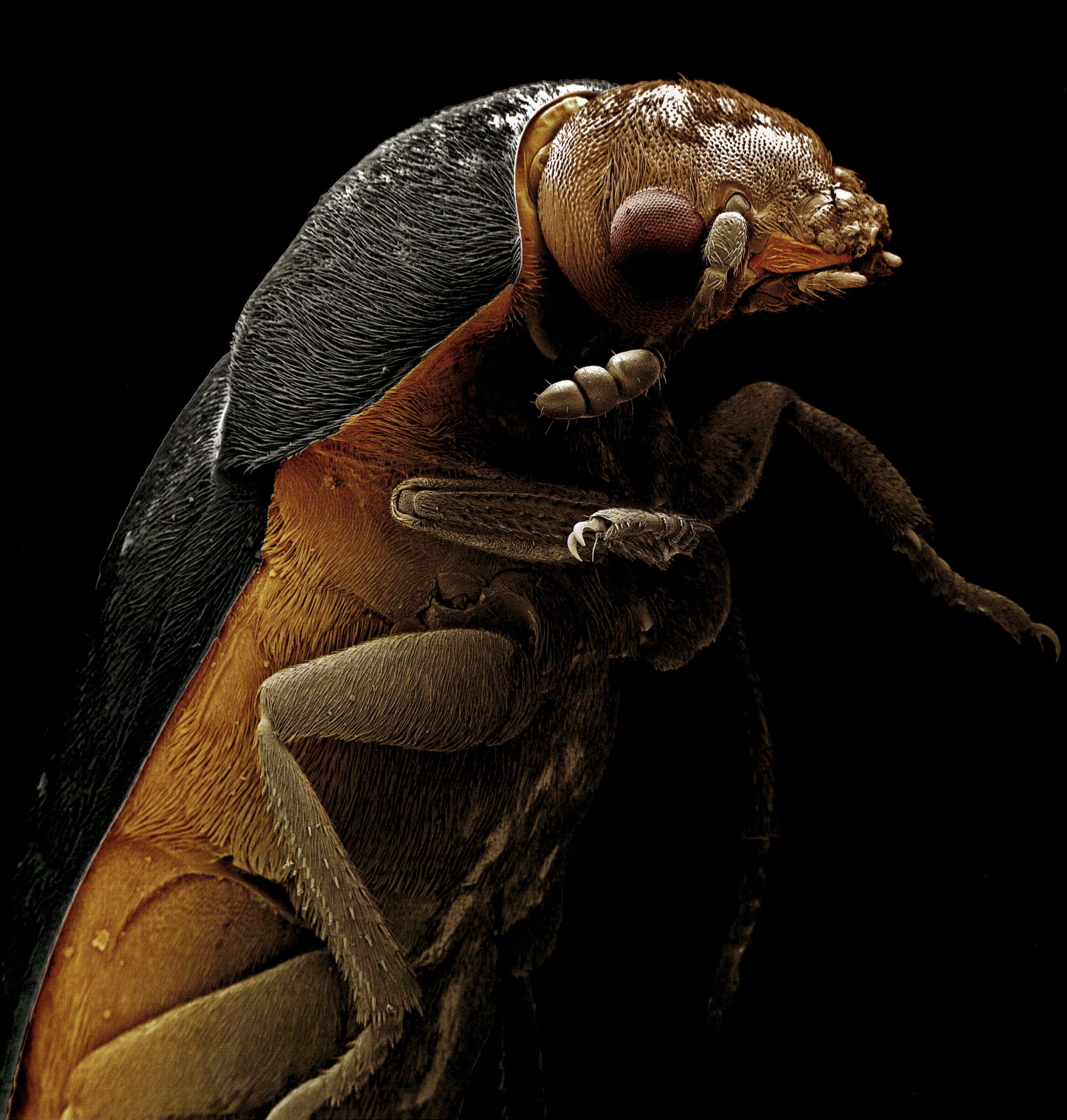

larvae feed on their hosts by sucking out their body fluids. The adult female *Theocolax* also probes the wood tunnels from the surface but is small enough to actually enter the woodborer tunnels to hunt for larvae on which it will deposit its eggs. The wasp larva pupates within the infected wood and after emerging the adult is capable of escape by boring its own exit tunnel. Another parasite is a mite, *Pyemotes ventricosus*, which is also parasitic on the larvae of many insects that infest straw, grain, hay, and grass seed. It will also bite us given the chance, sometimes causing an allergic skin reaction known as 'grain itch', and can cause asthma or other allergies.

BEETLES

Apart from the woodborers, there are several species of beetles and weevils that invade our homes to feed on stored food products, clothing, carpets and soft furnishings. In the kitchen, it is always advisable to keep dried foodstuffs in well sealed plastic or glass containers, to keep vagrant beetles out, and prevent the spread of insects introduced accidentally with newly purchased goods. Grain, cereals, flour and nuts are all targets for rice weevils, granary weevils, biscuit beetles and flour beetles. Typically, these insects are very small with bodies less than 5 mm long, and reddish brown to black in colour. There is also a whole range of species, including the larder beetle, leather beetle, fur beetle and carpet beetle that, between them, will eat almost anything. The larder beetle, for example, attacks various products of animal origin, including bacon and other fatty meats, but its larvae will chew pupation-holes in wood, cork, paper, textiles, mortar and even soft metals.

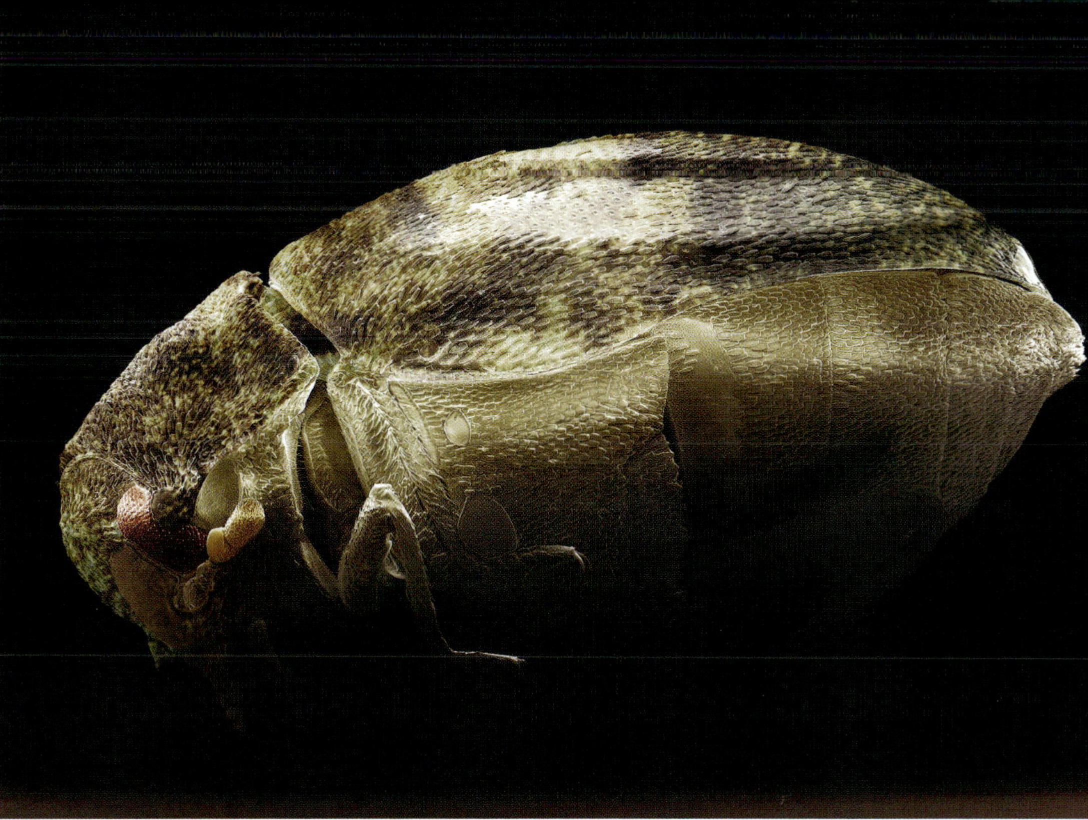

Varied Carpet Beetle, *Anthrenus verbasci* (above)

The adult varied carpet beetle feeds on nectar and pollen in flowers. It lays its eggs in old birds' nests, and sometimes enters houses to deposit them on carpets or any objects containing natural fibres or animal fur.

Magnification: x60
Scanning Electron Micrograph

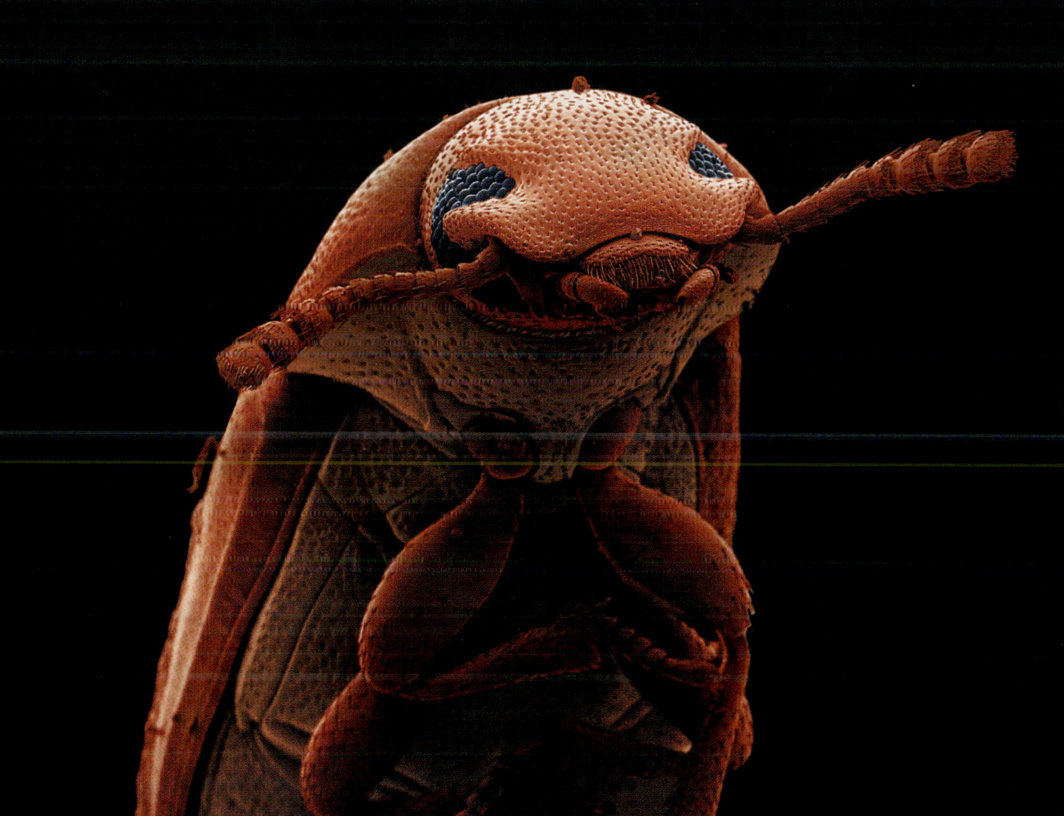

Leather or Hide Beetle, *Dermestes maculatus* (left)

Adult leather beetles and their larvae like to feed on raw animal skins. Before pupating, the larvae often bore into wood weakening structural timber.

Magnification: x55   *Scanning Electron Micrograph*

Confused Flour Beetle, *Tribolium confusum* (right)

Confused flour beetles infest milled grain products such as flour or cereal products, but can also survive on food scraps in cracks and crevices in cupboards or pantries.

Magnification: x50   *Scanning Electron Micrograph*

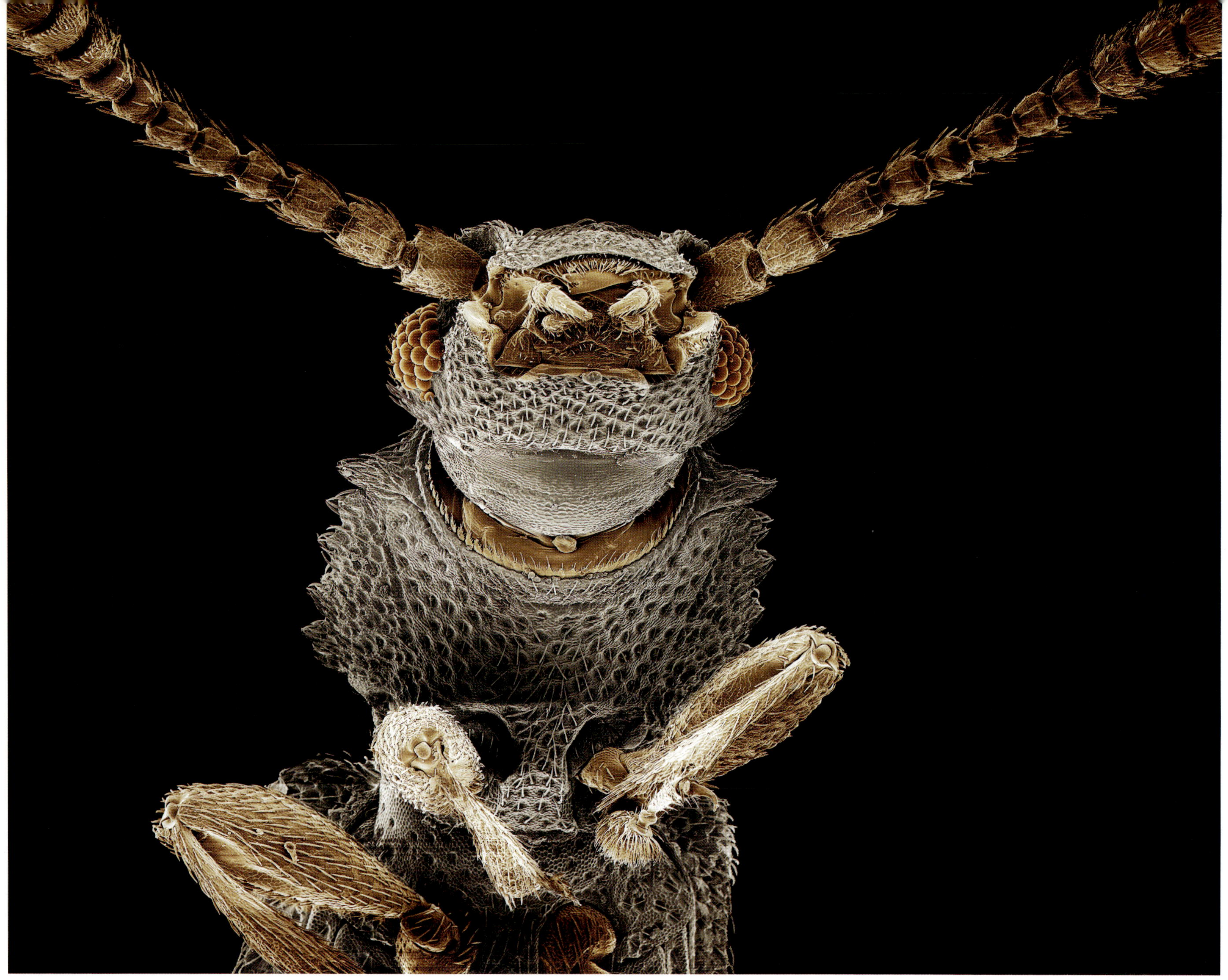

Merchant Grain Beetle, *Oryzaephilus mercator* (above)

Recognised by the saw-tooth projections on either side of the thorax, the merchant grain beetle infests stored foodstuffs such as grain, dried fruit, or nuts.

Magnification: x90   *Scanning Electron Micrograph*

Bluebottle, *Calliphora sp.*(right)

Bluebottles, or blowflies, are large stout insects, with a dark blue metallic sheen on the body, which is covered in bristly hairs. The adult flies feed mainly on carrion and other decaying animal matter, secreting saliva onto the surface of solid food to liquefy it before sucking it up the proboscis.

Magnification: x40   *Scanning Electron Micrograph*

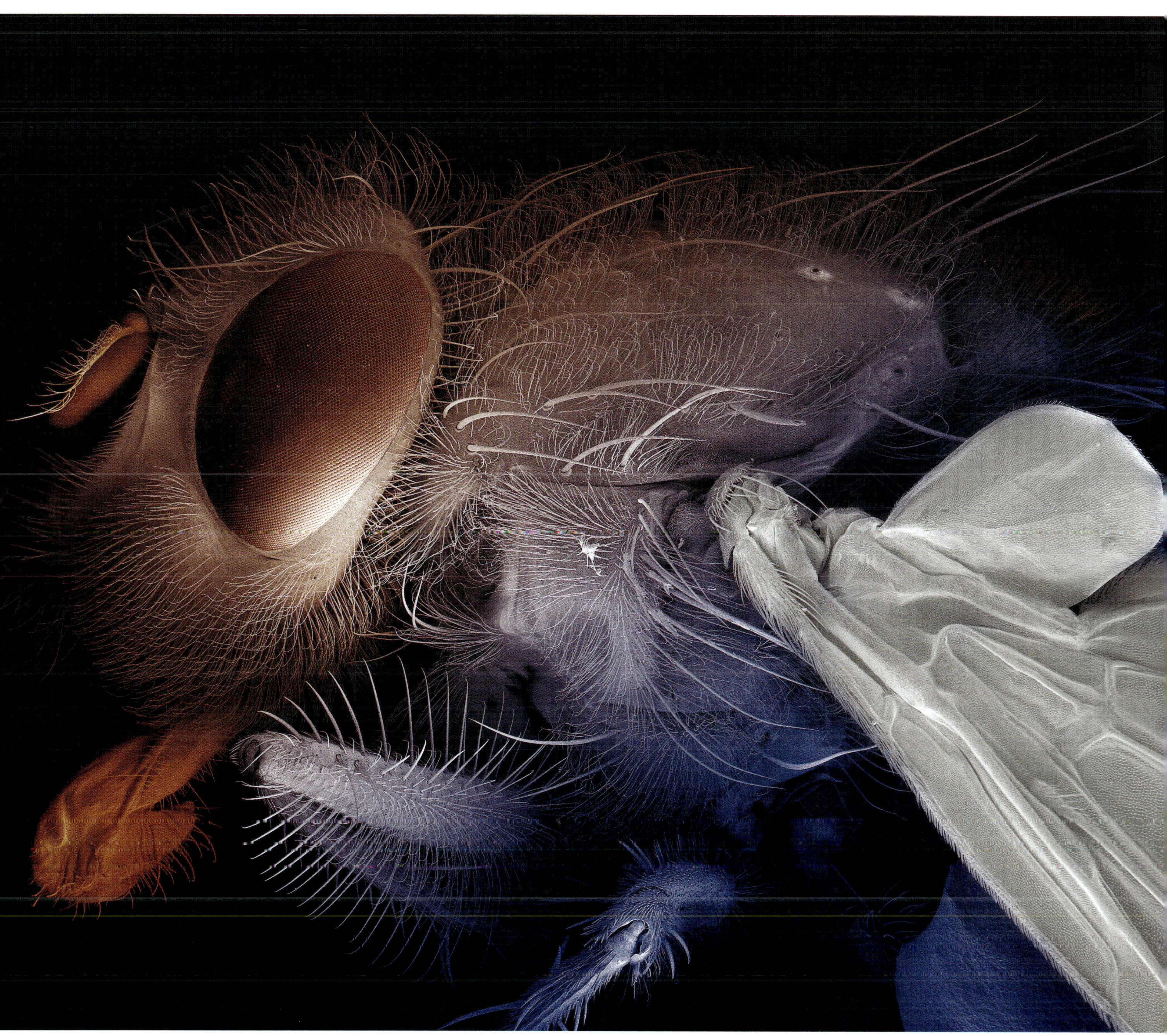

Plenty of the creatures that share our homes depend on us for food, but among the most irritating household pests are flies. They enter and exit the house with ease and speed through open windows and doors, tending to focus their attentions on the kitchen where they look for exposed food or traces of it on unclean surfaces. The problem with flies is that they frequent and feed indiscriminately on any solid food, which could be putrefying material as well as food stored for human consumption. Whatever they eat, flies first liquefy it by regurgitating digestive juices and their stomach contents on to it. Since flies pick up pathogenic organisms on their feet and bodies, or in their gut, which are transferred when they land, they are potential vectors of many diseases, for example dysentery, gastroenteritis, typhoid, cholera, tuberculosis, and intestinal worms. Throughout history, flies have been associated with poor sanitation and disease, but they have also proved to be useful. Since some fly larvae, or maggots, consume dead animal tissue but not living flesh, certain species have been used by field surgeons on battlefields, to clean infected wounds, and are still used to help heal gangrenous conditions.

The common housefly, *Musca domestica,* is highly active with a flight range of about 8 kilometres. The adults are 6-8 mm long with a wingspan of 13-15 mm; coloured grey with dark stripes, and when at rest the wings are held open in a delta shape. They spend the winter hibernating either as pupae or adults, in crevices that are sheltered from the weather. Under warm conditions, however, houseflies can remain active through the winter months and even reproduce. They reach a population peak in August and September. During her adult life of 1-3 months, a female is capable of laying 4-5 batches of 100-150 eggs, which are deposited in moist decaying matter such as household refuse, compost or dung. White maggots hatch in 8-48 hours, feeding greedily on the food that surrounds them. When mature, they pupate and the adults emerge a few days later. The lesser housefly, *Fannia canicularis*, is only slightly smaller but can be distinguished from the common housefly because when at rest its wings are folded along its back.

Long ago, before the months were named after Roman emperors, and gods, the month we now call July was 'worm' month in certain parts of Europe. The worms concerned were blowfly maggots, a reminder of what a problem it must have been to keep meat fit for human consumption during the heat of the summer. The blowflies most commonly seen in the house are the bluebottle, *Calliphora sp.*, and the greenbottle, *Lucilia sp.*. Blowflies are stout-looking, with a dark blue or green metallic sheen. They are most active during warm weather, breeding on carrion and other decaying animal matter. The eggs hatch in less than a day and the larvae burrow straight down into the food source, or in the case of *Lucilia* crawl on the surface where they liquefy food collectively. They grow very rapidly and once mature, they crawl away to a safe distance to seek out a suitable place in which to pupate. Adult blowflies emerge about a week later. Blowflies found indoors may originate from dead birds or mice brought in by domestic cats. Like other flies, blowflies are capable of carrying diseases.

PHARAOH ANTS

Pharaoh ants, *Monomorium pharaonis*, are tiny, the workers only 2 mm long and are reddish in colour. Native to Central and West Africa, they have spread to many parts of the world with international trade goods and have managed to establish themselves in northern Europe through the advent of central heating. In Britain, therefore, they can only survive in buildings where they make nests close to warm heating pipes in wall cavities or under floors. Due to their small size, these ants can penetrate the tiniest cracks and crevices, and since there are many queens, each occupying satellite nests, a colony can quickly grow from a few hundred individuals to several hundred thousands occupying entire buildings, making infestations extremely difficult to eradicate. The term 'pharaoh' ant was first used by the 18th Century biologist Carl Linnaeus, who associated the ants with the biblical plagues of Egypt.

A pharaoh ant colony consists of queens, males, immature worker ants, eggs, larvae and pupae. Throughout their lifetime, queens lay around 400 eggs in batches of 5-10, which hatch after about a week. Worker ants may remove developing larvae from a nest and carry them to form a new satellite nest elsewhere. The behaviour of the worker ants towards developing larvae determines whether they will become worker or queen ants. They feed on almost any foodstuffs and, as they forage, ants move out from the nest site laying down scent trails as they go, inducing other workers to follow. If a trail leads successfully to a food source, large numbers of workers reinforce the scent trail to form a major route. In a short time, a complex network of scent trails are established between the mother nest site, satellite colonies and food sources.

Pharaoh Ant, *Monomorium pharaonis*

Native to tropical Africa, pharaoh ants have managed to establish themselves in Britain, where they can survive in heated buildings. Very small, they can penetrate the tiniest cracks and crevices, and since there are many queens, each occupying satellite nests, a colony can quickly grow from a few hundred individuals to several hundred thousands, and occupy entire buildings.

Magnification: x230   *Scanning Electron Micrograph*

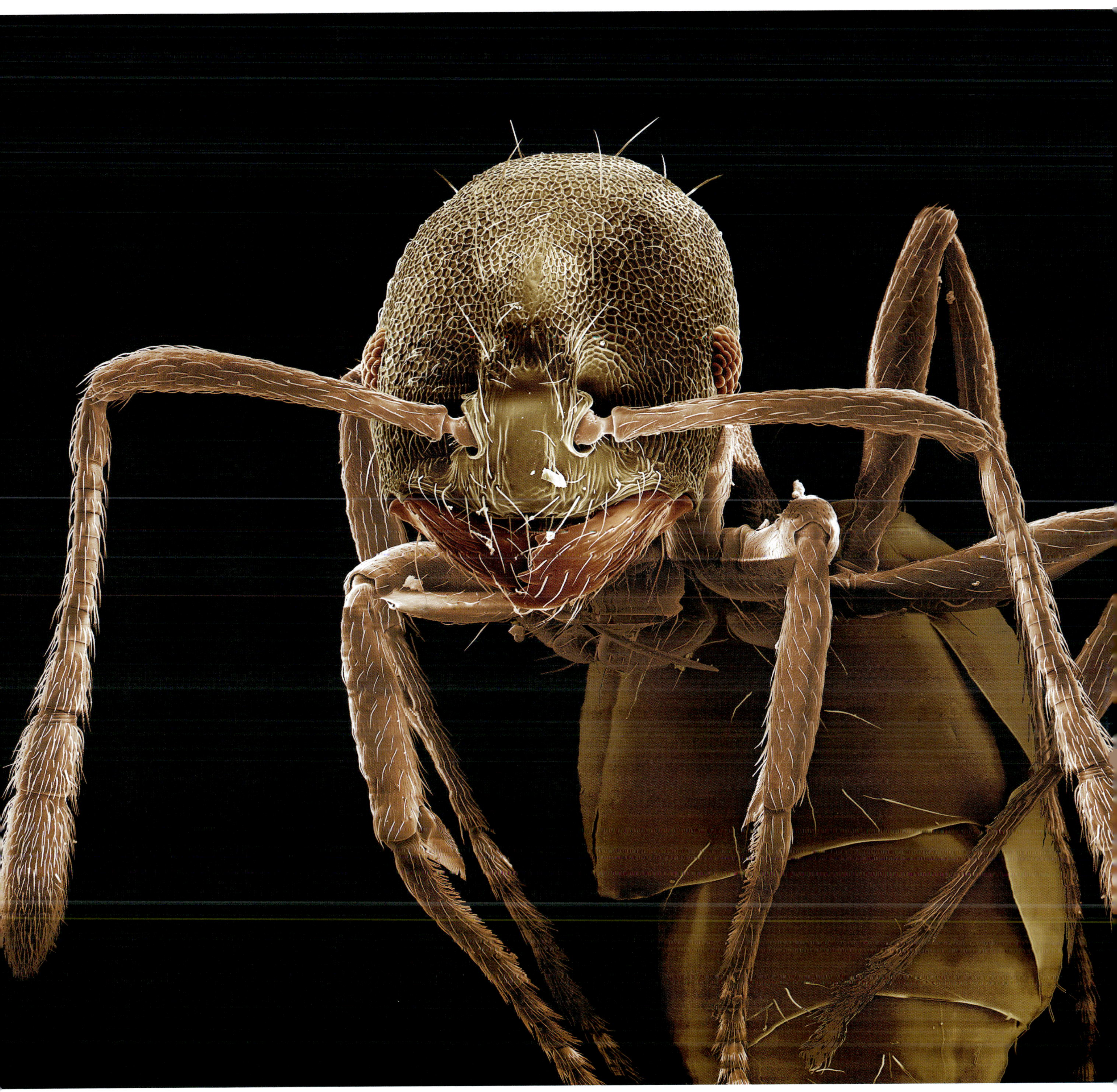

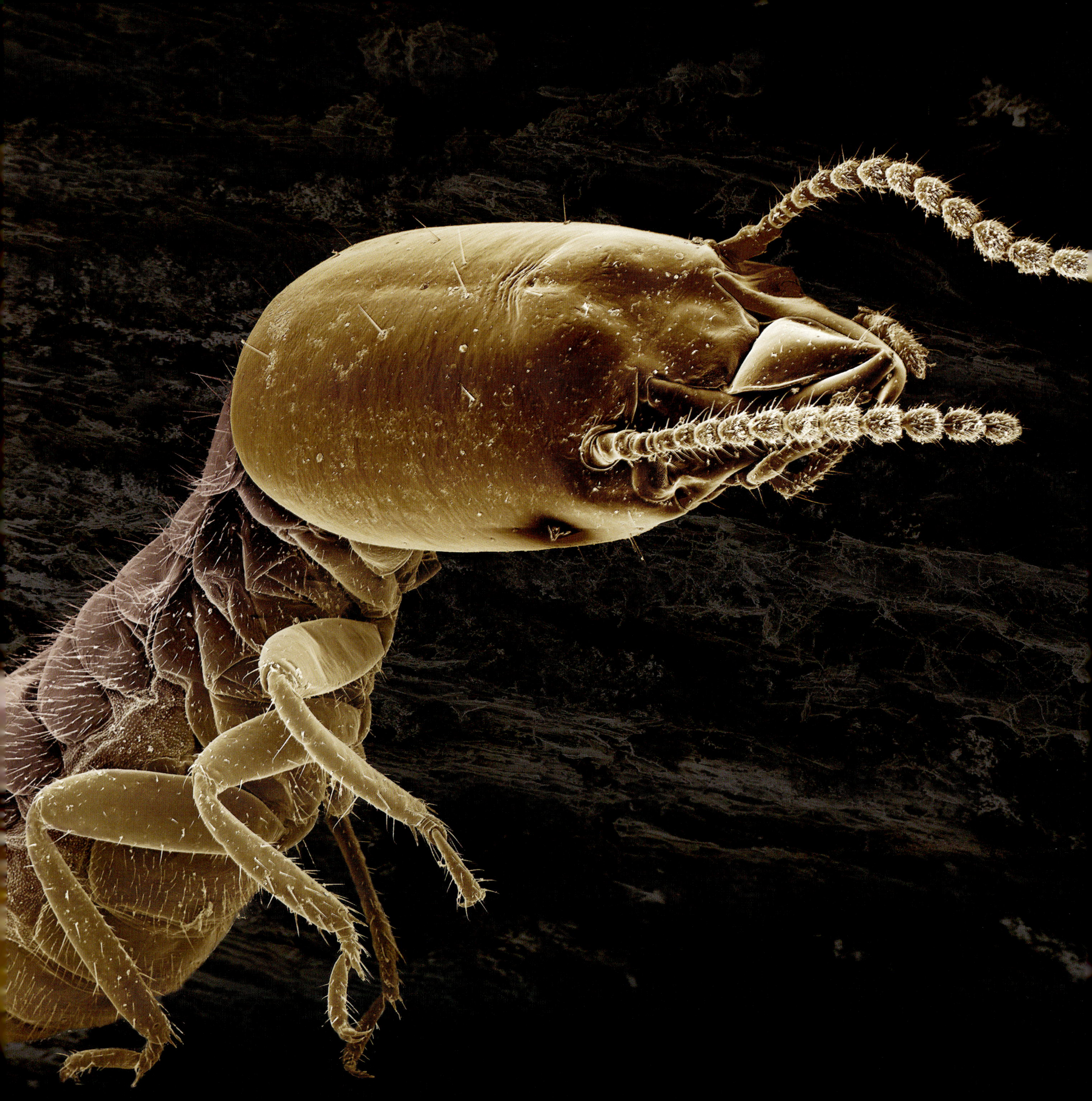

One of the effects of globalisation is that non-native creatures are arriving in Britain with increasing regularity, hidden in fruits, vegetables, flowers or wooden products. With current trends in climate change bringing warmer summers and milder winters, some of these visitors are not only able to survive, but establish breeding populations. One of these arrivals is the termite, a major threat to timber buildings in warmer climates. Most tropical termites would never be able to survive here but there are a few species, including *Reticulitermes lucifugus* and *R. santonensis*, which are more tolerant of cooler conditions. They have certainly become firmly established in Europe; they were known in the south of France as early as the 18th Century, spreading along the Mediterranean coast and have since steadily moved northwards, reaching Paris by 1945, taking advantage of city warmth and the abundance of food. There have been a few isolated cases of established termite colonies in Britain. Discovered during renovation work on houses, some of these have comprised thousands of individuals and even spread from building to building. If our climate continues along the current warming trend, these discoveries will become more frequent and familiar. If so, termites will present a significant

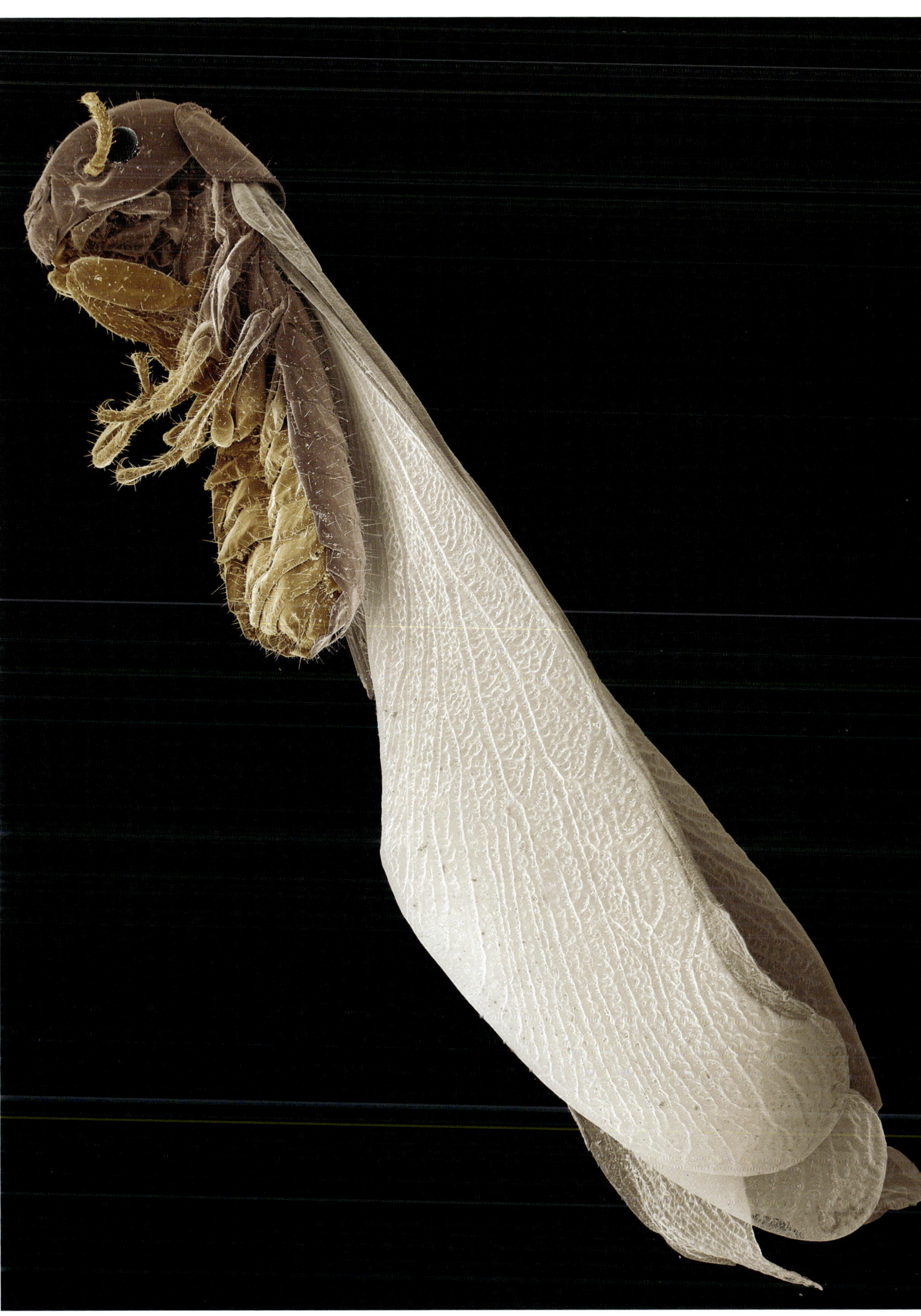

Termite, *Reticulitermes lucifugus*

Termites are highly organised social insects, living in colonies comprising thousands of individuals, with a queen, winged alates, soldiers and workers. Each has a specific role to perform for the benefit of the colony. The task of a soldier termite (see left) is to scout for food sources and to guide worker termites to it, as well as protecting them as they forage. Seasonally winged, or alate, termites (see right) are reproductive forms, which disperse en masse when climatic conditions are right to establish new colonies.

Magnification: x95
Scanning Electron Micrograph (left)
Magnification: x33
Scanning Electron Micrograph (right)

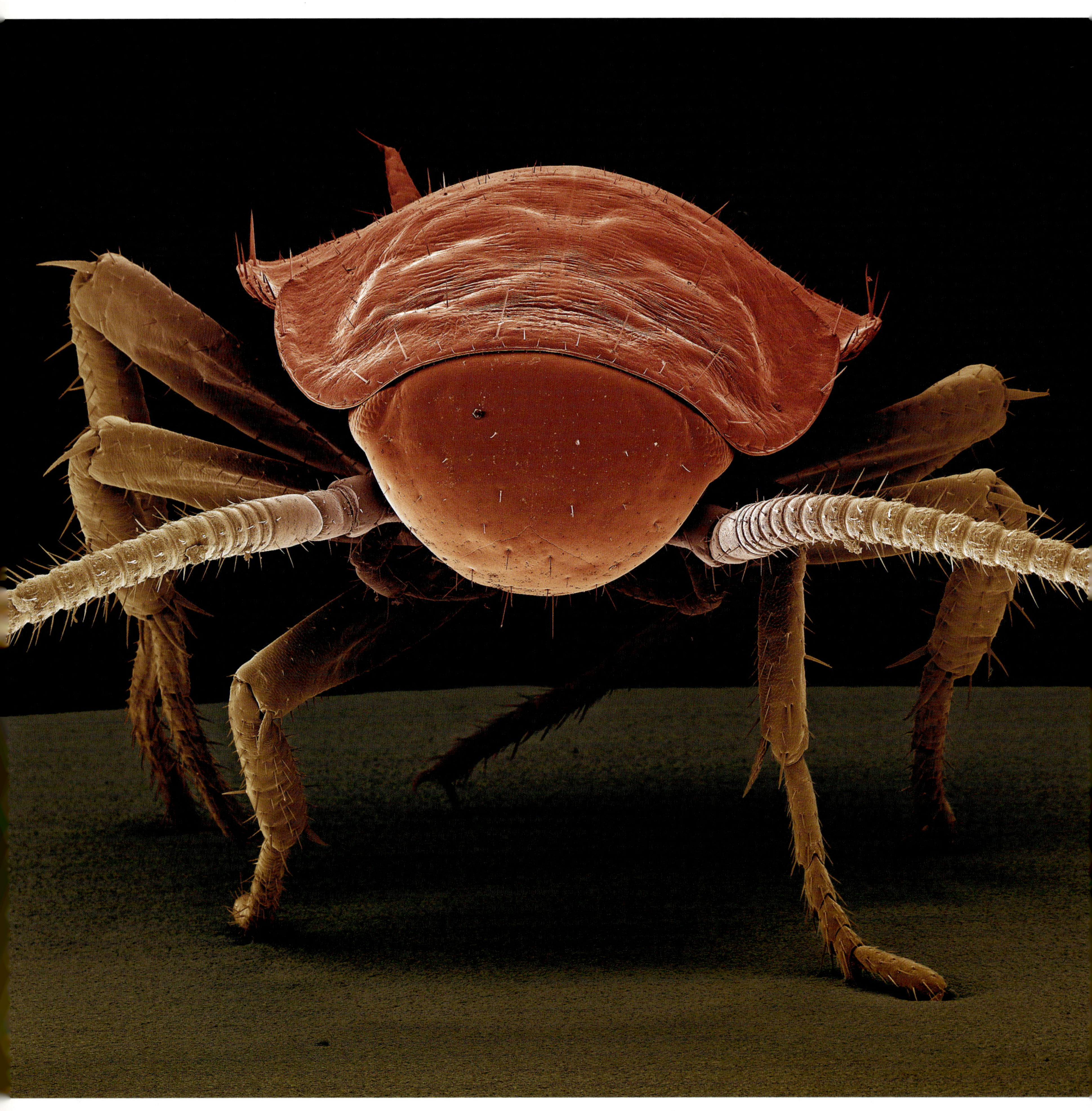

threat to housing infrastructure, eating not only timber, but vast quantities of paper, cardboard boxes, plants, cotton, and even chewing the plastic insulation of electric cables.

A nest of *Reticulitermes* comprises a number of interconnected subterranean chambers divided into cells constructed of semi-digested cellulose. The queen is confined to a special chamber and, as the colony expands, eggs and larvae are carried along tunnels to more distant chambers where new reproductive nuclei are established. Colonies of the more robust *R. santonensis* are far larger than those of *R. lucifugus*, and the termites more aggressive, attacking not only dead wood but living trees and can cause havoc in a garden by eating the roots of many herbaceous plants.

COCKROACHES

Cockroaches are living fossils and have changed little for 250 million years. There are about 3,500 different kinds, the majority of which are tropical with only a few small species native to Britain. Most cockroaches have no significant impact on us, but there are a few foreign species that arrive frequently with imported goods, especially foodstuffs, which manage to establish colonies in heated buildings where they can raid kitchens. These include the oriental cockroach, *Blatta orientalis*, German cockroach, *Blattela germanica*, American cockroach, *Periplaneta americana*, smokey-brown cockroach, *Periplaneta fuliginosa*, and brown-banded cockroach, *Supella longipalpa*. They vary from 1-4.5 cm long, are light to dark brown or mahogany in colour and have a flattened oval shape, with spiny legs, and long, filamentous antennae. Most of them have wings, though these are rarely used for flight and are sometimes even vestigial. A female will carry her eggs in a capsule, or ootheca, for a few hours before she drops or glues it to a suitable surface near a food supply. After hatching, the nymphs, which look like small pale versions of the adults, will moult several times over a period of a few months before reaching adulthood.

Cockroaches can be common in commercial premises where food is stored, and infestations quickly spread through large buildings and blocks of flats via service ducts, drains, refuse chutes, loose-fitting doors and windows, or gaps where pipes pass through walls or floors. Gregarious and nocturnal, they spend the day hiding in warm, moist, dark places, behind cupboards, or in cracks and crevices, leaving their signature behind in liquid excreta, faecal pellets, discarded egg cases, nymphal skins and an unpleasant odour. Apart from foodstuffs, they eat almost anything including leather, wallpaper and wallpaper paste, bookbindings, soiled septic dressings, faeces, and even each other if there is no other food available. Cockroaches cause health problems, including amoebic dysentery *(Entamoeba histolytica)*, gastroenteritis *(Escherichia coli)* and *Salmonella* food poisoning. Food may be spoiled with faeces and gland secretions, or by dead insects, and, in addition, allergic reactions may result from the inhalation of tainted airborne dust.

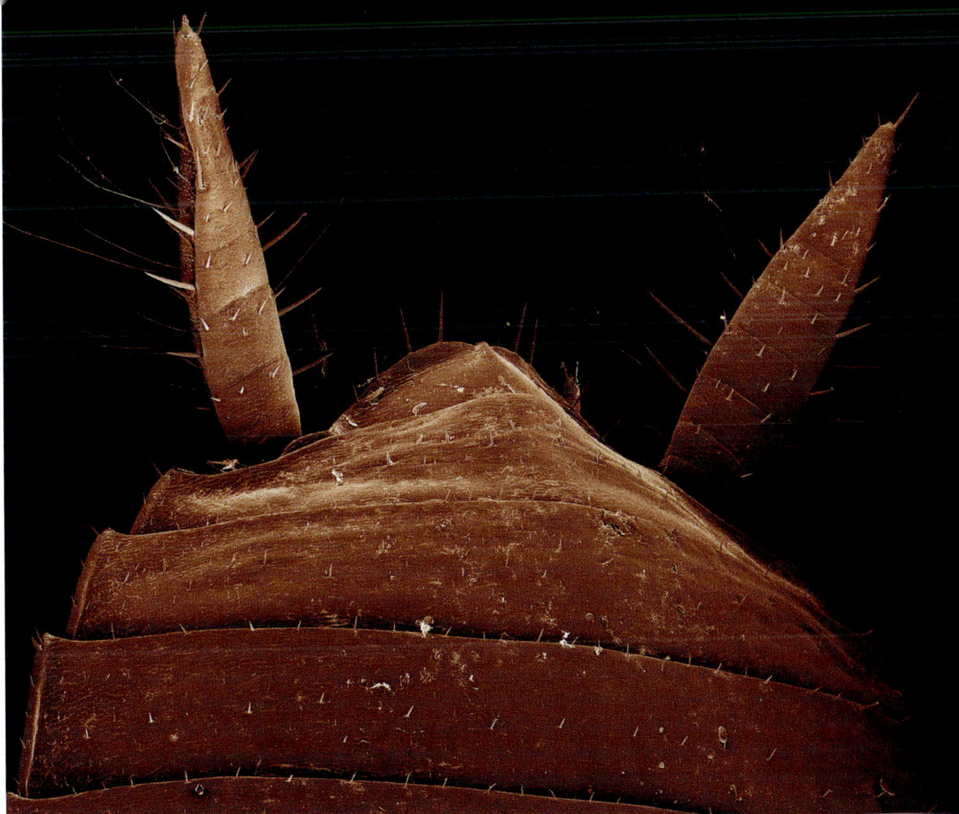

American Cockroach, *Periplaneta americana*

An immature or nymph cockroach (see left) resembles the adult except that it is wingless. In this frontal view, the small head is hidden under the large thoracic plate. On the tip of the abdomen (see above), there are two appendages called cerci that are sensory organs covered in hairs (see below) to detect air currents.

Magnification: x60   *Scanning Electron Micrograph* (left)
Magnification: x75   *Scanning Electron Micrograph* (above)
Magnification: x575   *Scanning Electron Micrograph* (below)

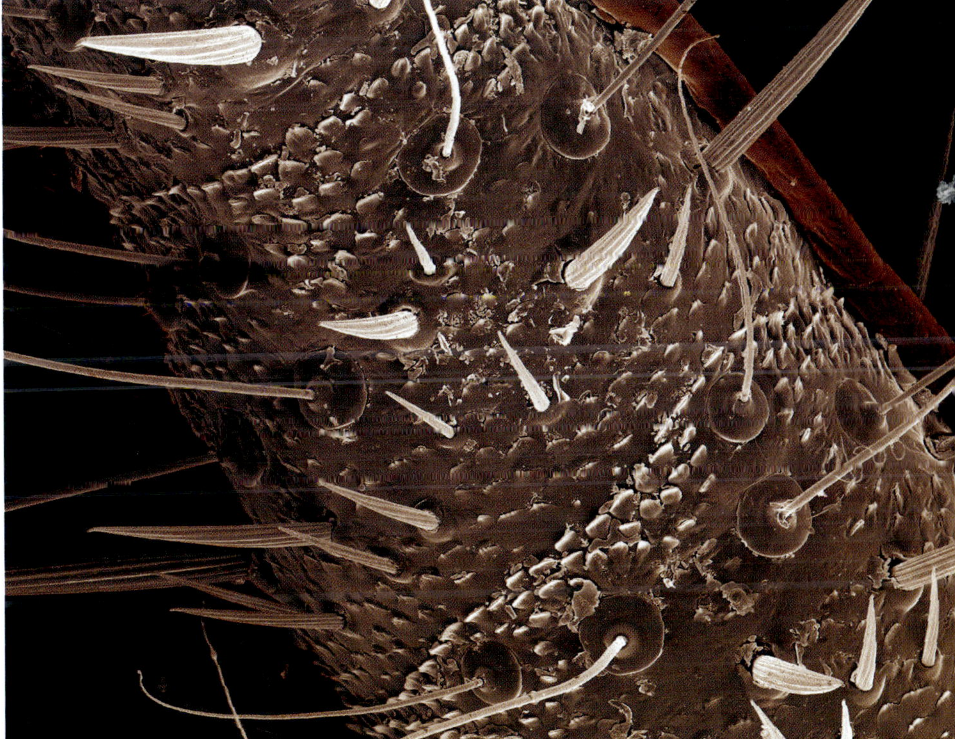

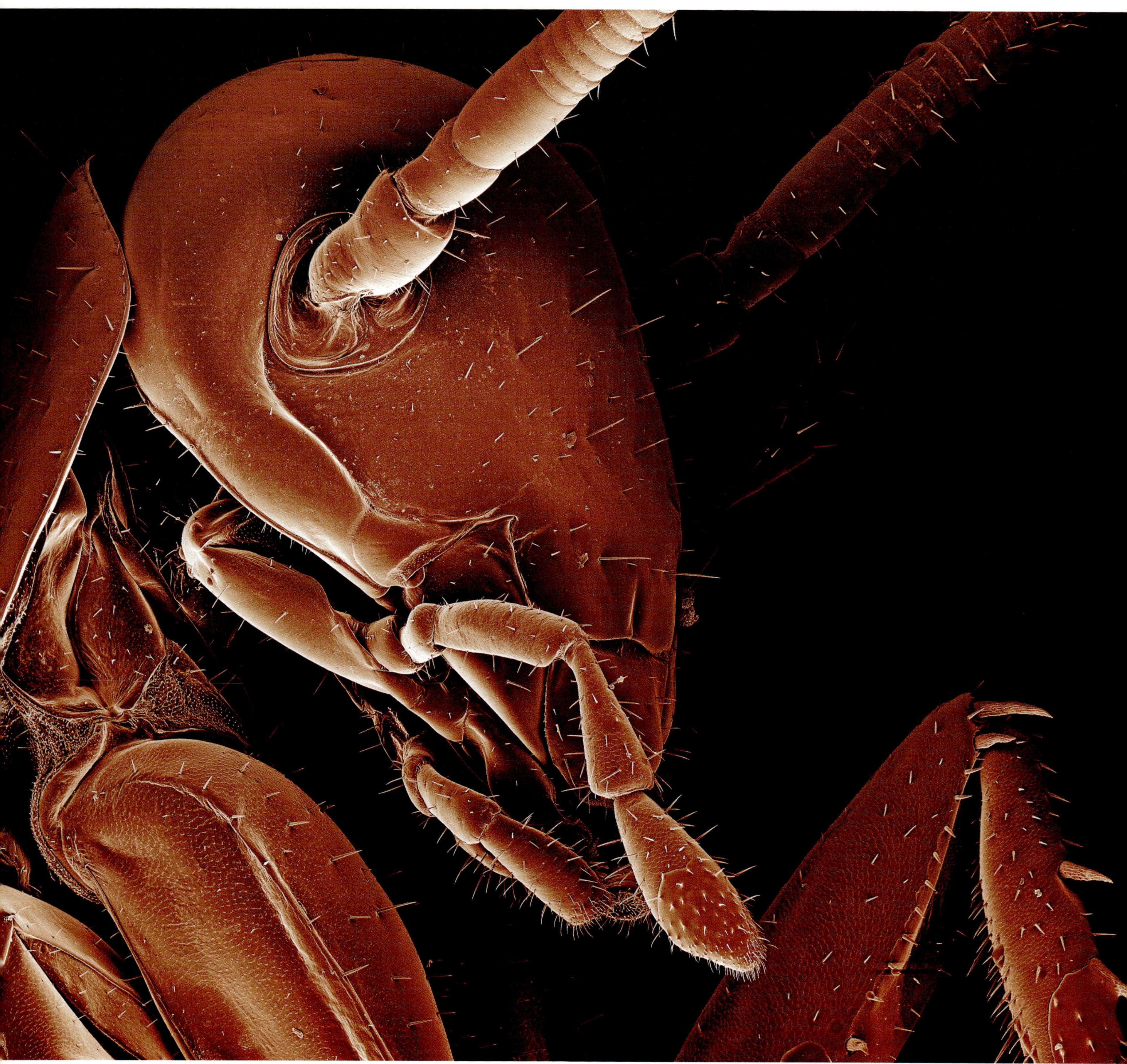

A fascinating insect that sometimes turns up in the kitchen sink is the silverfish, *Lepisma saccharina*. A small wingless insect 1-2 cm long, it is a living fossil, its form having remained little changed for about 300 million years, in fact insects of this type may well be the ancestors of all the different types of modern insects alive today. They crawled around primeval swamps at a time when the first vertebrates ventured out of water on to land, and survived through the age of the dinosaurs. Metallic silver or brown in colour, silverfish have three bristly filaments extending from the tip of the abdomen, hence the alternative name 'bristle-tail'.

In nature, silverfish are restricted to southern Europe and parts of Asia, where they live under rocks, bark and leaf mould, in bird nests, and ant nests. By exploiting damp habitats in households, silverfish have been able to increase their natural distribution into cooler latitudes including Britain. They

American Cockroach,
Periplaneta americana (left)

A close-up view of the head and mouthparts; in front of the eyes are two large antennae, below them are jointed appendages called maxillary palps, used to examine food, and behind these are mandibles, used for chewing.

Magnification: x120
Scanning Electron Micrograph

Silverfish, *Lepisma saccharina* (right)

A link with the past, this primitive insect has changed little in form since the Palaeozoic Era, some 300 million years ago. It is frequently found in damp kitchen cupboards, and occasionally in the sink.

Magnification: x115
Scanning Electron Micrograph

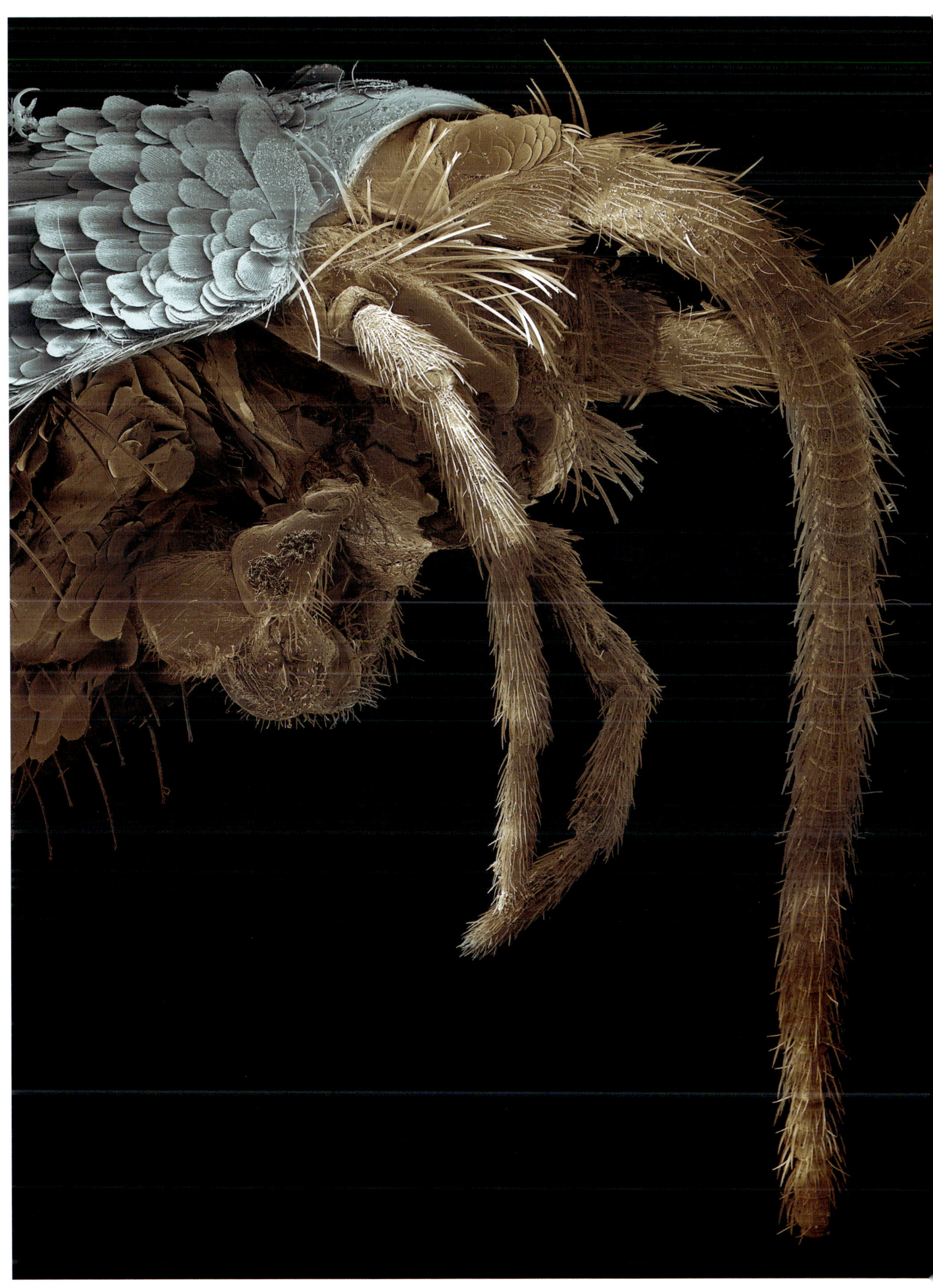

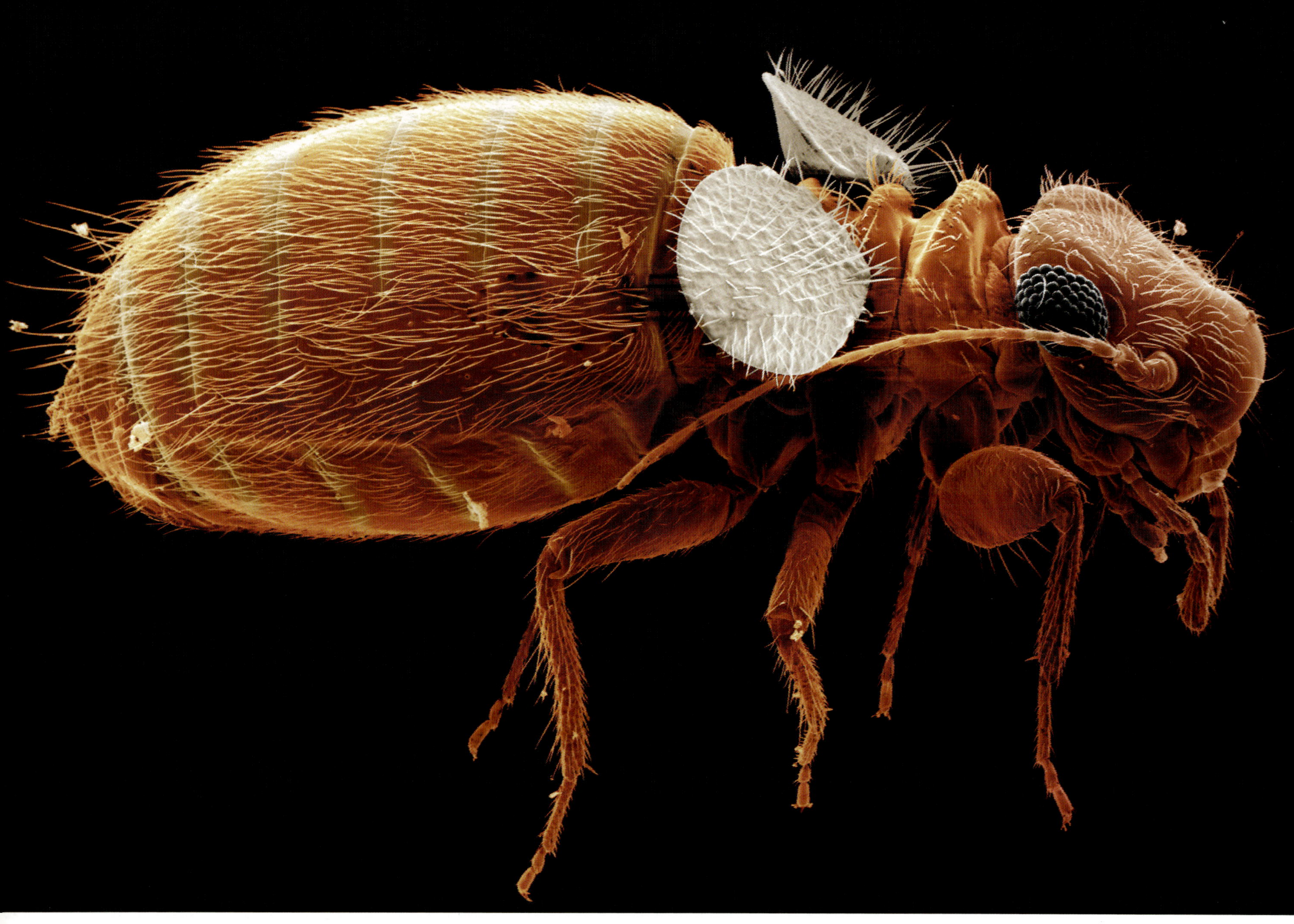

require high humidity in order to survive, so in houses they tend to frequent kitchens and bathrooms, and since they are nocturnal, they are most commonly seen scurrying for shelter in a dark crevice when the light is turned on.

Interesting as these animals are, they do cause some damage, feeding on flour, starch, paper, gum, glue, cotton, linen, rayon, and silk, sugar, moulds and breakfast cereals. They can however, survive for long periods without food. During her lifetime, which might be three or four years, a female will produce less than a hundred eggs, so they are normally present in small numbers. A large infestation would therefore indicate that a house has been infested for a long time. The eggs are tiny and deposited in damp, dark crevices.

Psocid, *Lepinotus sp*

In nature, psocids live on tree bark, stone walls and old birds' nests, feeding on algae, fungi or lichen. They are winged and called bark lice. There are several wingless species that infest the home, however, among them the book louse, *Liposcelis bostrychophila,* which feeds on paper, glue, paste and mould. Psocids are not lice, but soft-bodied insects. The adult psocid, *Lepinotus sp.* **(**see above), which has vestigial wings, and its nymph (see right) were both found living in a raffia tablemat.

Magnification: x150   *Scanning Electron Micrograph* (above)
Magnification: x150   *Scanning Electron Micrograph* (right)

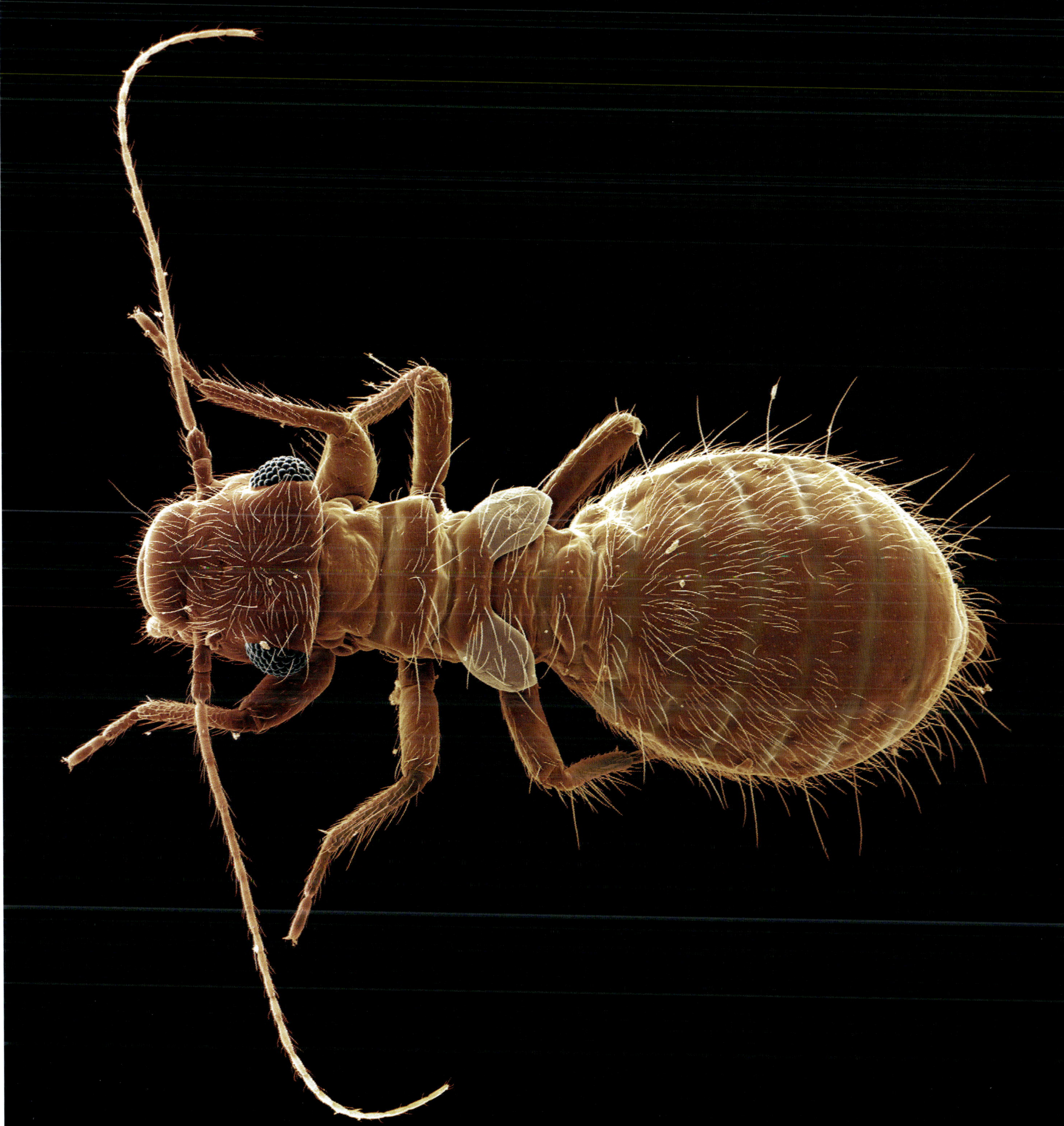

BOOK LOUSE

Book lice are not lice at all but small, soft-bodied, wingless insects belonging to a group known as psocids. The most common book louse, *Liposcelis bostrychophila*, is about 1.5 mm long, with a broad head, protruding eyes and long filamentous antennae. It infests our homes, hiding in cracks and crevices, feeding on paper, glue, paste, mould, flour or cereal products, damaging books, stamp collections, wallpaper, natural fibre tablemats, and tainting foodstuffs. It is the principal psocid pest species in Britain and in Europe has a very wide distribution throughout the world. The females deposit eggs singly or in small batches. The nymphs are almost colourless and difficult to see until they mature into brown adults. The majority of psocids in nature live on tree bark, leaves, fences, stone walls, or in old birds' nests feeding on algae, fungi or lichen. Often called bark lice, they are winged and range from 1 mm to 1 cm long.

MOTHS

There are several species of moths that invade our homes to feed on paper, fabrics and stored foods and each tends to be quite specialised in both its habits and food preferences; thus, there are clothes moths, meal moths, flour moths, rice moths, dried currant moths and so on. The adults of some of these species do not feed at all, and it is the caterpillars that do all the damage. Like all caterpillars, their appetites are voracious, consuming clothing, carpets, soft furnishings, grain, cereals, flour, nuts and other stored foods. The adult moths are generally drab in colour, small, about 5-8 mm in length, and their caterpillars are pale-yellow or dirty white. Most of these moth pests have been distributed worldwide with the help of international trading.

In nature, many of these moths live in the nests of birds and small mammals, where the caterpillars feed on dried plant and animal material; they are among the few insects that are able to digest the keratin (a tough, fibrous protein) of hairs and feathers. In the home, they have a wide choice of feeding material, but prefer natural fabrics such as wool, cotton, or silk, especially if soiled by food, sweat or urine. Caterpillar presence can be detected as surface damage on the pile of carpets, or tell tale holes in garments. The worst culprit is the common or webbing clothes moth, *Tineola bisselliella*, whose caterpillars will consume almost anything from vegetable matter and fabrics, to fur and feathers. The case-bearing clothes moth, *Tinea pellionella*, can also be a serious pest and its caterpillar can be recognised by its habit of constructing a tubular case for itself out of general debris. Caterpillars of the tapestry moth, *Trichophaga tapetzella*, which in nature often lives in owl pellets, specialises in targeting coarser materials such as canvas or sacking. Another moth pest is the brown house moth, *Hofmannophila pseudospretella*.

The frequency of moth infestations has declined considerably over the last fifty years, mainly due to the drier atmosphere in our homes through central heating and also the more widespread use of man-made fibres in fabrics. However, the risk of damage to articles of clothing stored for long periods and carpets in undisturbed places is still there so it is always worthwhile to periodically check for any signs before an infestation develops. The best control for clothes' moths is prevention: it used to be common practice to put white, marble-sized mothballs in drawers and wardrobes where clothing was kept, as they contained the chemicals naphthalene and paradichlorobenzene, which sublime to produce a gas that is toxic to insects. However, it was also toxic to us, and our pets, and should not now be used. Cedar wood balls provide a natural and far less toxic alternative to mothballs, and frequent washing or dry cleaning kills any eggs or larvae. Soft toys can be cured of moths by placing them in the freezer for a few days.

There are many different moth species whose caterpillars attack stored foodstuffs. Familiar examples include the flour or meal moth (*Pyralis farinalis*), the Mediterranean flour moth (*Ephestia kuehniella*), and the Indian meal moth (*Plodia interpunctella*), which was first recorded in Britain in 1847. Food exposed in open packets and containers attracts the moths and dry foodstuffs stored for long periods are often the source of infestations. Meal and flour moth caterpillars can chew their way into paper packets, cardboard boxes and through thin polythene bags. If an infestation is discovered in food stored in glass or tough plastic containers with tight-fitting lids, it was most likely already infested when purchased and easily eradicated by carefully discarding the entire contents.

Common Clothes Moth, *Tineola bisselliella*

The presence of an adult clothes moth in the home indicates that there are caterpillars infesting clothing, carpets and soft furnishings for it is the larvae that are responsible for any damage, feeding mainly on fibres and materials of animal origin such as wool, silk or fur.

Magnification: x110   *Scanning Electron Micrograph*

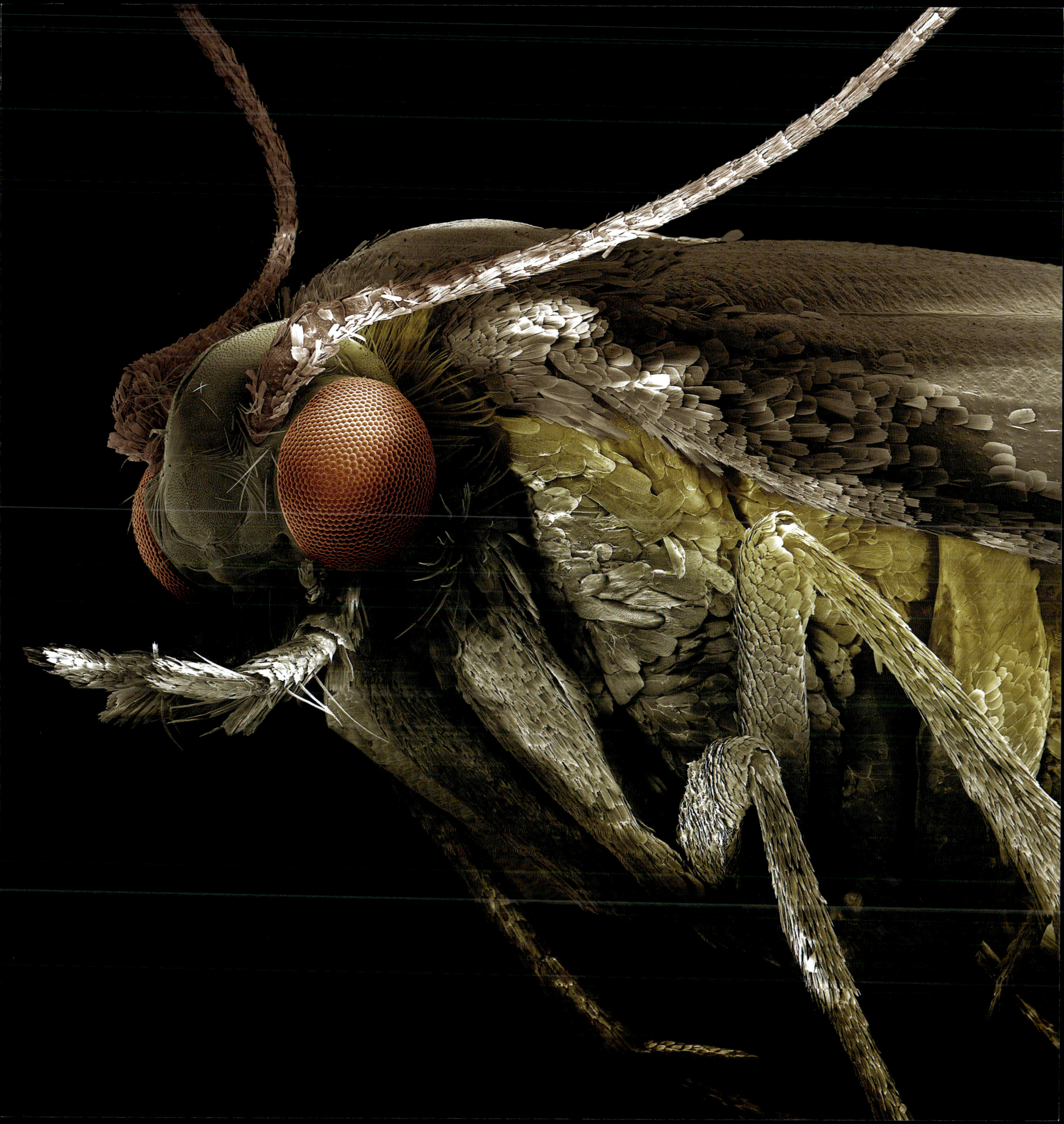

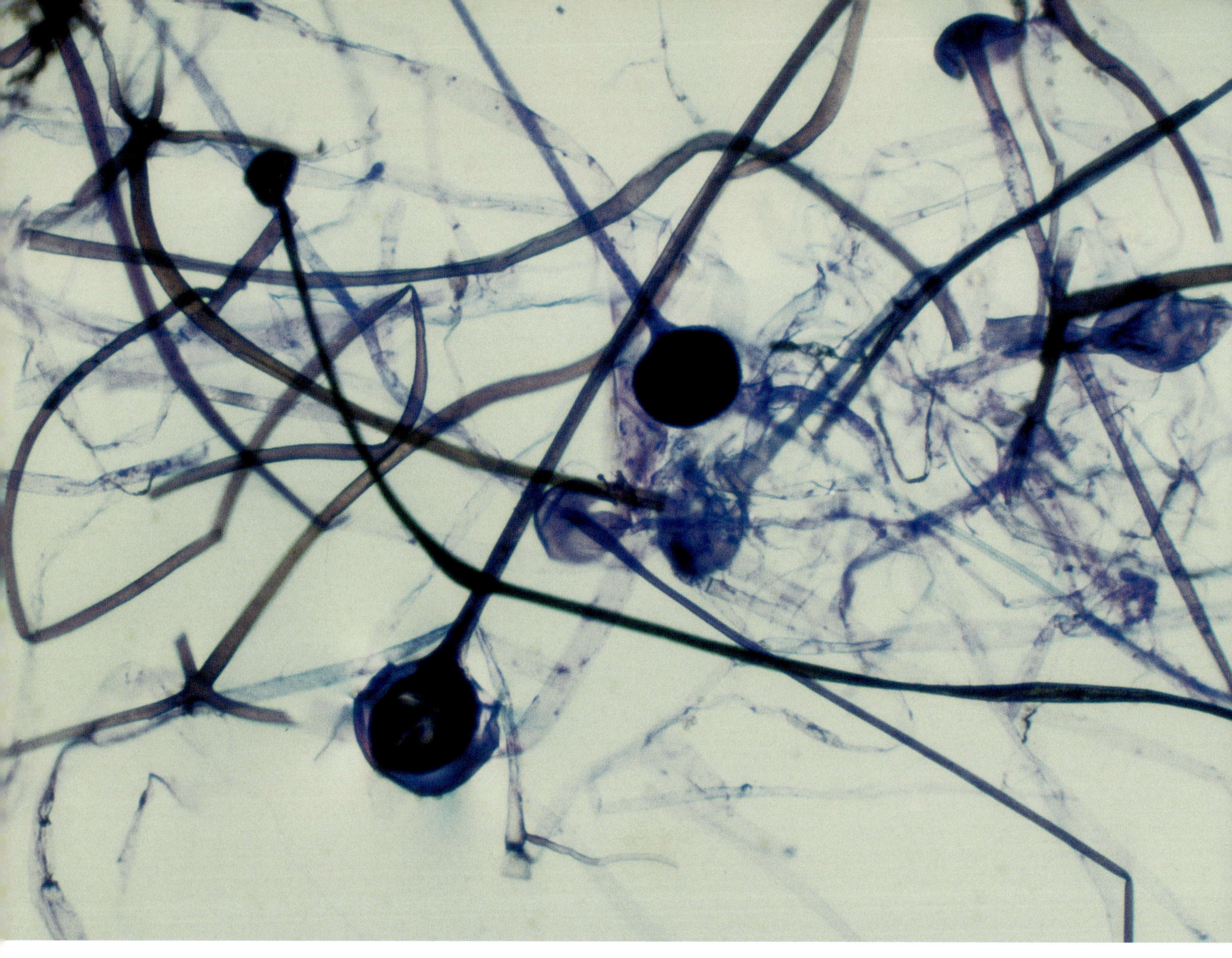

Aspergillus, *Aspergillus sp.*

The fungus *Aspergillus* includes over 185 known species, of which some 20 species cause invasive infections, often respiratory, in humans. They are also commonly found as mould growing on food, for example the **Common Bread Mould,** *Aspergillus niger,* (see above), a light microscope photograph showing filamentous hyphae with fruiting bodies, or conidiophores, which produce spores by asexual reproduction. A scanning electron microscope view of the **conidiophores** (see right) of another *Aspergillus* species shows the individual ripening spores. The spores are everywhere, in the air we breathe and on every surface we touch.

Magnification: x1750   *Light Micrograph* (above)
Magnification: x5000   *Scanning Electron Micrograph* (right)

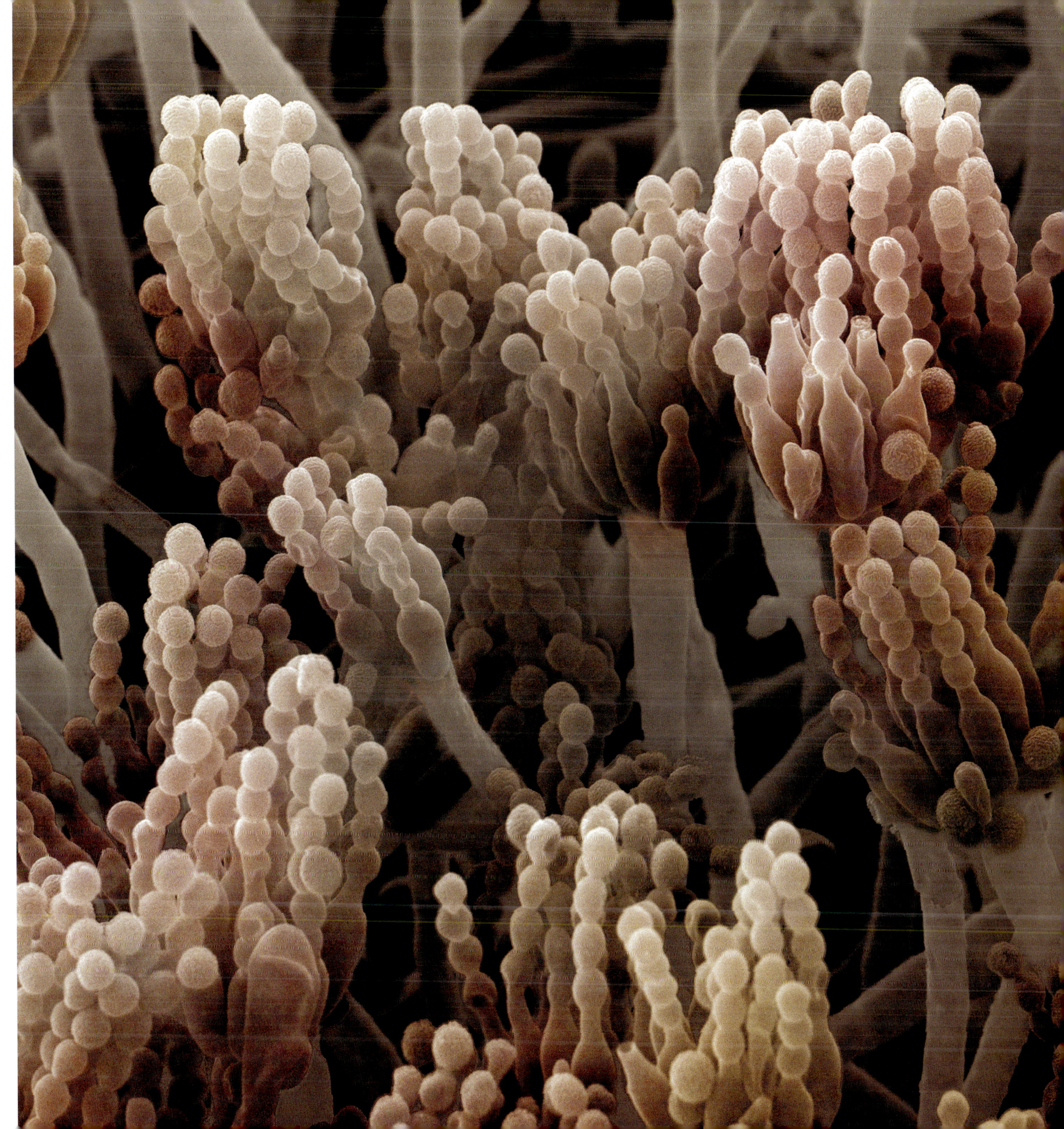

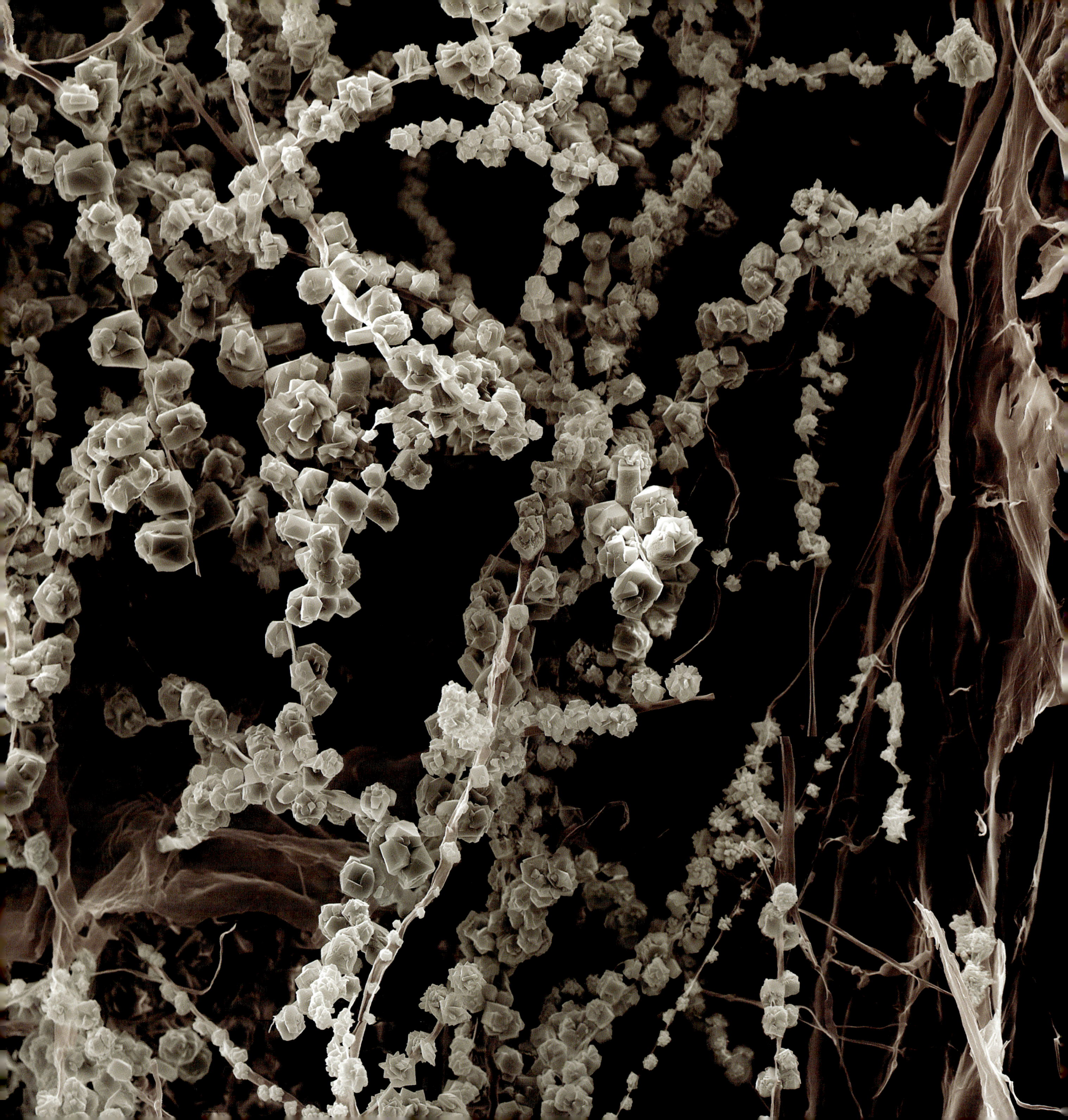

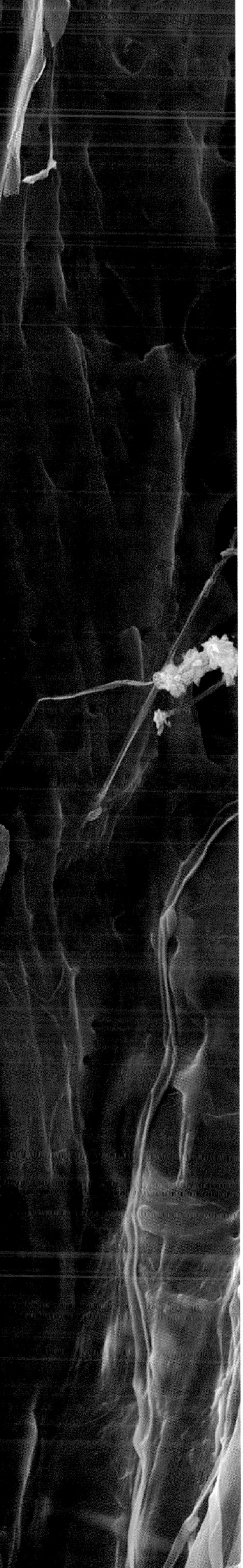

FUNGUS and MOULD

We may enjoy houseplants but these, together with vegetables and fruits that we carry into the home, can be an entry route for many kinds of creatures, including moulds and fungi. Fungal spores also waft into the house on air currents, and given the right atmospheric conditions will germinate, resulting in moulds on breads and jams, those vegetables that have been left in the rack for too long, and both wet and dry rot which can cause serious problems within the fabric of a building.

MOULD

Moulds are fungi of the class *Deuteromycetes* that appear as masses of unsightly grey, green or black hyphae growing on damp surfaces, such as walls or ceilings, giving off an obnoxious musty smell. Moulds spread by releasing millions of tiny spores into the air, which are so numerous they are everywhere, in every room of every house and probably every breath we take. These spores are dormant until, given the right conditions, they can germinate. They need moisture to grow and the resulting mould is usually found in damp, poorly ventilated areas of homes, such as bathrooms, where there is frequent condensation, although it can also grow as a result of rising or penetrating damp in walls. Mould needs very little nutrient and will grow anywhere with sufficient moisture. In the long term, clothes, furnishings and carpets can be destroyed. Even a small contamination can be a serious health risk causing asthma or other respiratory diseases. When inhaled, airborne mould spores can produce allergic reactions similar to hay fever. Mould also attacks foodstuffs resulting in dangerous toxins, which can, in turn, give rise to food poisoning.

WET ROT

Wet rot is caused by a number of fungi, among them the cellar fungus, *Coniophora puteana*, which is identified by dark brown threads spreading over the surface of the affected wood. Wet rot is a natural process of timber decay in the presence of high levels of moisture. In a house, the problem is almost always due to a structural defect, where the wall adjacent to the affected timber is damp, or the surface paint is damaged allowing the wood to absorb moisture.

DRY ROT

Dry rot is caused by a wood-destroying fungus, *Serpula lacrymans*, found in most temperate parts of the world. In nature, it affects forest timbers helping to break it down to recycle the wood, but dry rot is best known for its ability to destroy timbers in ships and buildings, and causes widespread structural damage. Despite its name, dry rot will only affect timber that is damp, often where the timber is in contact with, or embedded in, wet brickwork or masonry, approximating the conditions in which it naturally thrives. The fungus breaks down the cellulose in the wood and produces water, growing outwards, across and even through bricks and mortar, to other timbers in unventilated conditions, carrying the moisture with it. The decayed timber crumbles into cube shapes as it dries and eventually disintegrates. The spreading mycelium, or body of the fungus, is white and fluffy, and emits a musty odour. Where damp conditions prevail, fruiting bodies develop. These grow into a pancake shape with a yellowish colour that darkens as it matures to produce a dust of rust brown spores. The fungus cannot survive in temperatures over 25° Centigrade especially if the atmosphere is dry.

Fungus, *Sporothrix schenckii*

Sporothrix is a fungus found worldwide, in soil and on decaying wood where there is plenty of moisture. Young growth is white, later becoming a dirty candle wax colour, with a leathery surface. It produces spores either in rosette-like clusters or directly attached to the hyphae (see left). Contact with this fungus on scratched or broken skin can cause sporotrichosis, a chronic subcutaneous infection that spreads through the lymph system. It begins with a small swelling, followed by additional bumps across the body that may develop into open lesions. It is sometimes called 'rose-handler's disease', since the fungus may grow on rose thorns and be introduced into the body via thorn pricks.

Magnification: x2150   *Scanning Electron Micrograph*

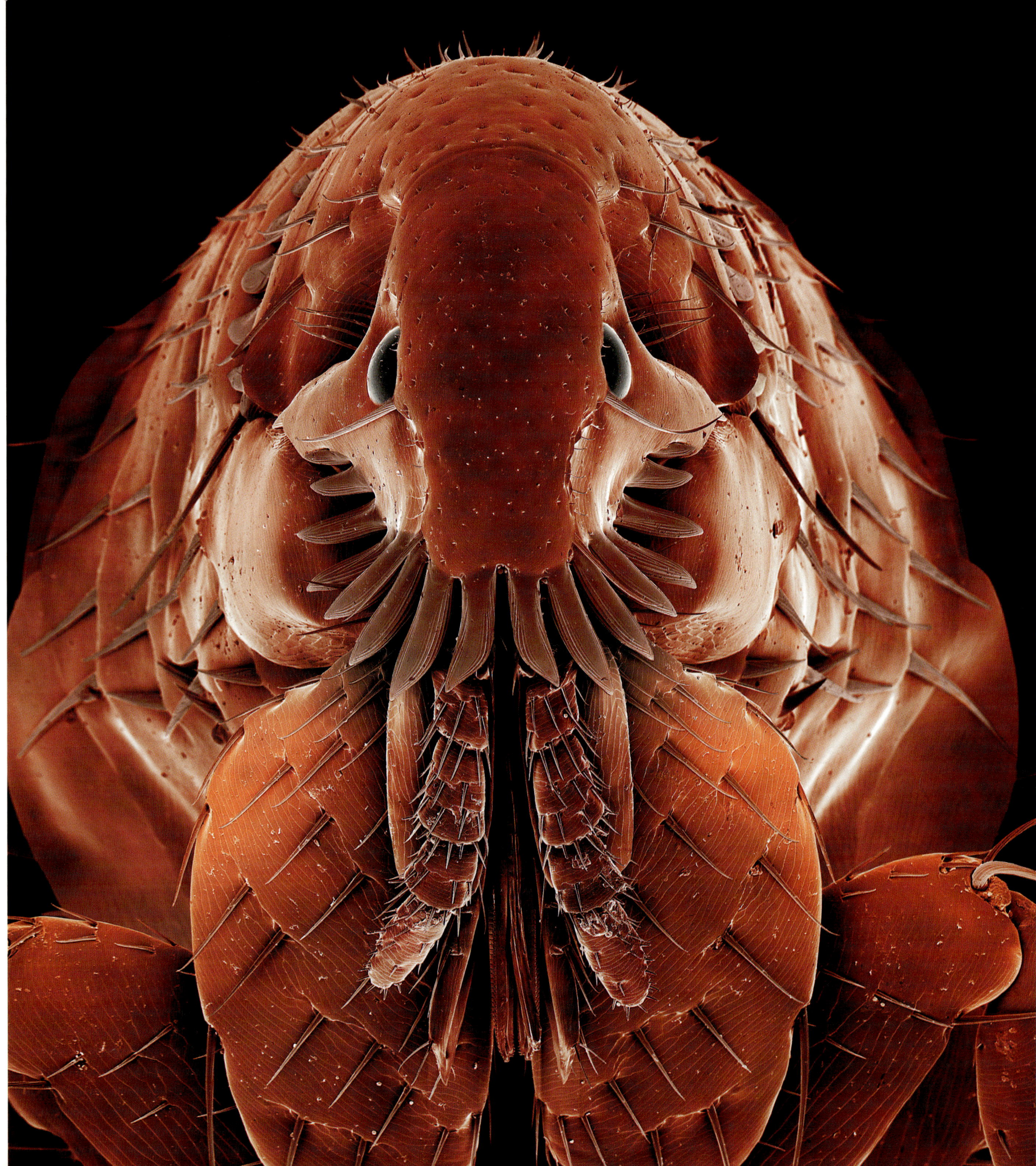

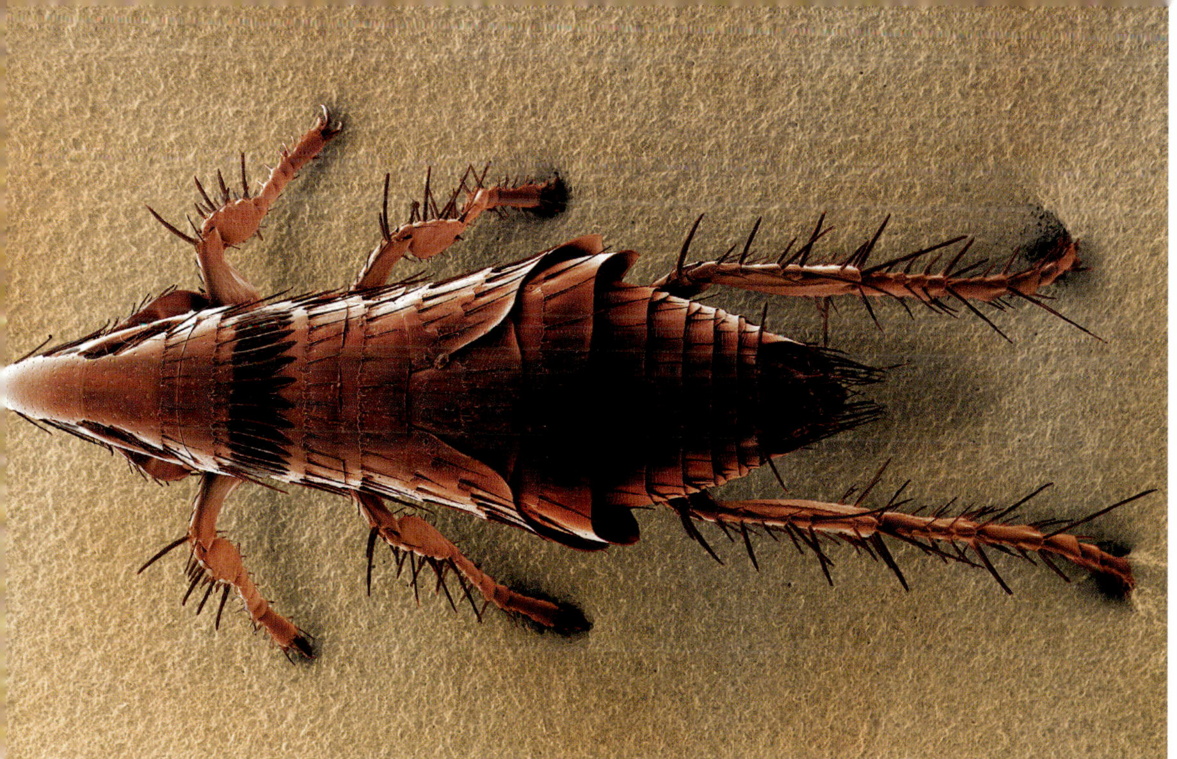

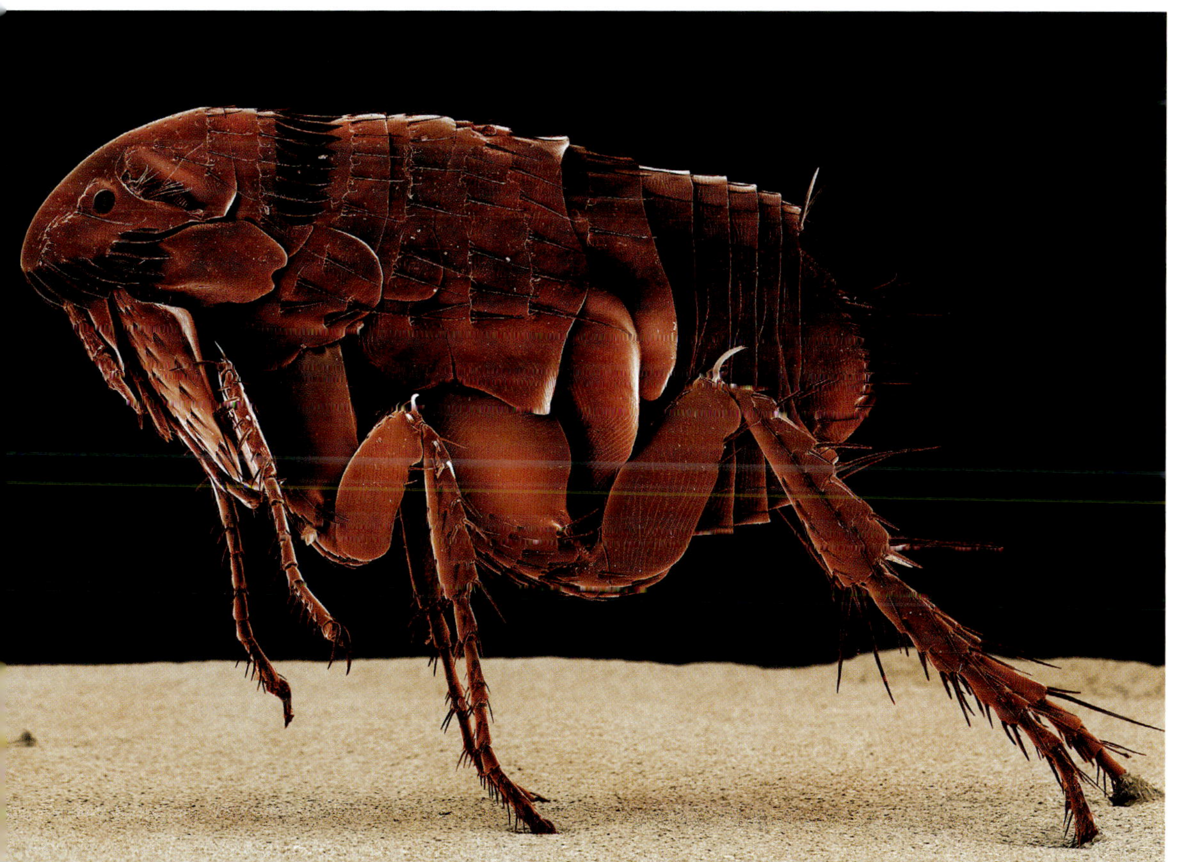

Cat Flea, *Ctenocephalides felis*

Streamlined, armoured and covered in bristles, the cat flea is the most common flea pest found in the home. Long hind legs, used for leaping, make the flea one of nature's greatest athletes. Just behind the eyes are dark pits, which bear heat-sensitive organs.

Magnification: x250   *Scanning Electron Micrograph* (left)
Magnification: x40   *Scanning Electron Micrograph* (above)
Magnification: x40   *Scanning Electron Micrograph* (below)

CAT FLEAS

For those of us who have pets, the summer months can bring fleas into our homes, as hitchhikers on our dogs or cats. When it is warm, populations of fleas expand rapidly, making lives miserable for many wild and domestic animals. Although there are many species of fleas, it is the cat flea, *Ctenocephalides felis* that most frequently bites us. An infestation can occur with little warning for our beloved pet may harbour a few without us even realising it. Once in the home, fleas will hide deep within the confines of carpeting, rugs, soft furnishings and in pet bedding, making their elimination extremely difficult. They cannot breed without a blood meal from their primary host (in the case of a cat flea, a cat) but even so, cat fleas are masters of survival, happy to subsist on a diet of human blood or even stay alive for a year without feeding. The first sign of an infestation in the home is usually the presence of unexplained bites on our ankles, or the sight of a flea landing on us after leaping from the carpet, before scurrying across the surface of our skin looking for a place to feed. Fleas can jump vertically upwards to about 15 cm, about 75 times the length of their bodies, equivalent to a person jumping nearly a 150 metres up into the air. This is by no means a world record for insect jumps but it is very impressive and it makes the capture of individual fleas very difficult.

In the warmth of the home fleas reproduce rapidly and, as an infestation deepens, pets experience uncontrollable itching. The complete metamorphosis from egg, through larva and pupa to adult can take as little as two weeks. Following a blood meal on a cat, a female can produce up to 20 eggs per day, depositing them in the fur, but eggs are often scratched or shaken off. Eggs hatch after about two days and the larvae migrate into floor cracks or carpet pile, under rug edges, furniture or beds. The pale, bristly, blind larvae do not suck blood, but feed on digested blood from adult flea droppings, dead skin, and other organic debris. The larval stage lasts from one to three weeks before pupating. The adult emerges either within a few days, or it can remain in its cocoon for a year or more, waiting for a stimulus to emerge, triggered by the warmth or vibration of a potential host nearby.

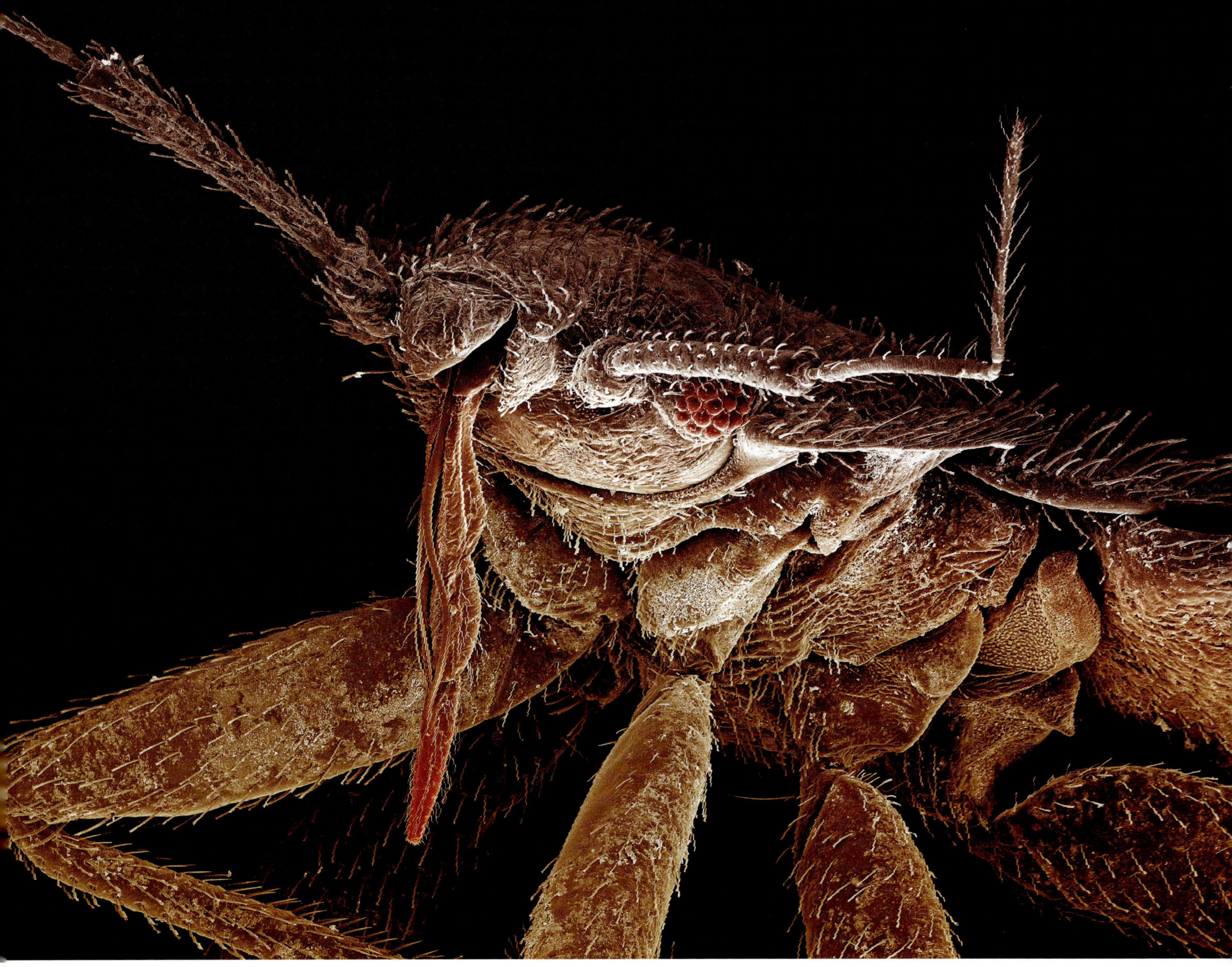

Bedbug, *Cimex lectularius*

The piercing sucking mouthparts of the bedbug (see above) are used to penetrate human skin and suck blood. Mating (see right) is achieved by traumatic insemination; the sharp sex organ of the male is inserted into the female through her abdominal wall at any convenient spot, injecting sperm directly into her body cavity.

Magnification: x100   *Scanning Electron Micrograph* (above)
Magnification: x150   *Scanning Electron Micrograph* (right)

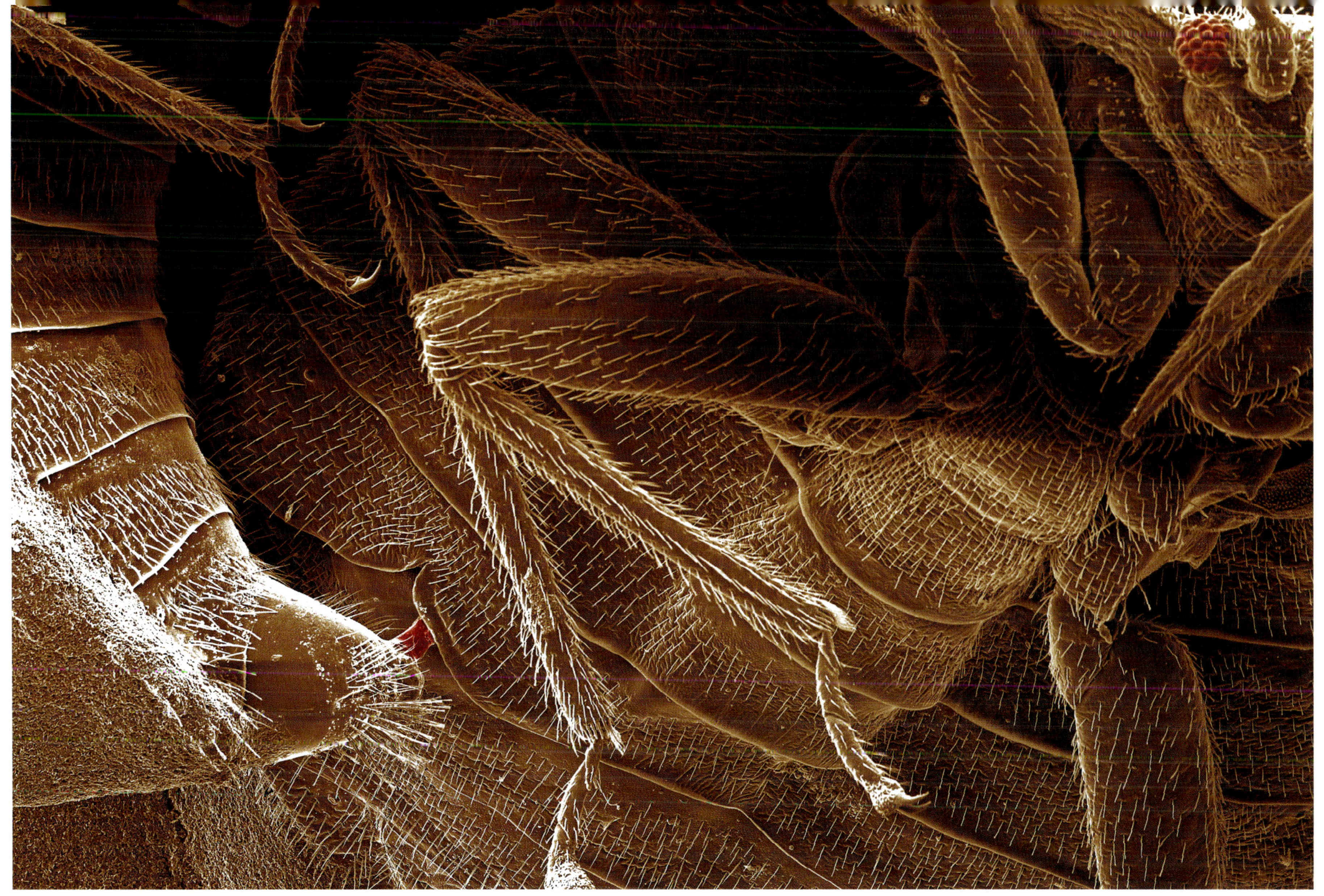

BEDBUGS

If the idea of mites living in our beds to feed on dead skin is repugnant, then how about bedbugs? These tiny masters of concealment, up to 5 mm long, are voracious blood feeders, and they do it while we are asleep. During the day, they hide in the folds of bed sheets, seams of mattresses, cracks in furniture, or somewhere close by, perhaps behind loose wallpaper, under floorboards or in wall crevices. They are very adept at this since their oval-shaped bodies are quite flat, making it easy for them to squeeze into tight places. But at night, they emerge to feed, attracted to the warmth and carbon dioxide emanating from a potential host. After locating a suitable feeding site, the victim's skin is held by the forelegs of the bug and pierced by the mouthparts, simultaneously injecting saliva, which contains an anticoagulant to keep the blood flowing. Although it may take several minutes to complete feeding, the victim is normally unaware of being bitten.

Bedbugs, *Cimex lectularius*, 4-5 mm in length, are found worldwide and have been associated with human habitation for centuries; they are even mentioned in classical Greek writings from the time of Aristotle. Although they are usually associated with substandard housing and poor hygiene, this is not always the case; they can be introduced on clothing or luggage so even the cleanest hotels are prone to infestation given the frequent and varied nature of paying guests. Their presence can be detected by dark spots of dried excrement found along mattress seams or spots of blood on bed linen. An adult female will lay two to three tiny eggs every day; these stick to rough surfaces and the nymphs hatch within ten days, growing to adulthood in six to eight weeks. They can survive for long periods without feeding if neglected, and an infestation will grow rapidly, usually accompanied by a sweet musty smell. Bedbugs are not proven vectors of any human diseases, but they cause a great deal of discomfort, distress, sleep deprivation and embarrassment to those who come into close contact with them.

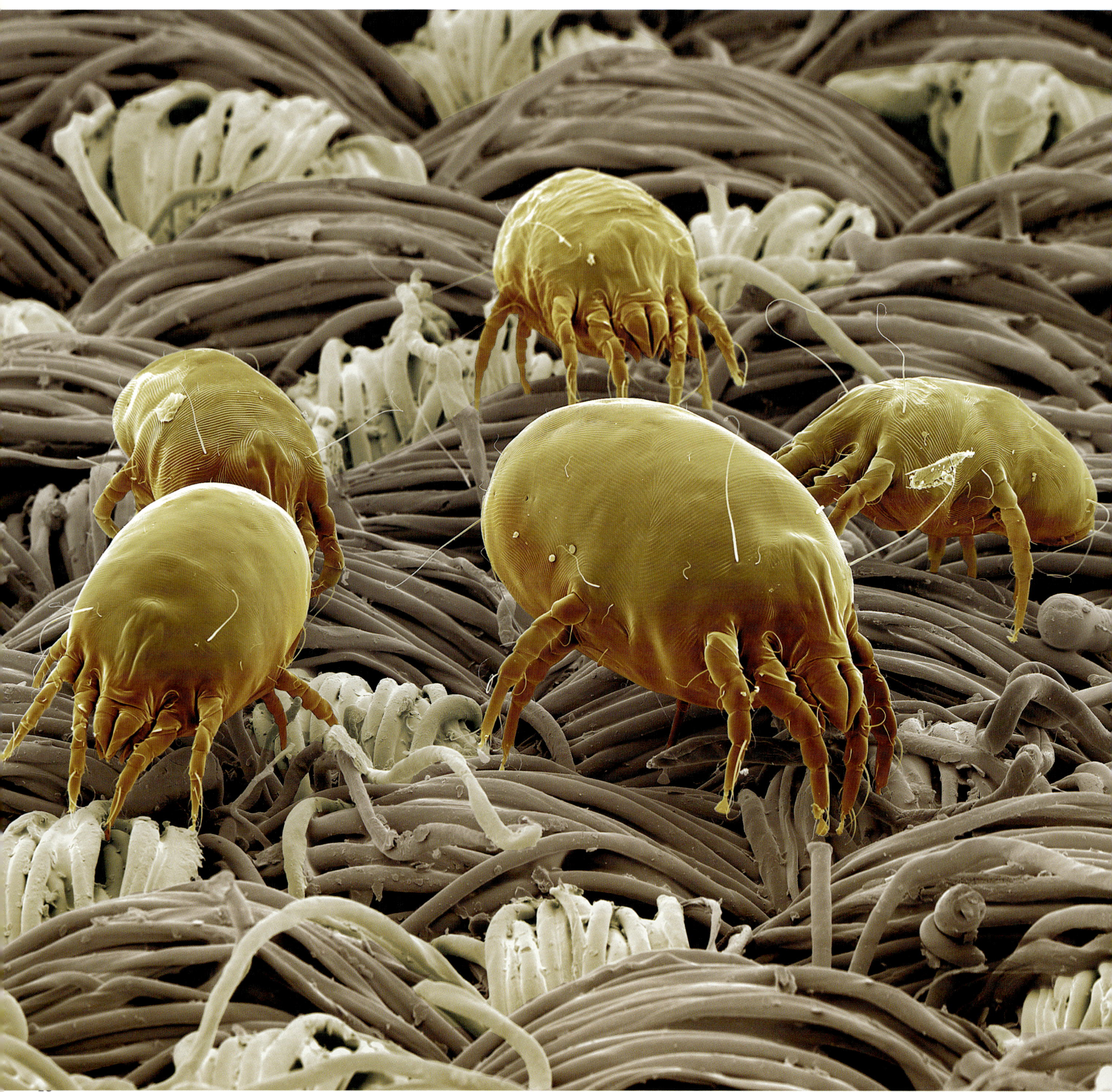

MITES

Several species of mites of the family *Acaridae* attack foods such as grain, flour, dried fruits, cheese, dried milk, or sugar if they have been stored for a long time under damp conditions. In serious cases, the surface of infested foods may visibly heave due to the enormous number of mites present, together with other mite debris; dead bodies, shed skins and droppings in and around the foodstuffs. The most common food mites are the grain mite (*Acarus siro*), the cheese mite (*Tyrolichus casei*), the mould mite (*Tyrophagus putrescentiae*), and the flour mite (*Tyroglyphus farinae*). These creatures are barely visible to the naked eye, being less than 0.5 mm long and more or less translucent.

There is another mite that frequents our homes and prefers a much closer relationship with us. Much of the dust in our living space originates from the flakes of dead skin that we continuously shed from our bodies. We produce huge volumes of it, helping to fill the vacuum cleaner bag. Dust mites, *Dermatophagoides pteronyssinus*, feed on this and other organic debris, making themselves comfortable in our beds, in carpets and soft furnishings. They probably occur in most houses, sometimes reaching huge numbers, in fact an old pillow may contain thousands of living and dead mites, along with faecal pellets. Their presence can contribute to skin allergies or asthma, due to potent proteinases, or digestive juices, from the mite gut. The mites themselves are tiny, less than 0.2 mm long, and almost translucent, so they are effectively invisible. The extent of any infestation depends on standards of cleanliness, and can be kept to a minimum by frequent and thorough vacuum cleaning, including bedding.

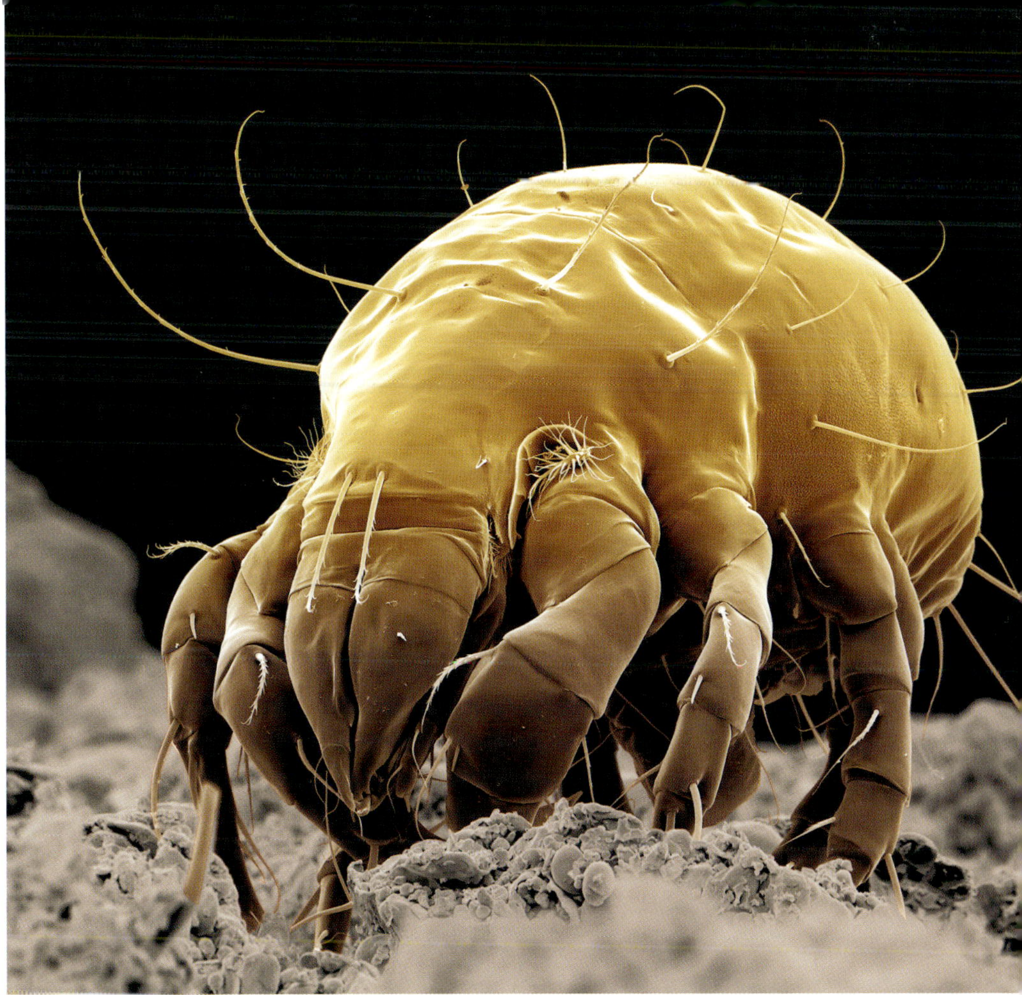

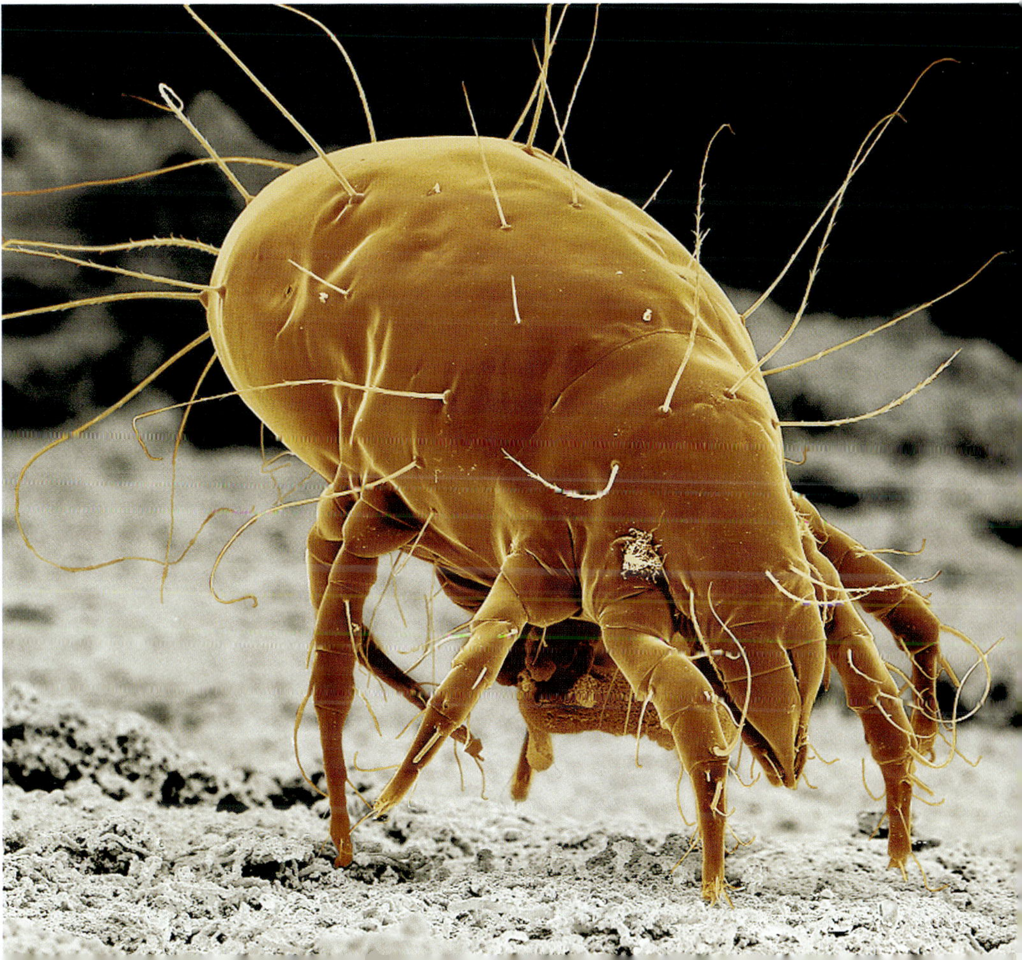

Dust Mite, *Dermatophagoides pteronyssinus* (left)

A group of dust mites forage for flakes of human skin on the surface of a bedsheet. Live mites, their faeces, moulted skins and dead bodies can account for a significant proportion of the volume of an old pillow, and have been implicated in causing skin allergies and asthma.

Magnification: x350
Scanning Electron Micrograph (left)

Several species of mite can infest stored foodstuffs in the home, among them the **Flour Mite,** *Acarus siro,* (above right), and the **Cheese Mite,** *Tyrolichus casei* (below right).

Magnification: x250
Scanning Electron Micrograph (above right)
Magnification: x250
Scanning Electron Micrograph (below right)

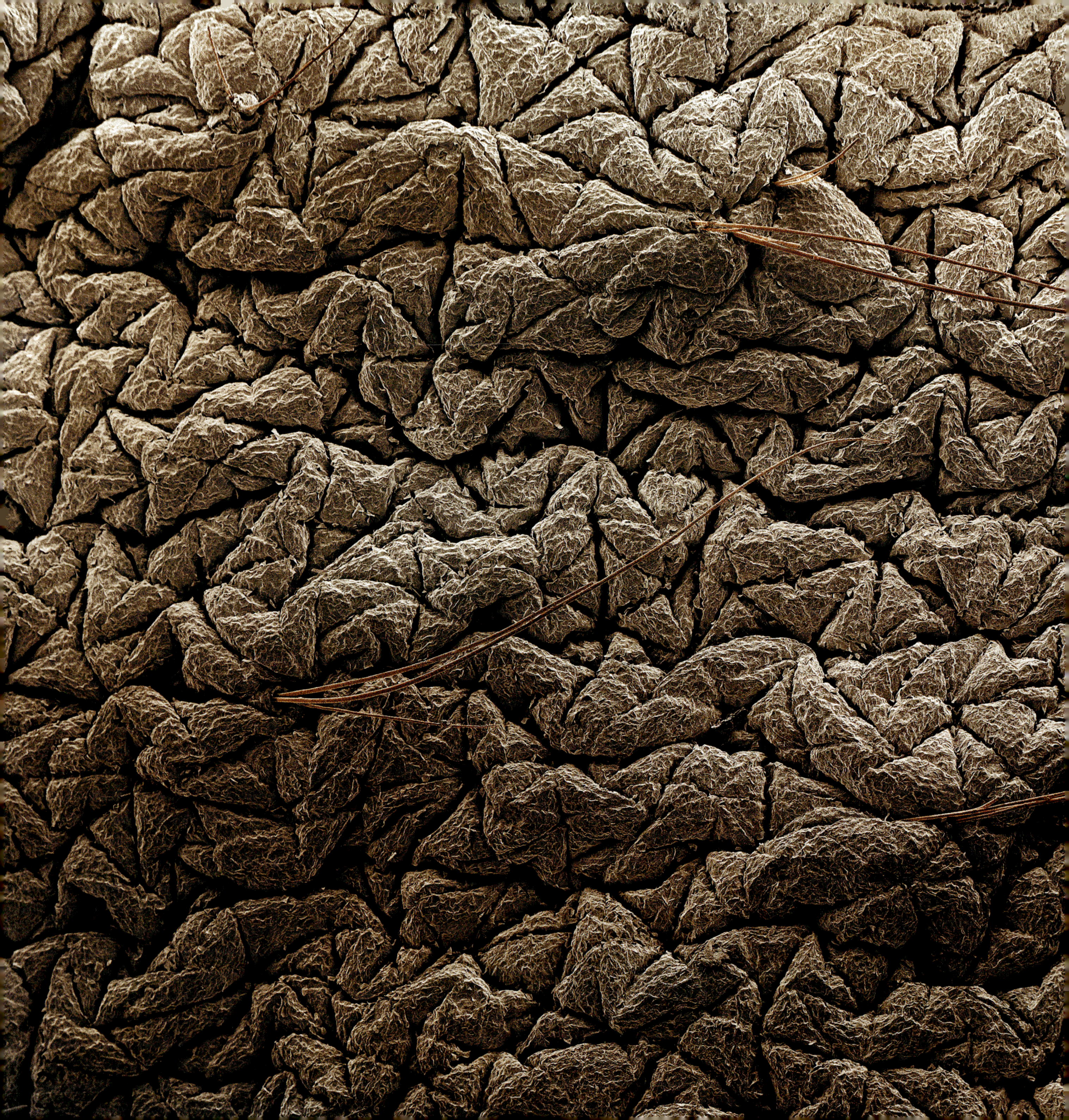

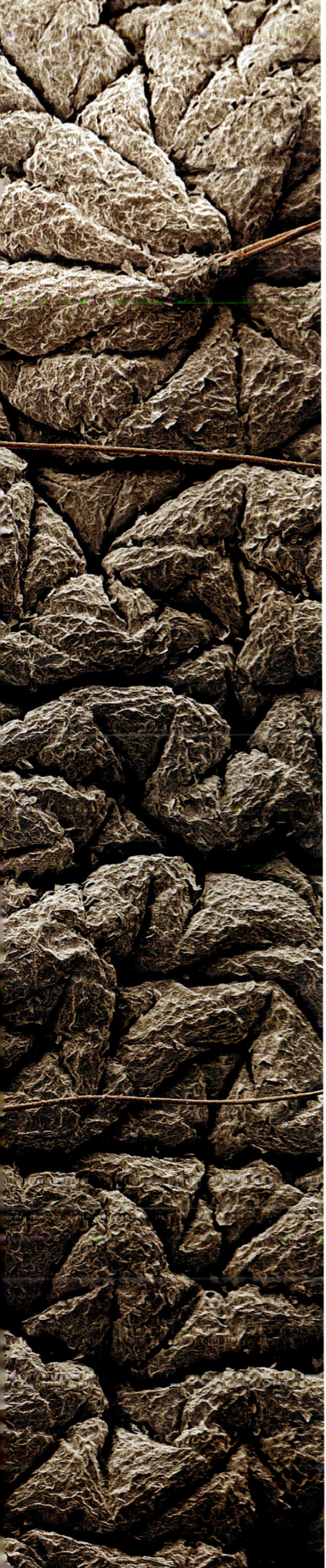

INTIMATE COMPANIONS

It is a hostile place. Its crumbling, cracked surface features broad expanses of desert, dense forests, and deep canyons; from time to time, it is inundated by flash floods, burned by the sun, or buried in the dark. At times, it even changes shape due to sub-surface upheavals. But this is not an alien world. Take a powerful magnifying glass, and look closely at your skin. Now just imagine you are a microscopic creature trekking across it, prospecting for a new home; the surface of the human body is a craggy, challenging environment of extremes that would require very special adaptations to enable survival. The interior of our bodies is even more inhospitable – dark, sometimes air-less, and deluged by various unfriendly fluids and gases. Despite this, and perhaps surprisingly, our bodies are heaving, both inside and out, with a myriad of tiny, uninvited guests, and yet we are rarely aware of most of them.

Our bodies may not seem very habitable, but in fact they present a huge variety of lodging opportunities. The relationships we have with the guests who come to live on us or in us are symbiotic – we provide them with a home and, in return, they may either offer us some benefit, or cause us harm. Some of our most intimate companions should be welcome, for they help our bodies to function more efficiently. A normal human intestine, for example, might contain some 100,000,000,000,000 (10^{14}) bacterial cells - 10 times more than the number of cells that make up our entire body. Some of these bacteria are only temporary visitors, but altogether they comprise some 800 different species, and a healthy digestive system needs this valuable cocktail of bacteria to aid digestion and assist in the production of vitamins. This is symbiosis in its purest form, and is called mutualism, since it benefits both us, and our tiny guests.

Some of the creatures that live on the surface of our bodies may not be welcome, but they do not necessarily cause us any harm. Many skin bacteria, by virtue of their antibiotic effect, help to protect us from infection by more harmful strains. Tiny mites on the surface of the skin may from time to time feed on dead skin cells, and while we may not even be aware of their presence, the mites derive great benefit. This form of symbiosis is called commensalism. However, the most harmful form of symbiotic relationship is parasitism. Parasites attract our attention the most since they are not only of no benefit to us, but we can be significantly disadvantaged or even seriously harmed. In some cases, unless steps are taken to neutralise them, these uninvited guests can cause pain or even life-threatening symptoms. A successful parasite should be a specialist in the art of survival, and should not kill its host before it has been able to reproduce. From the parasite's point of view, it is important to keep the host well enough to continue to be a source of food, shelter and to offer the right conditions for breeding. The study of parasites is important because of the impact it has on human health and wellbeing. While one kind of parasite might be harmless on its own, it may be acting as a convenient carrier for another that does us harm. They often have complex life cycles, using one or more vectors that help to distribute them far and wide. Today, with travel to distant places easier and more frequent than ever before, many parasites that were previously confined to certain parts of the world have become international hazards. Outbreaks of potentially serious tropical diseases are becoming commonplace in Britain and Europe as people returning from trips abroad unwittingly arrive home carrying a few unwanted guests.

Human Skin

The largest organ of the human body, the skin provides a protective barrier for the delicate organs within. Insulating against heat and cold, and wear and tear, it is soft, pliable, waterproof, and self-repairing. It oozes oils and other fluids to help its maintenance and is constantly regenerated by new cells growing from inner layers, while older cells on the surface are constantly worn away.

Magnification: x1100   *Scanning Electron Micrograph*

Of all the nuisance creatures on Earth, the mosquito is at the top of the list. In Britain, there are two distinct kinds, *Culex*, the most common, and *Anopheles*. They both thrive in the warmth of summer, irritating us with their whining flight and making us itch when they bite. Both male and female mosquitoes feed on nectar and fruits, but the females supplement their diet with blood, in order to provide protein for their eggs. Adult mosquitoes live for just a few weeks; a male is attracted to a female by the tone of her wing beat using feathery antennae that act as sensory receptors, and mating normally occurs on the wing. Once mated, the female immediately searches for a blood meal, following sensory cues such as carbon dioxide and body heat to home in on a victim. Her mouthparts form a long, serrated proboscis for piercing the skin, and her saliva contains a mild local anaesthetic and an anti-coagulant, which are injected into the wound as she feeds. The introduction of these foreign proteins generates an immune reaction in our bodies that results in the familiar inflamed, itchy spot at the site of the bite. After feeding, the female seeks out a resting place to digest the blood meal, after which the eggs are laid on the surface of any available standing water, such as a pond or water butt. *Anopheles* eggs are laid singly, while those of *Culex* are deposited in batches of 70-100, in the form of a floating raft. After hatching, *Culex* larvae hang just below the surface of the water breathing air through snorkels, while those of *Anopheles* lie parallel to the surface breathing through special openings. If disturbed they dive in a wiggling motion then float back to the surface. The larvae feed on algae, and after 7-14 days turn into comma-shaped pupae, the least active stage of the lifecycle. After 2-4 days, the pupa metamorphoses into an adult mosquito. The adults emerge by splitting the pupal case, resting for a few minutes on the water surface to dry and expand the wings before flying away.

There are around 3,500 species of mosquitoes worldwide, many of them relatively harmless. However, since a few are carriers of serious diseases such as malaria, mosquitoes could be considered the most potentially dangerous animals known to mankind. Malaria is certainly nature's biggest killer of humans; at current estimates, it affects some 300 million people globally, and kills up to 1.5 million each year. It is only transmitted by *Anopheles* mosquitoes, of which there are around 60 known species proven to carry the disease. Today, there is no indigenous malaria in Britain, but from Roman times until as recently as the 19th Century, the fatal disease was rife around the swampy Fenlands of East Anglia and other marshlands of southern and eastern England, where it was known as "marsh fever" or "ague". In the early 20th Century, it was brought back to Britain by infected soldiers returning from the First World War, and more recently by adventurous travellers, occasionally spreading through local communities for short periods, spread by the British species of *Anopheles* mosquito. Due to global warming trends, it is possible that malaria will soon be in Britain again to stay.

It was only in 1889 that the cause of malaria was identified as the protozoan *Plasmodium*, and only in 1897 was the *Anopheles* mosquito discovered to be its carrier. Exactly when the malaria parasite historically arrived in Britain is a mystery, but by 1649, a cure for the disease was being imported into Britain in the form of an extract from an obscure South American tree. It had been discovered for the outside world by Jesuit priest Agostino Salumbrino, who in 1630 was working in Peru where Quichua Indians were using the bark of the *Cinchona* tree to alleviate shivering, one of the symptoms of malaria. To prepare the medicine, the Indians dried the bark, ground it to a fine powder, and then mixed it with a liquid. In 1631, Agostino sent a small quantity of the potion to Rome to test in the treatment of the disease, which was rife in the Pontine marshes south of Rome and the Maremma area of Tuscany. In the years that followed, *Cinchona* bark became one of the most valuable commodities to be shipped from South America to Europe. The bark was shown not only to be an effective muscle relaxant, calming the shivering, but an effective cure for the disease. It contained the miracle drug quinine, which works by binding itself to blood proteins, forming complexes that are toxic to the malarial parasite. *Cinchona* bark continued to be the prime source of quinine until the Second World War, when American scientists found a way to synthesise the drug artificially.

There are four kinds of malaria that affect humans. The parasite may have evolved as a parasite of animals since it is so harmful to humans, but probably spread with our species as we moved out of Africa to settle elsewhere. It has a complex life cycle, requiring two different hosts and involving sexual and asexual stages,

Mosquito, *Culex pipiens*

A common and familiar biting insect in Britain, and the most widely distributed mosquito in the world, *Culex pipiens* carries a number of serious diseases in warmer climates, among them elephantiasis, encephalitis and the West Nile virus. Only the female bites and she does so in order to provide protein for her eggs.

Magnification: x30   *Scanning Electron Micrograph*

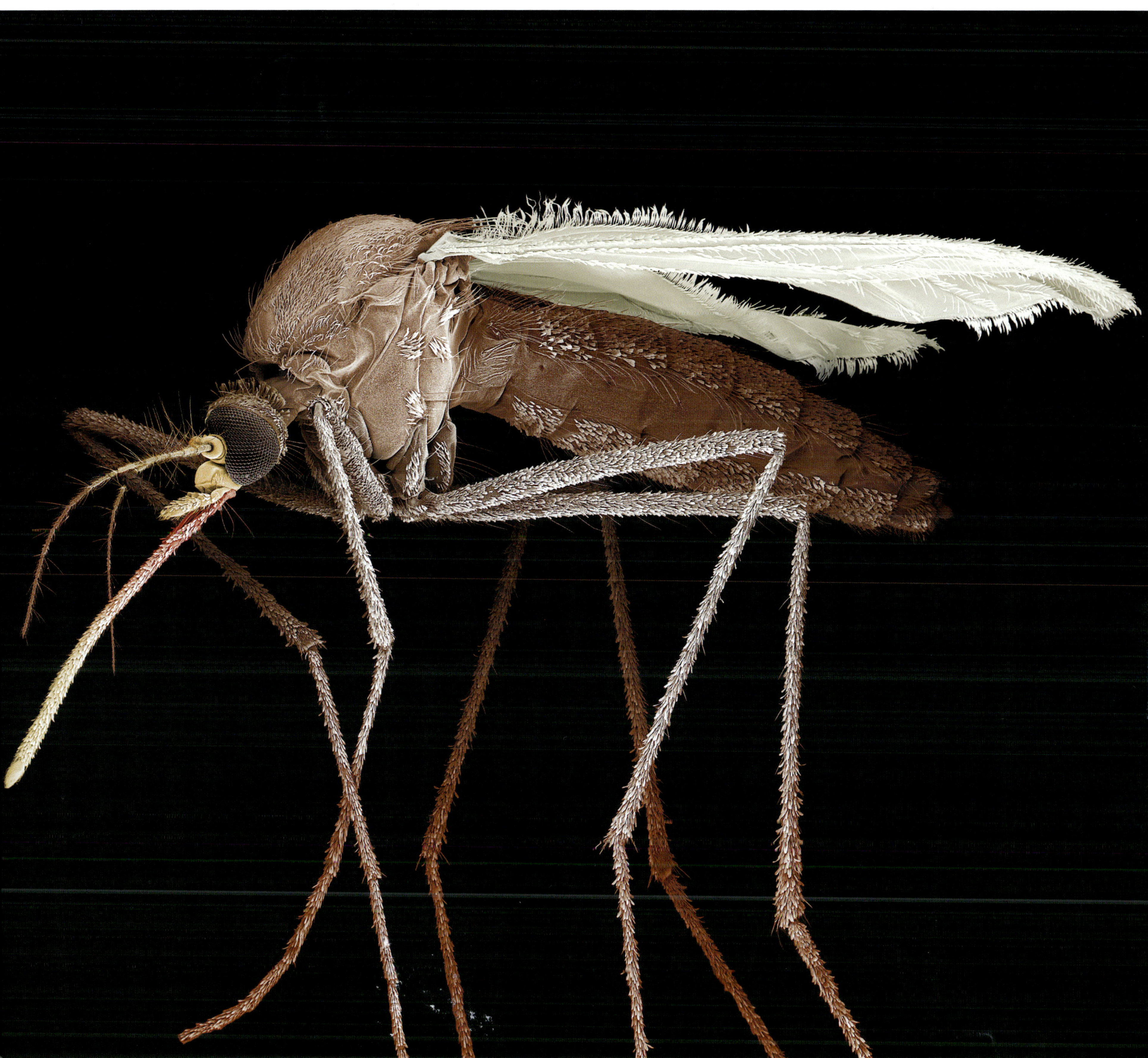

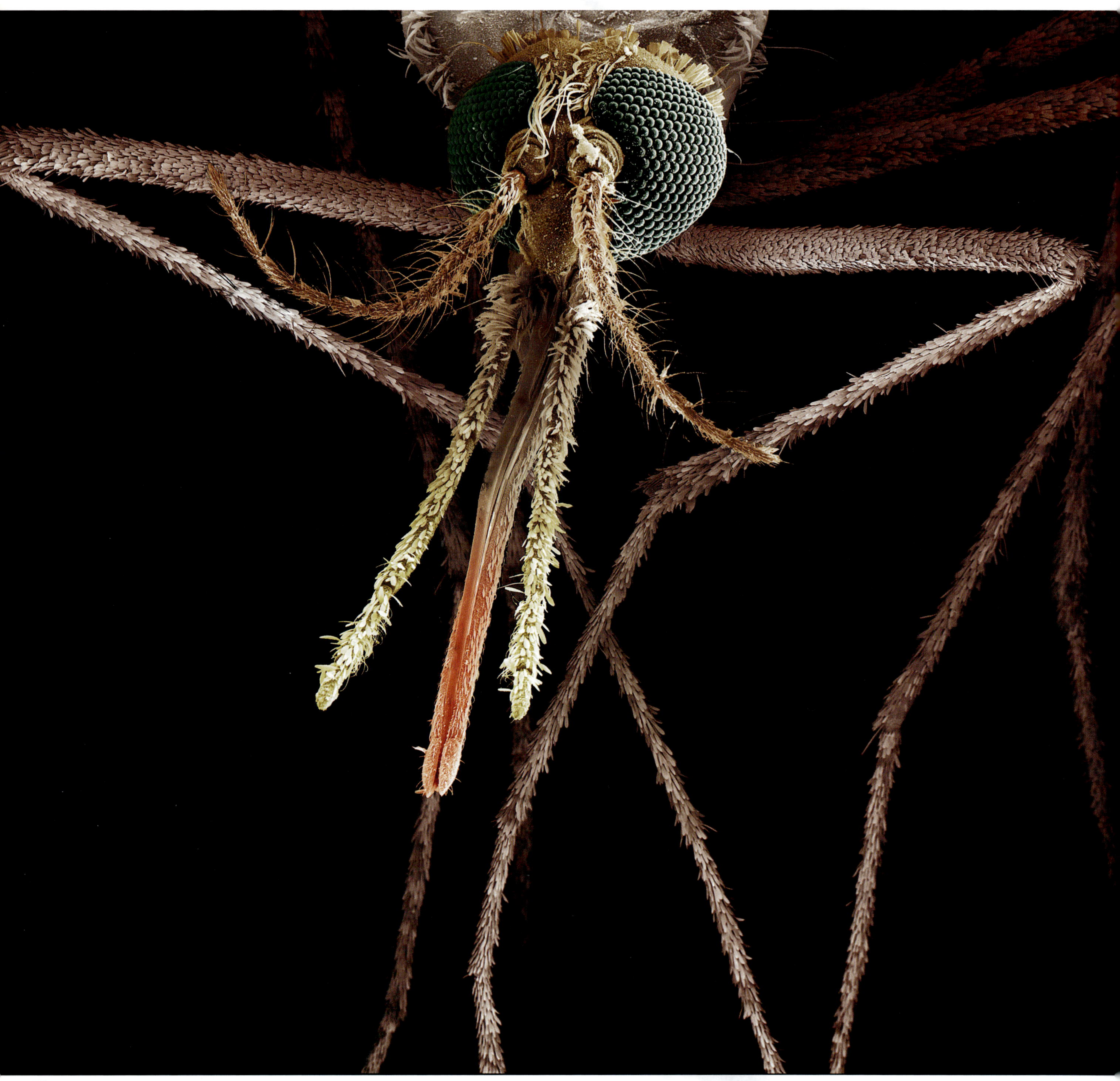

the sexual stage occurring in mosquitoes, and the asexual stage in humans. When an infected mosquito bites us, it injects saliva laced with active *Plasmodium* sporozoites, which make their way through the bloodstream to the liver. Once there they multiply and, after a week or two, they escape into the bloodstream to start invading and devouring red blood cells. This invasion repeats itself, and each time it induces fever as toxins are released from the bursting red blood cells. The periodic fevers occur with predictable regularity, and are accompanied by high body temperatures, intense headaches, delirium, and uncontrollable shivering. In cerebral malaria, the most deadly form of the disease, debris from destroyed red cells may obstruct blood vessels in the brain, often with fatal consequences.

BITING FLIES

Biting flies that attack us to obtain a blood meal occur in most parts of the world. Here in Britain, horse flies deliver a bite that can feel like a hot needle. Unlike insects that puncture the skin with needle-like mouthparts, horse flies have mandibles like tiny serrated scimitars, which they use to slice into the flesh, causing blood to seep out. They normally feed on nectar and sometimes pollen, but females need blood to obtain proteins to aid egg development. They have large compound eyes that serve as the primary sense for locating prey, and the flies are stealthy, usually waiting in shady areas for a potential victim to approach. They are attracted to large moving objects, and to exhaled carbon dioxide. When attacking humans they seem to target areas of the body outside our immediate visual range, such as the back or legs, where they can proceed with their task unseen.

In warmer parts of the world, biting flies come in all shapes and sizes, from sand flies and grassland flies so small they are

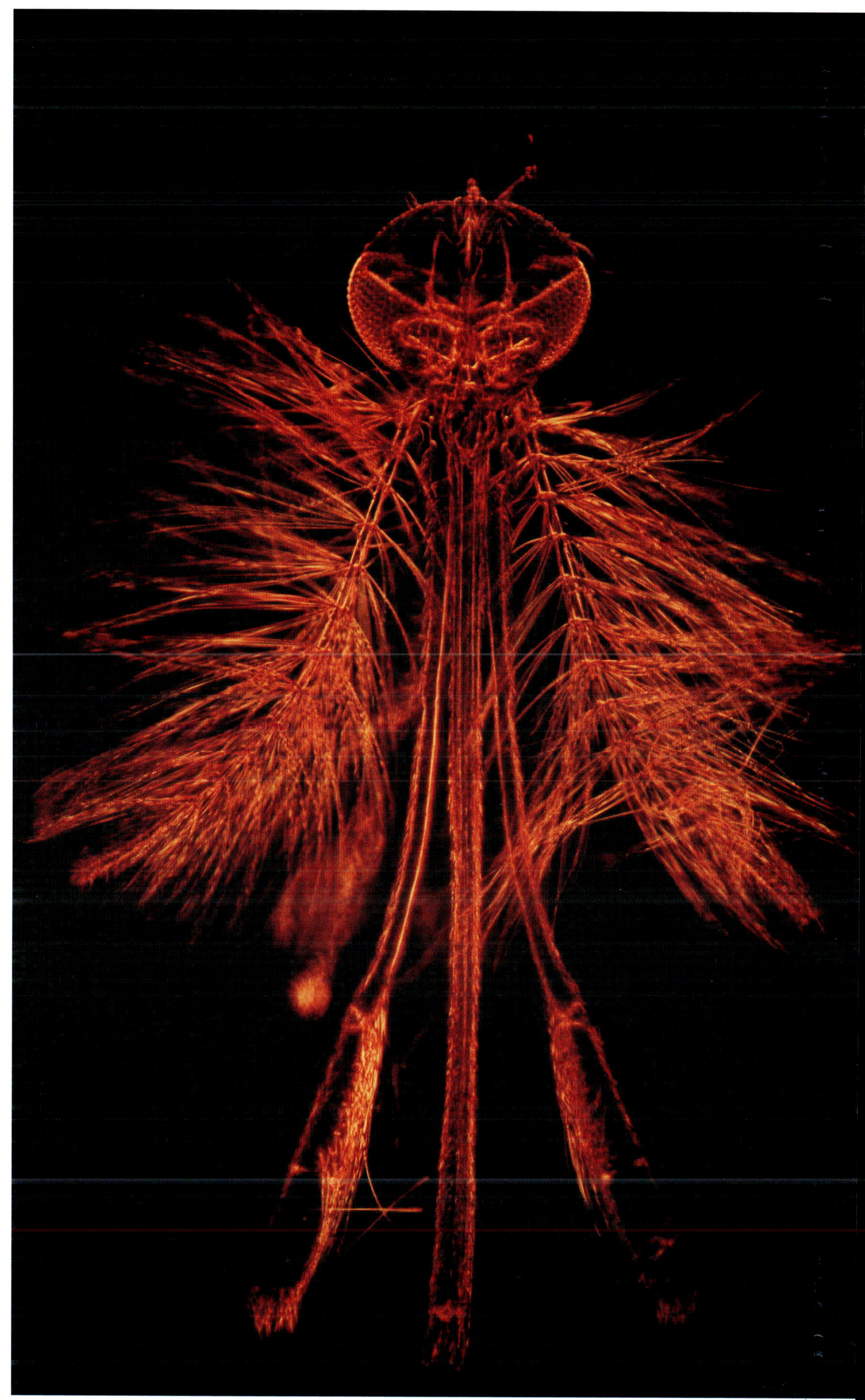

Mosquito, *Anopheles sp.*

Only **female *Anopheles* mosquitoes** (left) bite and transmit the malarial parasite *Plasmodium sp.*. **Males** (right) feed on nectar, and have bushy antennae to detect the sound and scent of females.

Magnification: x100
Scanning Electron Micrograph (left)
Magnification: x60
Dark Field Light Micrograph (right)

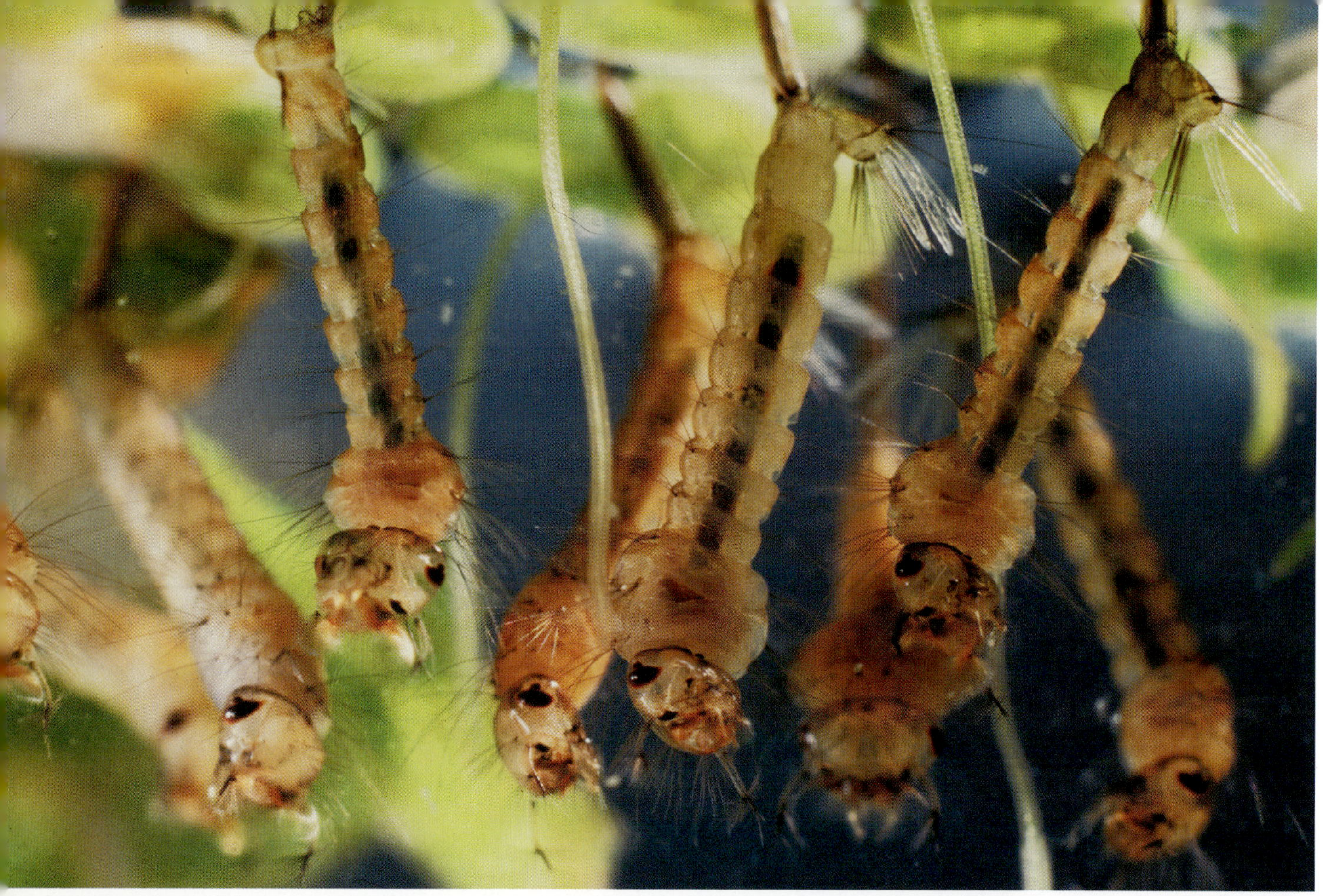

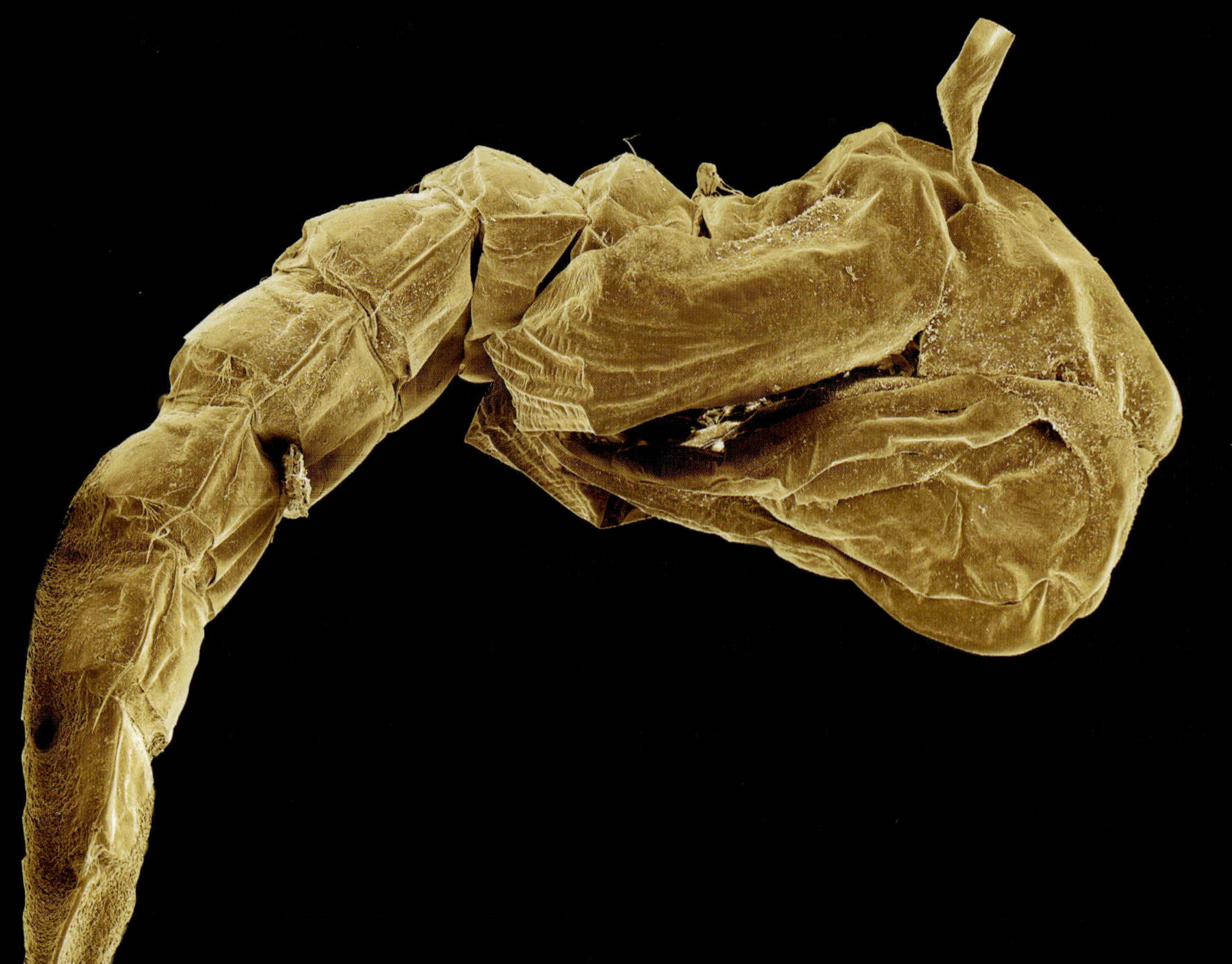

Mosquito larvae,
Culex pipiens (left)

Culex larvae hang from the surface of water, with a breathing tube, or snorkel, to breathe air. They wriggle to dive if disturbed. They feed on plankton or tiny particles of organic detritus filtered out of the water using their bushy mouthparts.

Magnification: x11
Macro-photograph

Mosquito pupa and larvae,
Anopheles sp.

The pupa (below left) breathes air through a pair of trumpet-shaped snorkels, one of which can be seen in the photograph. It hangs just below the water surface until the adult mosquito emerges by splitting open the outer cuticle. **The larvae** (right) breathe through special openings in the body wall.

Magnification: x65
Scanning Electron Micrograph (below left)
Magnification: x50
Scanning Electron Micrograph (right)

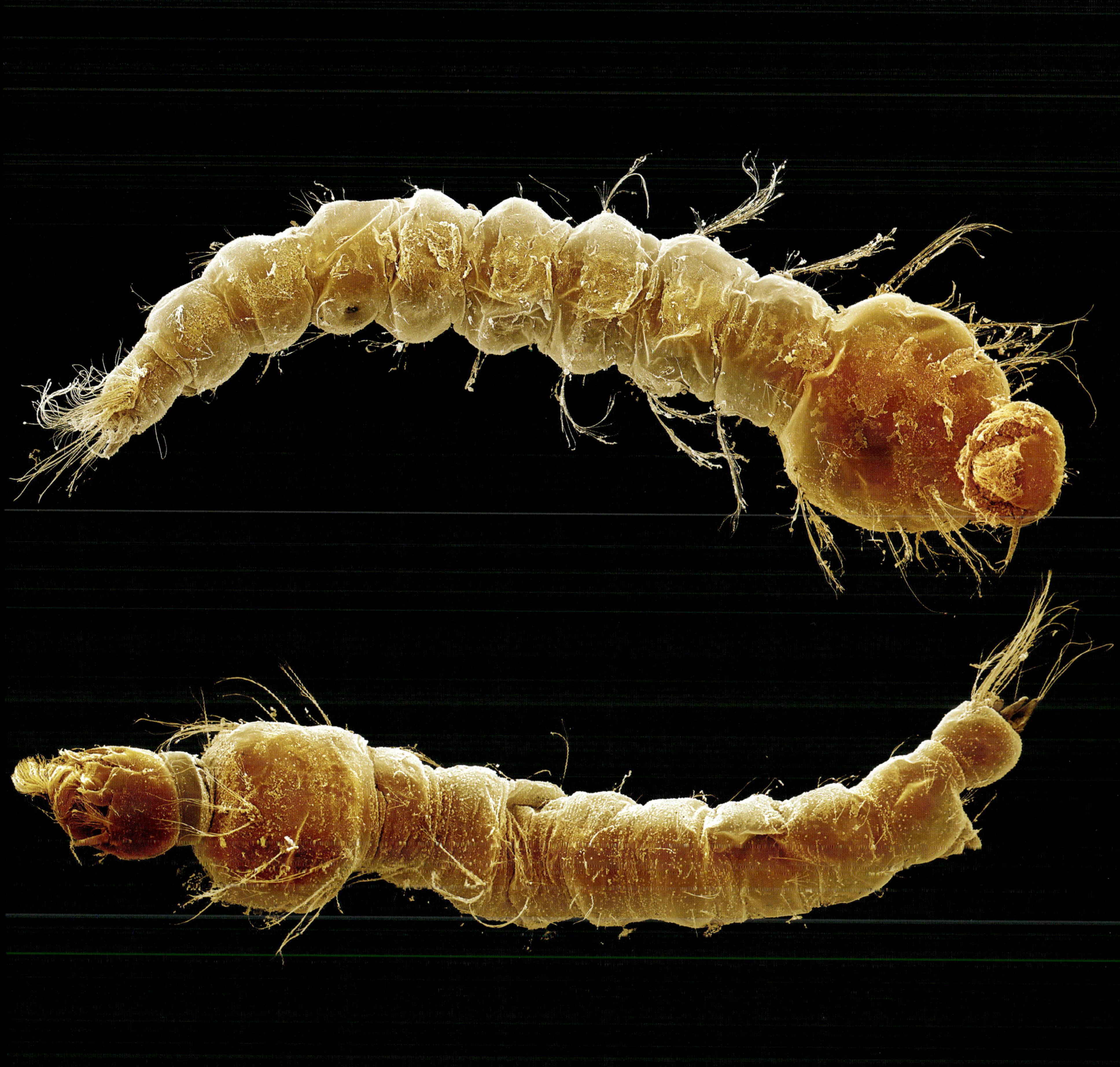

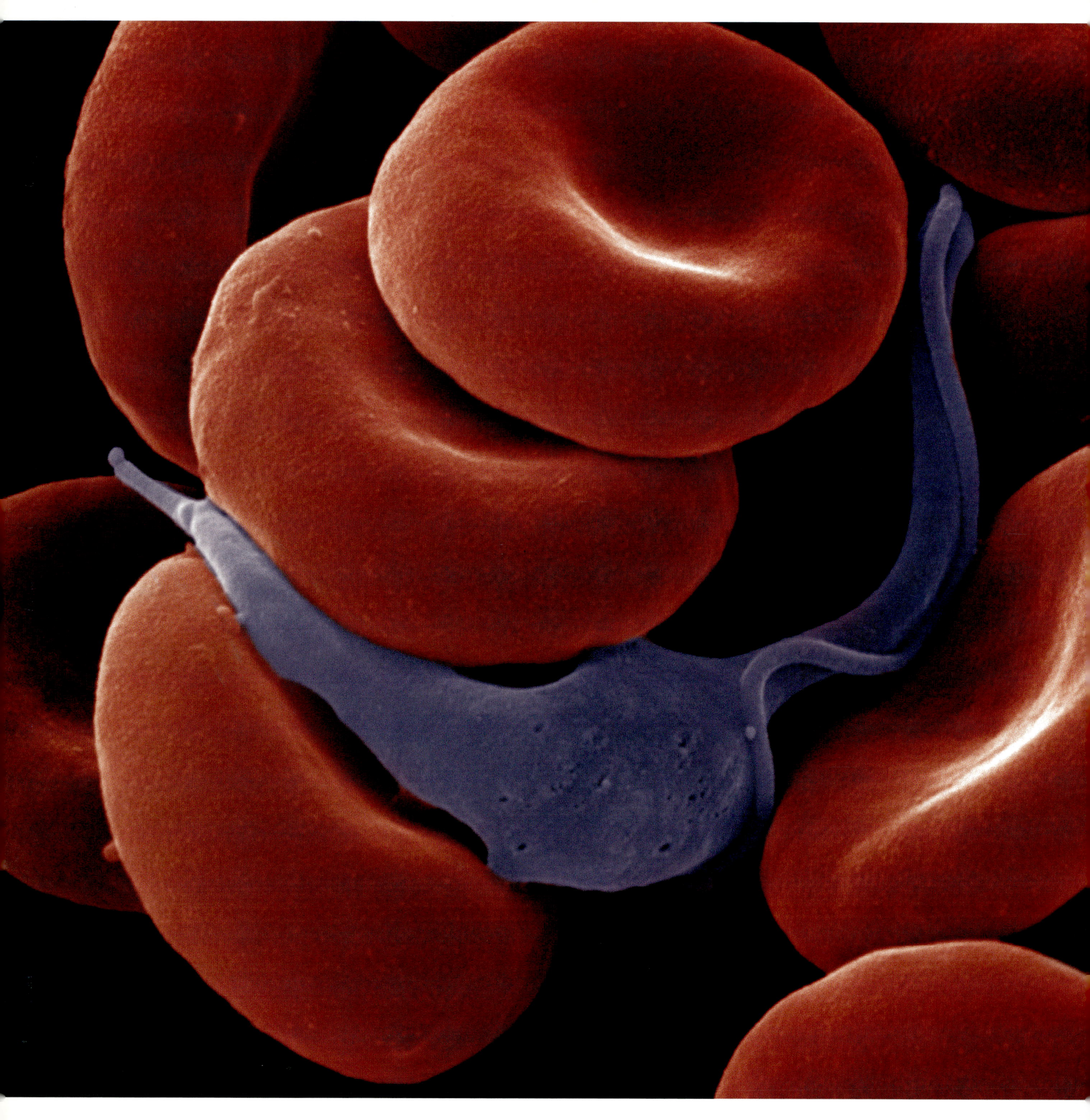

aptly named 'no see-ums', to larger, more infamous species such as the tsetse fly of sub-Saharan Africa, which carries the frequently fatal disease, sleeping sickness, which kills around 40,000 people each year. The natural targets for the tsetse are the grazing animal herds of the African plains, but they also attack humans. After being bitten by an infected fly, a red, painful swelling develops at the site of the bite. The disease is caused by the protozoan parasite *Trypanosoma*, which is injected with the fly's saliva then invades the blood stream, causing fever, headache, sweating, and enlargement of the lymph nodes. Parasites eventually invade the brain, causing mood swings, headache, fever, and drowsiness during the day, but insomnia at night. As the disease progresses, sleep becomes uncontrollable and the patient eventually becomes comatose.

FILARIAL WORMS

In addition to carrying malaria and a host of other diseases, for example yellow fever, dengue fever, epidemic polyarthritis, encephalitis, Rift Valley fever, and West Nile disease, mosquitoes are vectors for a disfiguring condition known as elephantiasis, caused by thread-like filarial worms such as *Wucheria bancrofti, Brugia malayi,* and *B. timori.* These tiny worms invade the lymphatic system, blocking lymph canals, causing inflammation, with corresponding thickening of the skin and underlying tissues. The condition most commonly affects the legs and feet, causing them to look not unlike elephants' legs, hence the name, but other areas of the body can be affected too, for example arms, breasts, ears and even genitals. Another species of filarial worm, *Onchocerca volvulus*, migrates to the cornea of the eye, causing the disease onchocerciasis, or river blindness. More than 1 billion people in approximately 80 countries live at risk of contracting filarial diseases from mosquitoes, and currently over 120 million people are already infected, with more than 40 million incapacitated or disfigured.

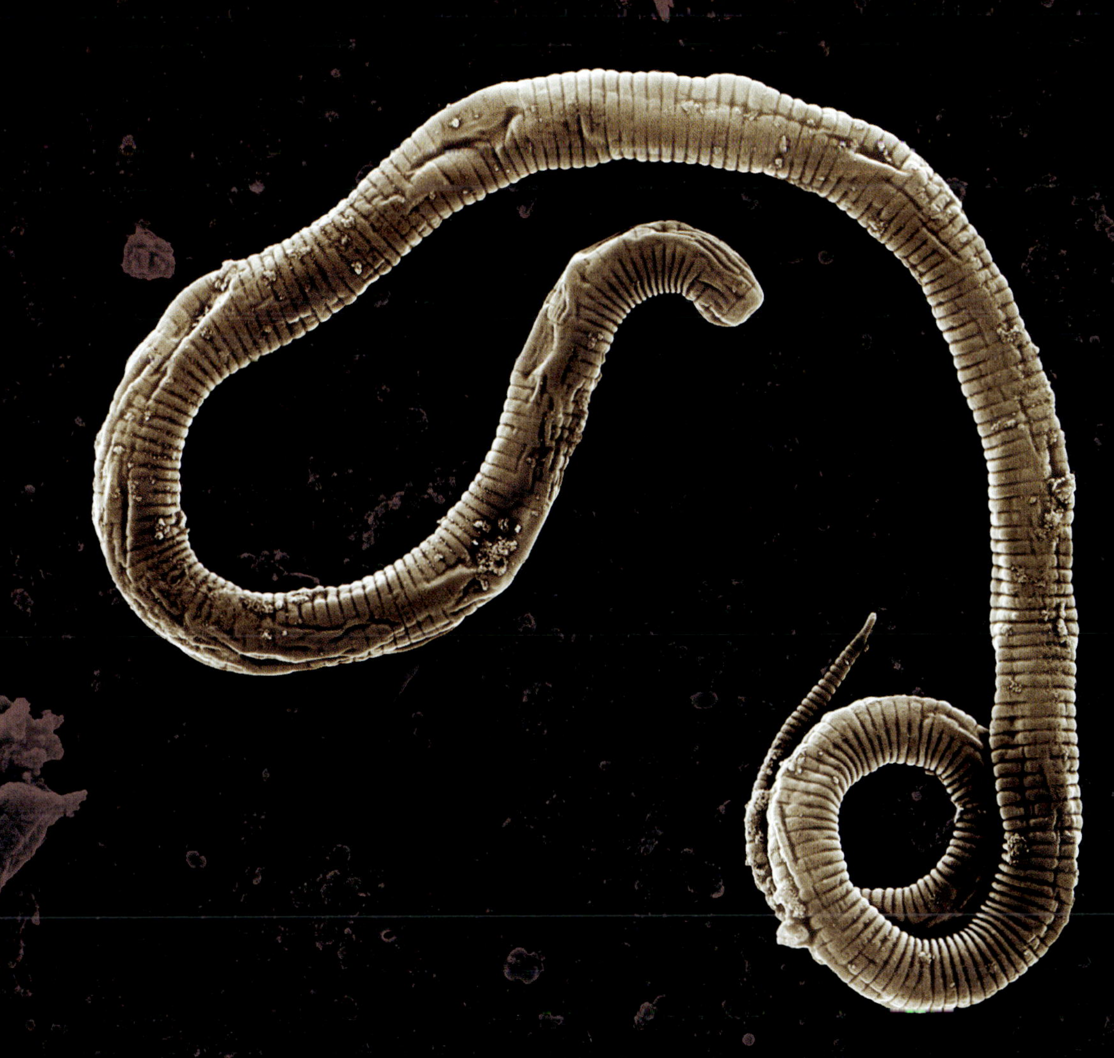

Micro-filarial worm, *Wucharia bancrofti*

Wucharia bancrofti is a micro-filarial worm transmitted by mosquitoes, causing elephantiasis. It lives and matures in human lymph canals but also circulates through the bloodstream. The surface of the worm is visible in the scanning electron micrograph (above) and a more transparent view is obtained in the light microscope photograph (below).

Magnification: x3300   *Scanning Electron Micrograph* (above)
Magnification: x1500   *Light Micrograph* (below)

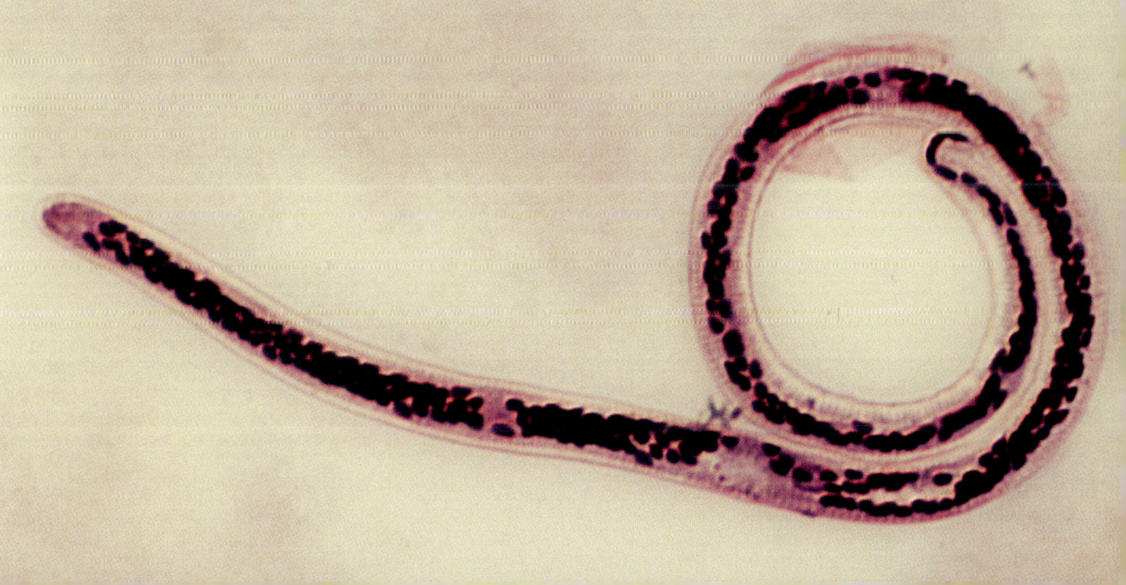

Sleeping Sickness Parasite, *Trypanosoma sp.* (left)

This *Trypanosoma* protozoan is in human blood, surrounded by red blood corpuscles. The parasite, carried by the tsetse fly in tropical Africa, causes sleeping sickness and is a major health problem.

Magnification: x32000
Scanning Electron Micrograph

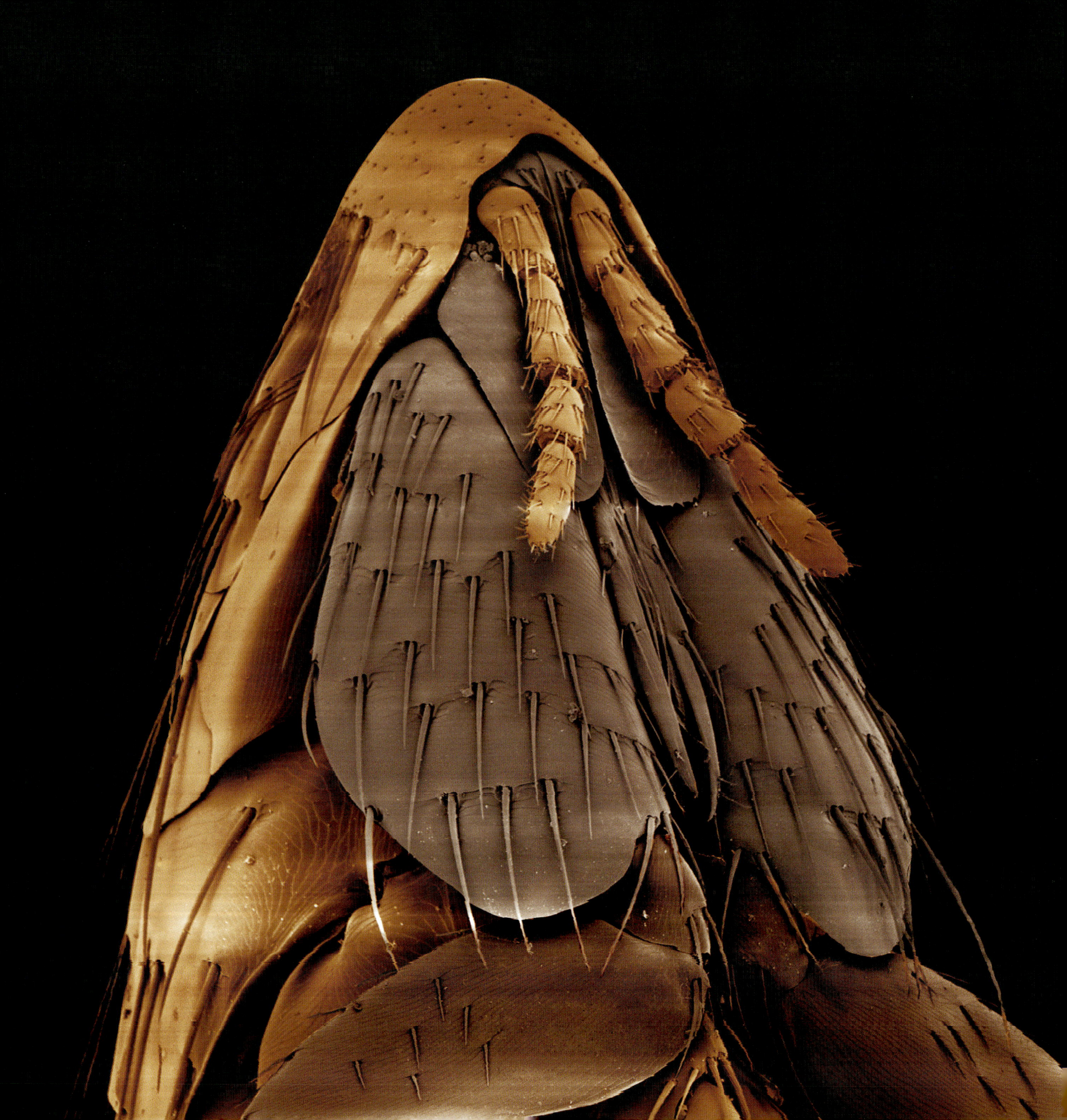

Our soft bodies attract many different kinds of blood-feeders. Fleas, like mosquitoes, are not only unpleasant companions but are also proven vectors of disease, capable of transmitting typhus, tapeworms, and other conditions to us and our pets. Rodent fleas can transmit murine typhus to humans from rats or mice. They are also frequent carriers of tapeworms that infest the digestive tract of cats, dogs, and us, for example the troublesome *Echinococcus multilocularis*, and *E. granulosus*, which produce life-threatening hydatid cysts. The Oriental rat flea, *Xenopsylla cheopis*, stands accused of spreading plague repeatedly through Europe with devastating results during the Middle Ages. The so-called 'Black Death', between 1347 and 1351, killed almost two-thirds of Europe's population, and simultaneous epidemics swept through the Middle East and parts of Asia around the same time. Worldwide, the Black Death pandemic killed some 75 million people, with the disease returning every few years like a haunting nightmare leaving death in its wake until the 18th Century. It is generally agreed that the Black Death was bubonic plague, caused by the bacterium *Yersinia pestis*, which still survives today in parts of tropical Asia, Africa and South America.

Fleas are particularly successful parasites since they are very agile, their long hind legs enabling them to leap on

Oriental Rat Flea, *Xenopsylla cheopis*

Typically an inhabitant of warm climates, the oriental rat flea has been successful in many temperate cities, infesting rats living in sewer systems and in seaports. It was the primary vector in the infective cycles of the Black Death and the Great Plague, which caused a series of devastating pandemics from the 6th Century onwards, killing millions of people worldwide. Infected rats were bitten by the flea, which then transmitted the bacterial agent, *Yersinia pestis*, to human hosts. The flea also carries other diseases, among them murine typhus. The scanning electron micrographs (left and right) show a male, while the dark field light microscope photograph (overleaf) shows a female.

Magnification: x450
Scanning Electron Micrograph (left)
Magnification: x150
Scanning Electron Micrograph (right)
Magnification: x200
Dark Field Light Micrograph (overleaf)

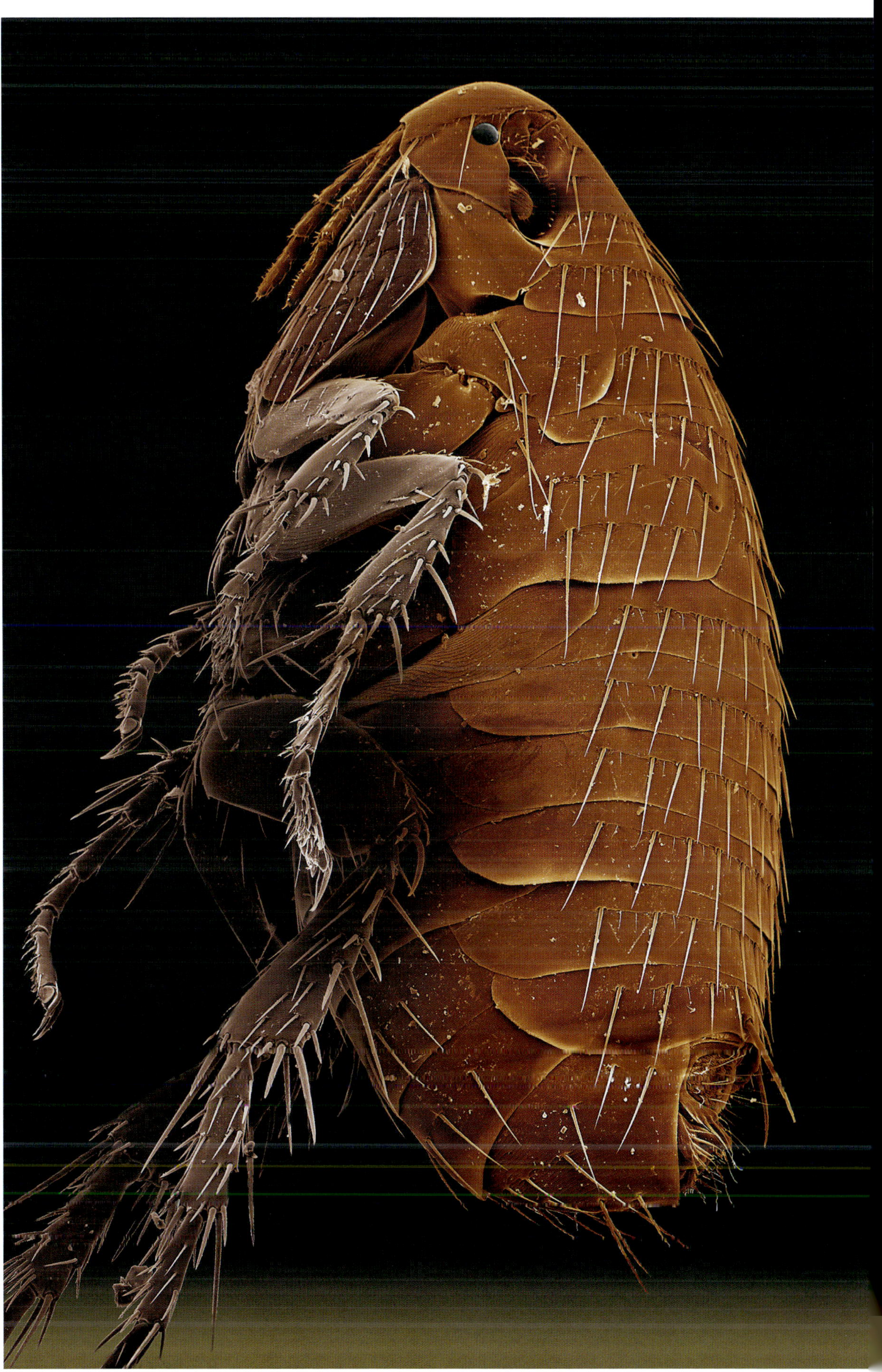

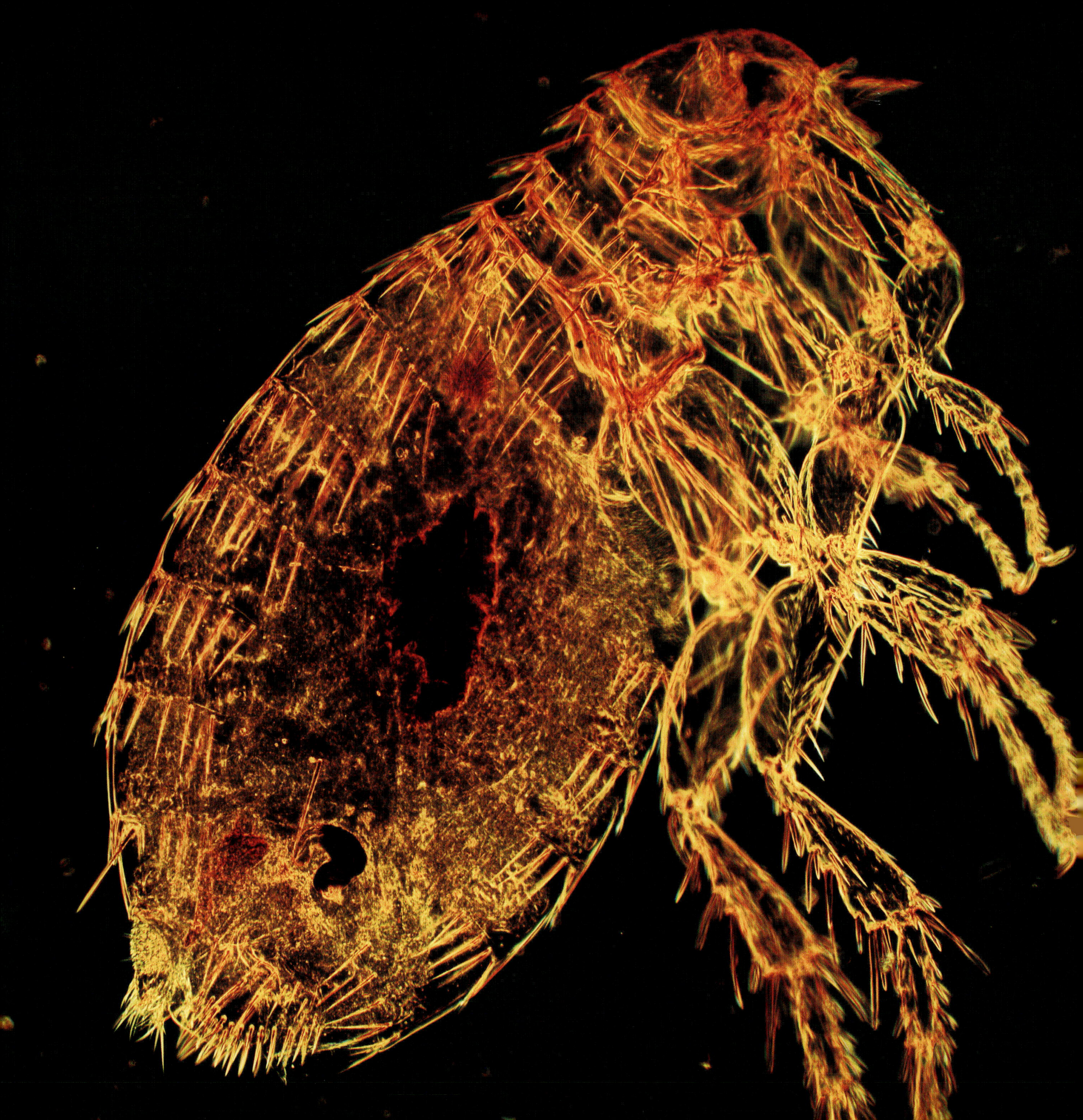

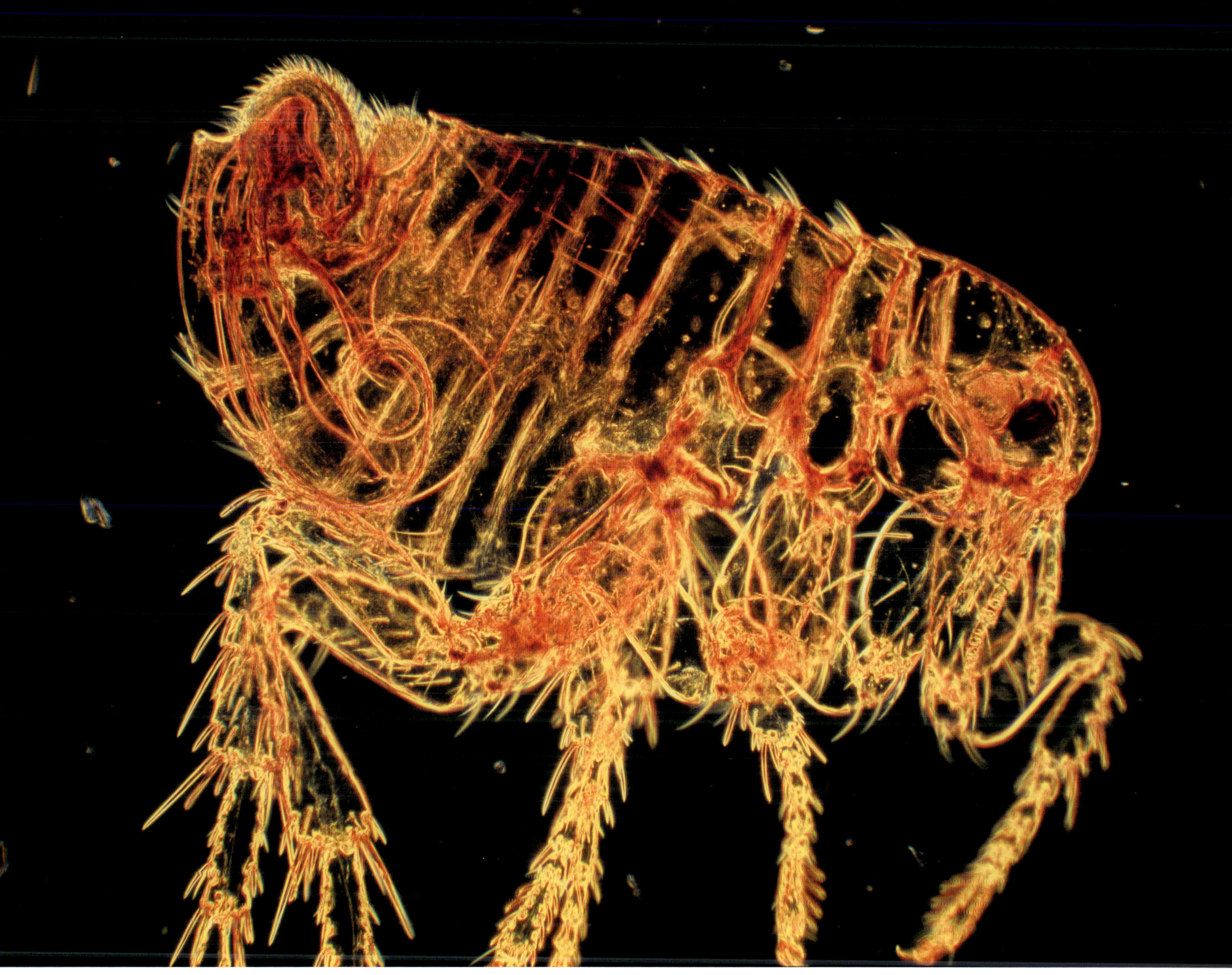

Human Flea, *Pulex irritans* (above)

Found worldwide, human fleas were extremely common in Europe before modern improvements in standards of hygiene. Previously the scourge of all classes of people, today they are often found on pigs. This specimen is a male.

Magnification: x100   *Dark Field Light Micrograph*

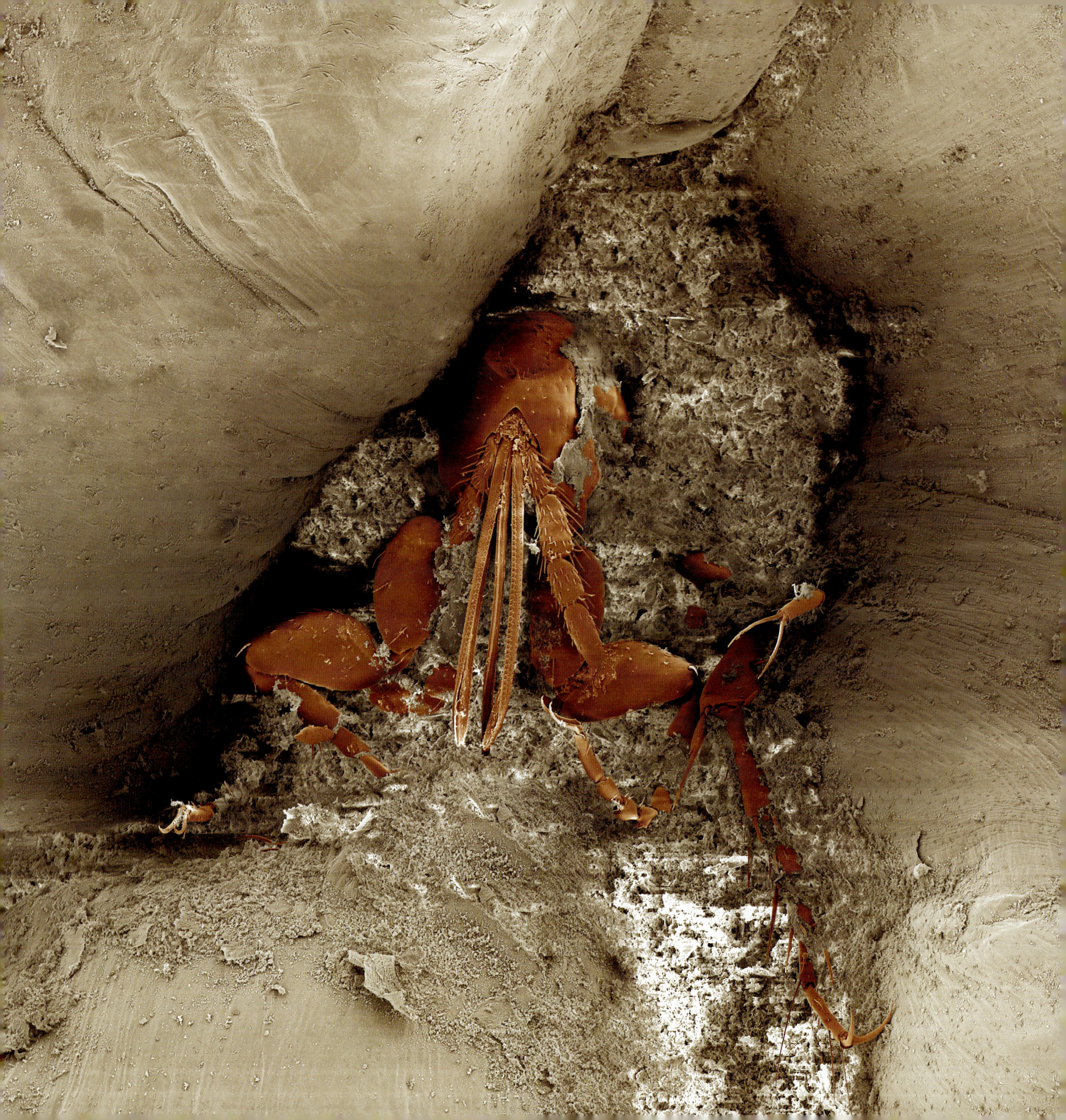

to a victim or to evade capture at lightning speed. The tiny, brownish-black insects, 1 to 4 mm long, are laterally compressed with small bristles, the perfect shape for scurrying at high speed through dense hair or fur. There are many species of fleas that will bite us but they tend to be host specific when it comes to breeding. Although cat fleas, *Ctenocephalides felis*, the most likely invaders of our homes today, happily bite us, they need feline blood in order to breed. Likewise, the human flea, *Pulex irritans*, needs human blood. Until recent history, human fleas were the persistent unwelcome companions to all classes of people. They were literally everywhere, and impossible to eradicate. From a time far back in history when people started to crowd together in towns and cities, dense populations and close living quarters made parasite and disease transmission easier. Before control of sewage or water purification, living conditions were filthy, infested with lice, fleas and rats, and diseases related to malnutrition and poor hygiene. In order to divest themselves of fleas, rough cloth or woollen blankets would be placed on the bed or floor as flea traps, and some women wore fur stoles known as 'flea cravats' to attract fleas from their bodies and clothing; in both cases, the idea was to trap the fleas for long enough to be taken outside and shaken.

Although we are the primary hosts for the human flea, it can also live on some domesticated animals, especially pigs. The worm-like larvae live in cracks and crevices in flooring, feeding on undigested blood from the faeces of adult fleas and other organic debris. Without eyes or legs, they move by using backward-projecting bristles and hooked projections at the tip of the abdomen to obtain purchase on a surface. They eventually spin silken cocoons to protect them during the pupal stage until the vibrations and warmth of a suitable host trigger their emergence as adults.

The chigoe flea, *Tunga penetrans*, at 1 mm long, is the smallest known flea, and is found in tropical West Africa, and South America. Frequenting areas around rural settlements, it lives in the soil or sand,

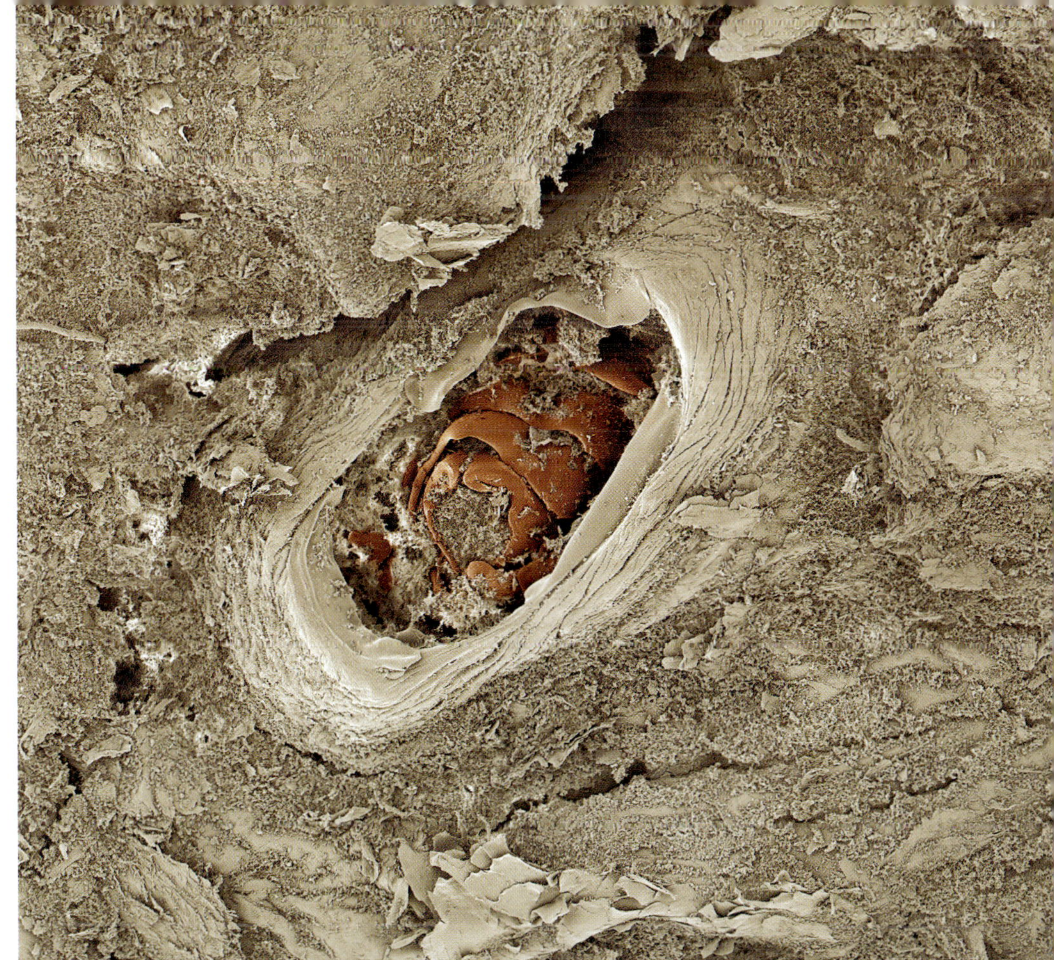

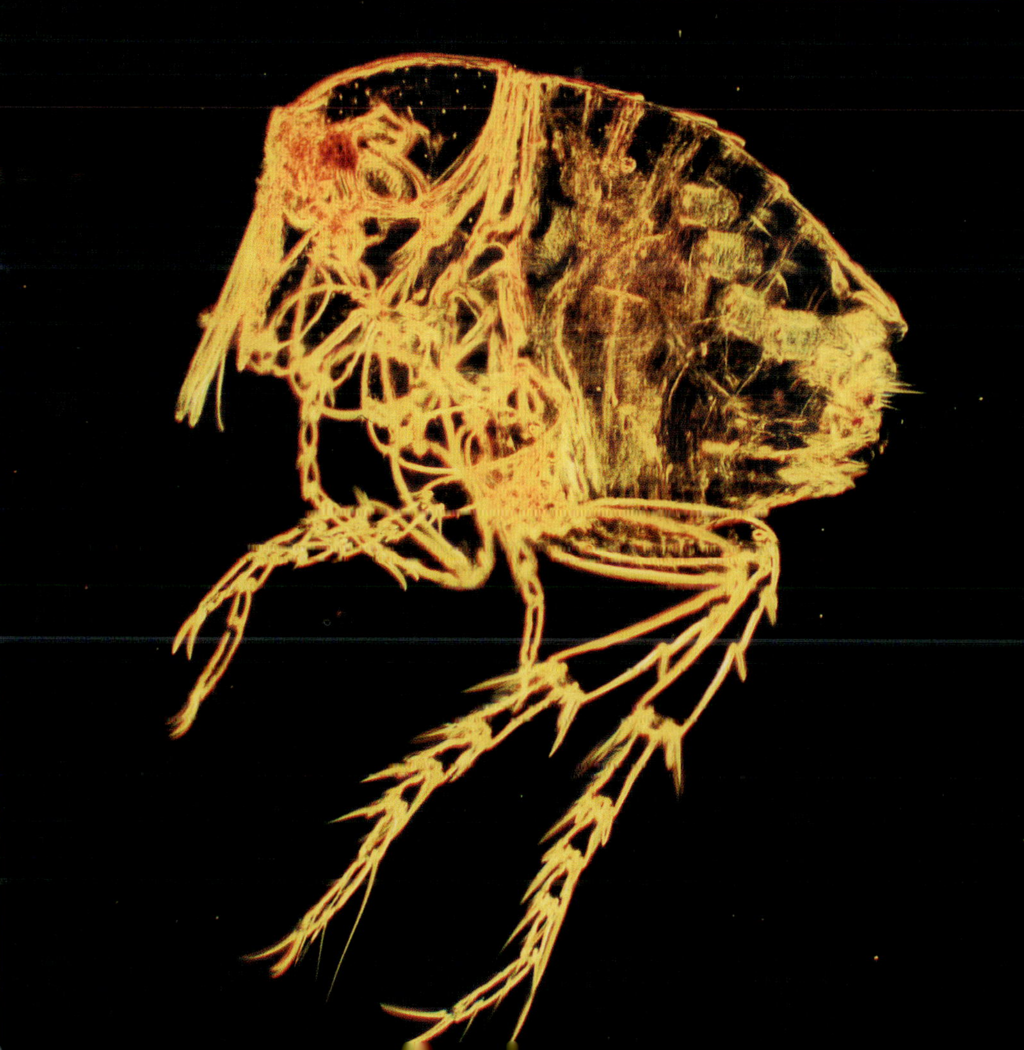

Chigoe Flea, *Tunga penetrans*

A female chigoe flea, her head just visible (see left), has burrowed into the skin of a human foot, where she remains embedded, her ovaries swelling and producing eggs. While she feeds on blood and tissue, she breathes through a lesion in the skin (see above right), which itches intensely. Once the eggs are ripe, the female releases them through the lesion, after which she dies. The chigoe flea is the smallest known flea and is found in tropical West Africa and South America, frequenting the soil or sand around human settlements.

Magnification: x200   *Scanning Electron Micrograph* (left)
Magnification: x200   *Scanning Electron Micrograph* (above right)
Magnification: x100   *Dark Field Light Micrograph* (below right)

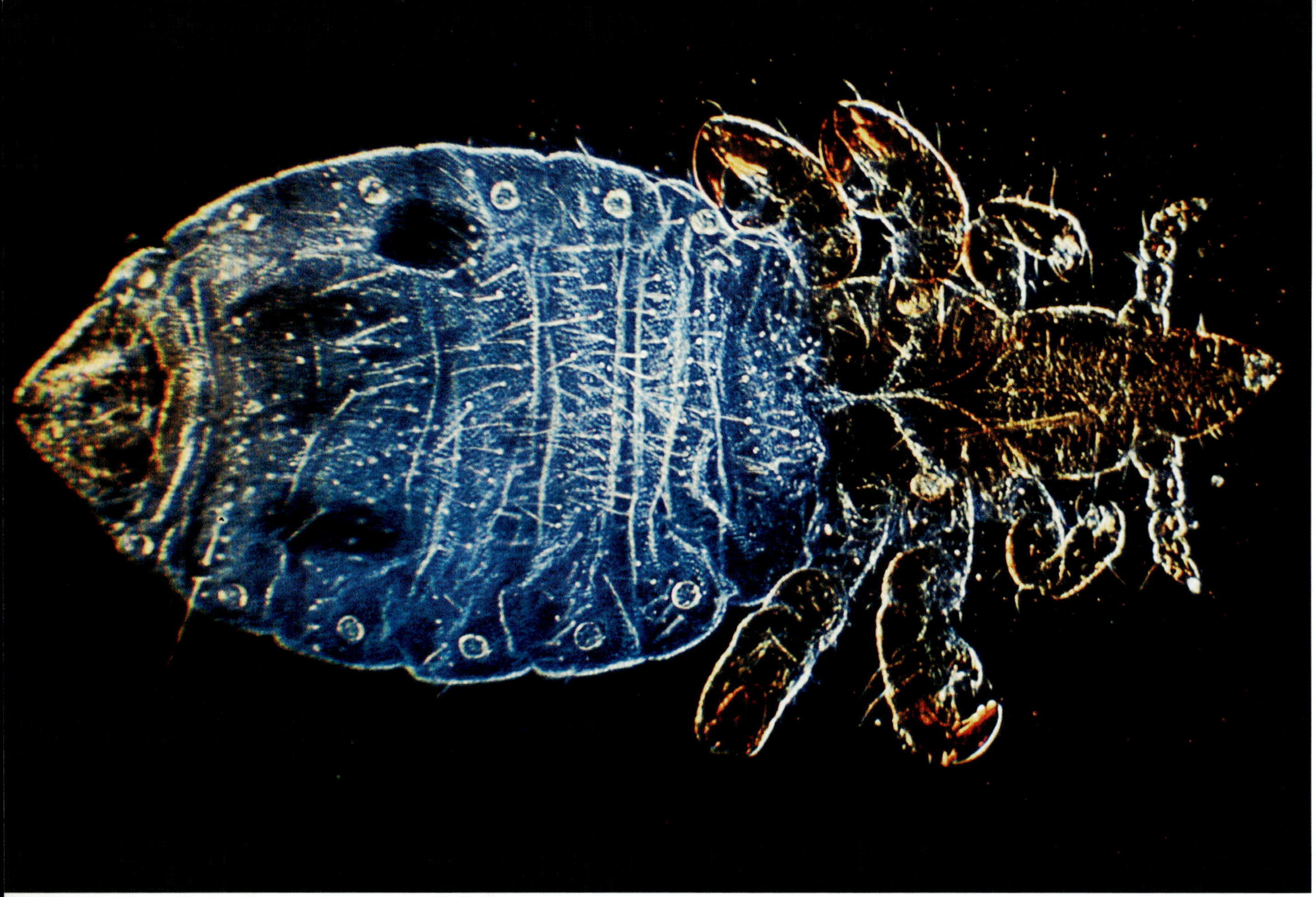

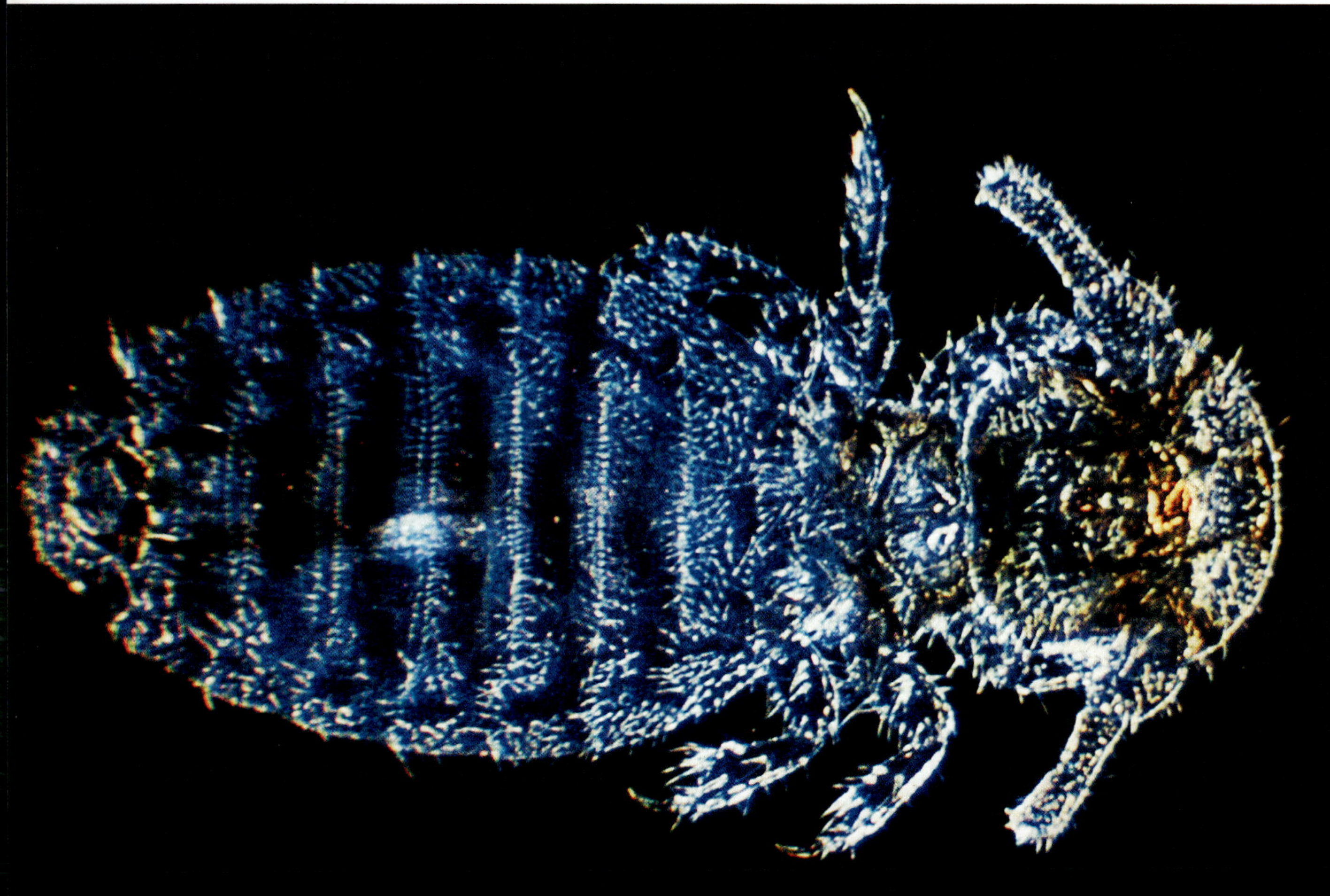

Human Head Louse,

Pediculus humanus capitis
(above left and right)

Powerful legs and long claws enable the human head louse to cling on to hairs and move about on the surface of the skin, biting and sucking blood.

Magnification: x80
Dark Field Light Micrograph
(above left)
Magnification: x250
Scanning Electron Micrograph
(right)

Bird Louse, *Order Mallophaga*
(below left)

There are nearly 3,000 known species of bird lice worldwide, each usually restricted to one species or a group of closely related birds. Living on the body, they feed mainly on particles of skin and feathers, occasionally sucking blood.

Magnification: x75
Dark Field Light Micrograph

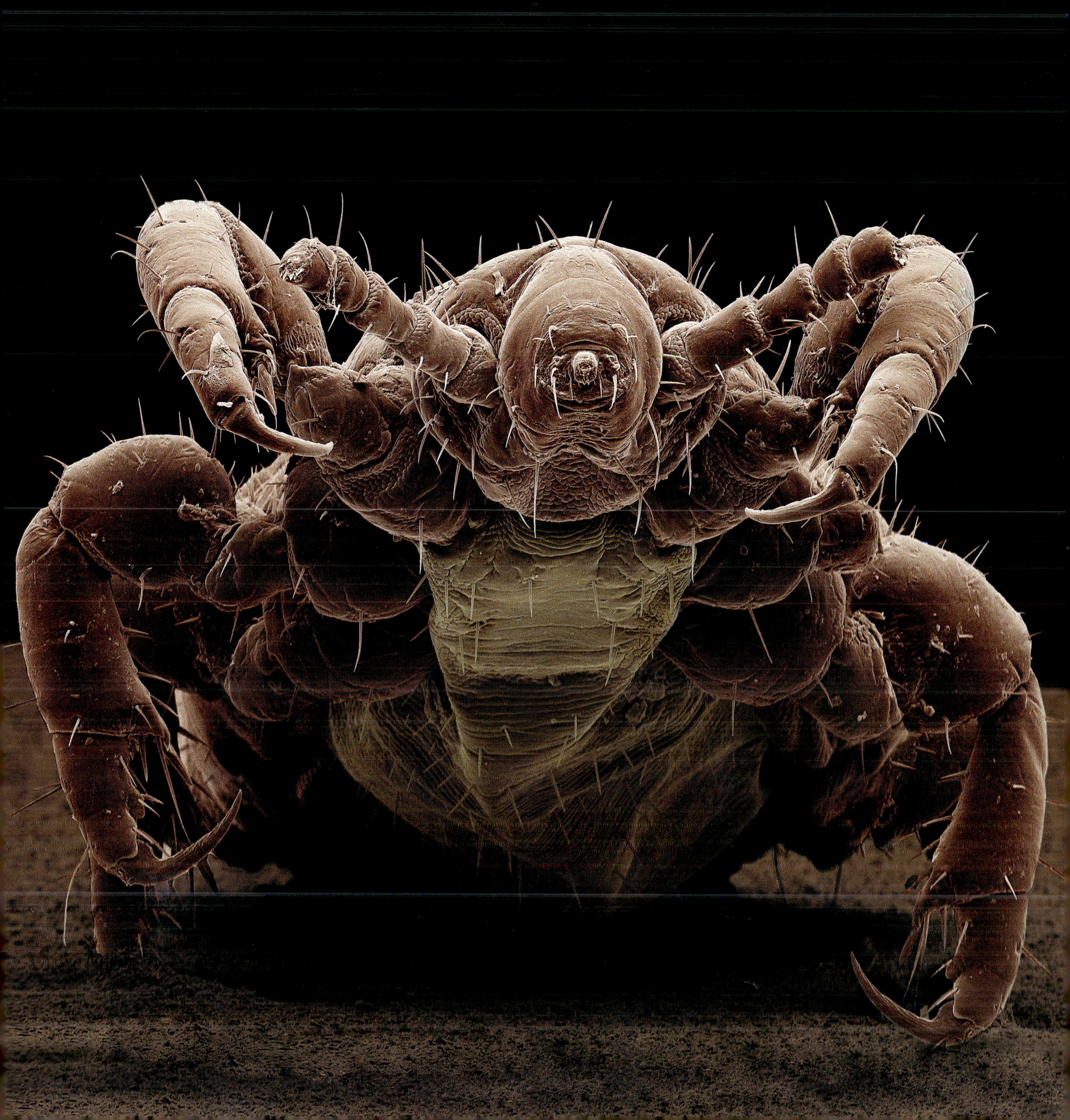

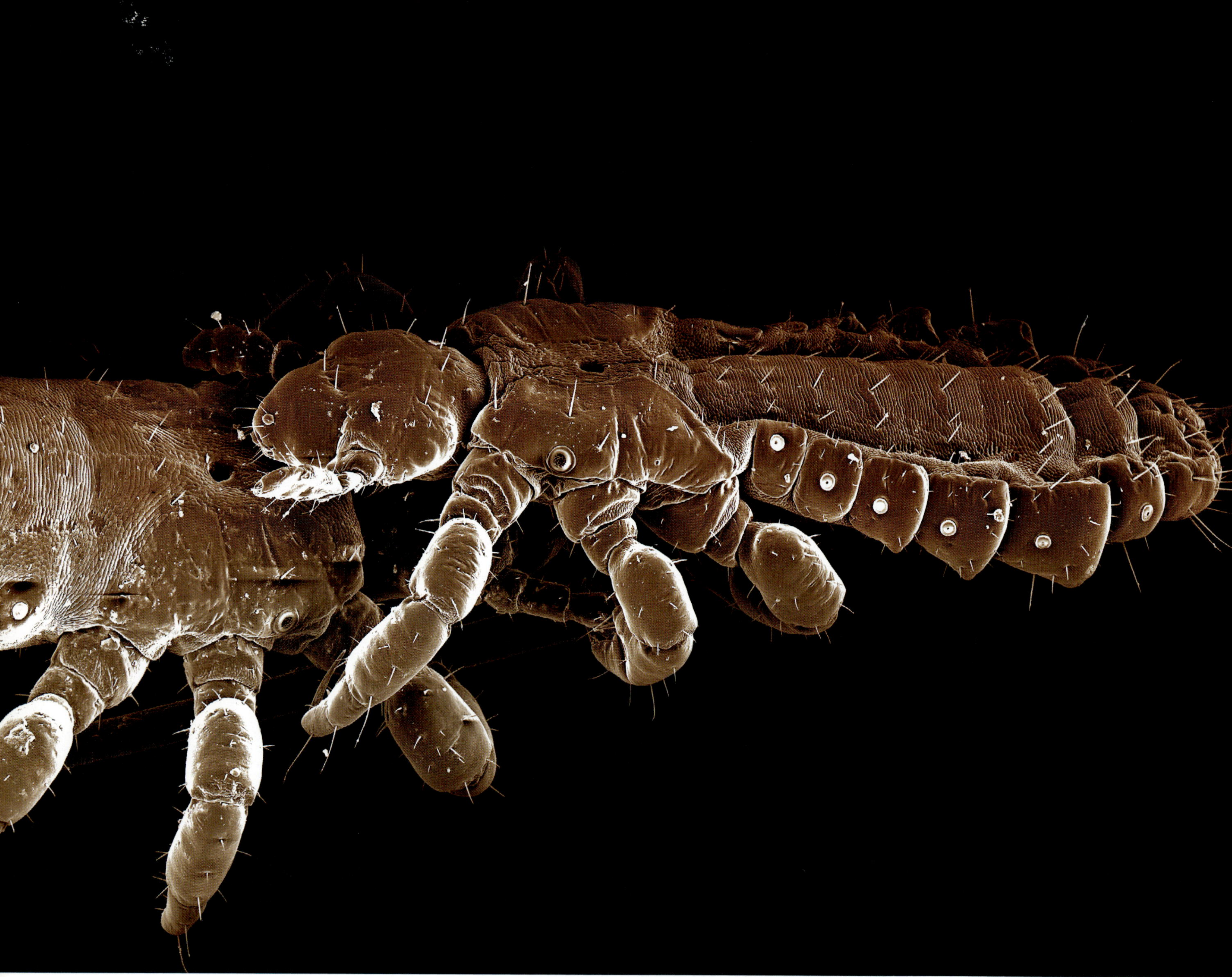

Human Head Louse, *Pediculus humanus capitis*

Two head lice, one crawling across the other (see above). The female glues her eggs, or nits, to the base of hairs (see right).

Magnification: x85   *Scanning Electron Micrograph* (above)
Magnification: x80   *Dark Field Light Micrograph* (right)

Human Crab Louse, *Phthirus pubis* (overleaf)

More squat and crab-like than the head louse, the crab louse is normally spread from person to person by physical contact. It feeds on blood and has well developed claws for clinging on to pubic hair.

Magnification: x230   *Scanning Electron Micrograph* (overleaf, left & right)

and attacks warm-blooded hosts, including us. As fleas go, the chigoe is a poor jumper, but sufficient for the female to leap from the ground on to the feet or ankles of a person walking close by. Wearing shoes affords some protection but not much in areas of severe infestations since hungry fleas can navigate around folds of clothing and squeeze under tight socks. Scurrying across the surface of the skin, the female chigoe seeks a suitable place, usually around toenails or between the toes, then burrows, often leaving the tip of its abdomen visible through a lesion, a tiny orifice that allows her to breathe while she feeds on blood and produces eggs. After a few hours, the site itches intensely and can be recognised as a small, pale spot with a central black dot. If left untreated, the female eventually releases the ripe eggs through the orifice on to the ground, after which she dies.

LICE

Like human fleas, human lice were also rife throughout history and caused their victims considerable and continuous irritation. There are three kinds of lice that frequent the human body: the head louse *(Pediculus humanus capitis)*, the body louse *(Pediculus humanus humanus)* and the crab louse *(Phthirus pubis)*. Infestations of any of these have reached epidemic proportions when people have lived closely together in situations lacking in hygiene, for example during the prolonged trench warfare in World War I. In World War II, widespread louse epidemics were prevented by the use of DDT. Today, large-scale lice infestations have been eliminated, so we are surprised when small outbreaks occur. Head louse infestations, however, still frequently appear among elementary school children; body lice frequent homeless people and others unable to care for themselves; and crab lice can be transmitted during sexual intercourse with an infested partner, through the sharing of personal articles such as towels or even toilet seats.

The presence of a louse on the body makes the skin itch and they are very contagious. Unlike fleas, lice can neither hop nor jump, but they can crawl quite swiftly. Wingless insects about 2 mm long, they have claws that grip and hold onto hair shafts. Their colour varies from pale to dark red, as the gut fills with blood. The female can produce up to six eggs a day, which adhere to hair shafts near the skin. Called nits, the eggs are yellow-white in colour and can be mistaken for flakes of dry skin or dandruff. Other signs of head lice infection are droppings, which appear as black specks on pillows. The eggs hatch within 10 days, and the nymphs resemble the adults except for size. The body louse is similar to the head louse but often hides in seams of clothing, where it deposits its eggs. During wartime, they were called 'mechanised dandruff' because they could frequently be seen running in and out of the weaves of clothing.

Body and head lice have been responsible for transmitting many diseases to humans, including typhus, impetigo, trench fever and relapsing fever. Historically, the disease typhus, caused by the bacterium *Rickettsia prowazekii*, was a common occurrence in confined populations such as cities under siege. Typhus can be fatal, sometimes killing more people than the wounds of war. The bacteria enter the body through the faeces or the vomit of the louse, or if the remains of a crushed louse are able to get into the skin through wounds caused by scratching. The symptoms begin about ten days after infection and usually include fever, pain, stiffness, and headache. After a few days, a rash appears all over the body, and the victim becomes delirious. Serious conditions such as gangrene or pneumonia may develop before the disease reaches the nervous system when it becomes deadly. In some epidemics, the death rate has been as high as 50-75% of those infected. During World War I, the disease killed about three million people, but by World War II, antibiotics helped to control its spread.

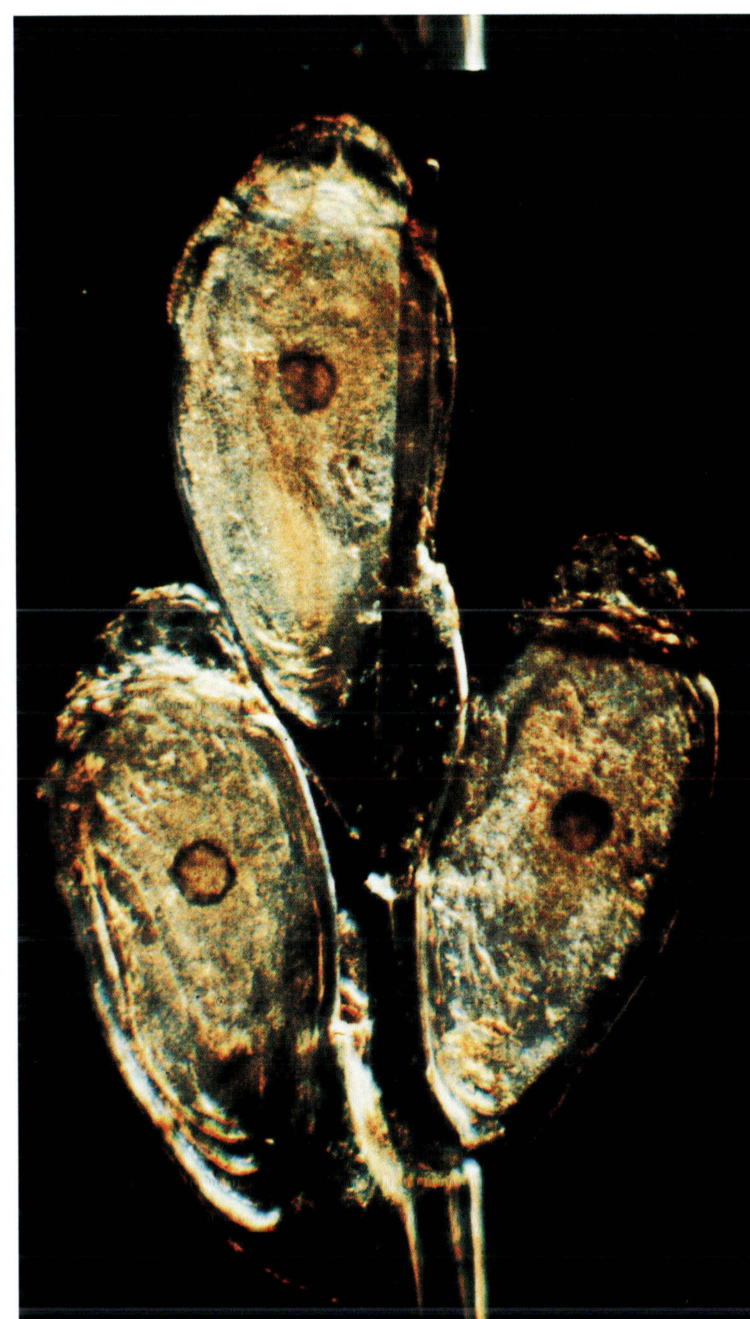

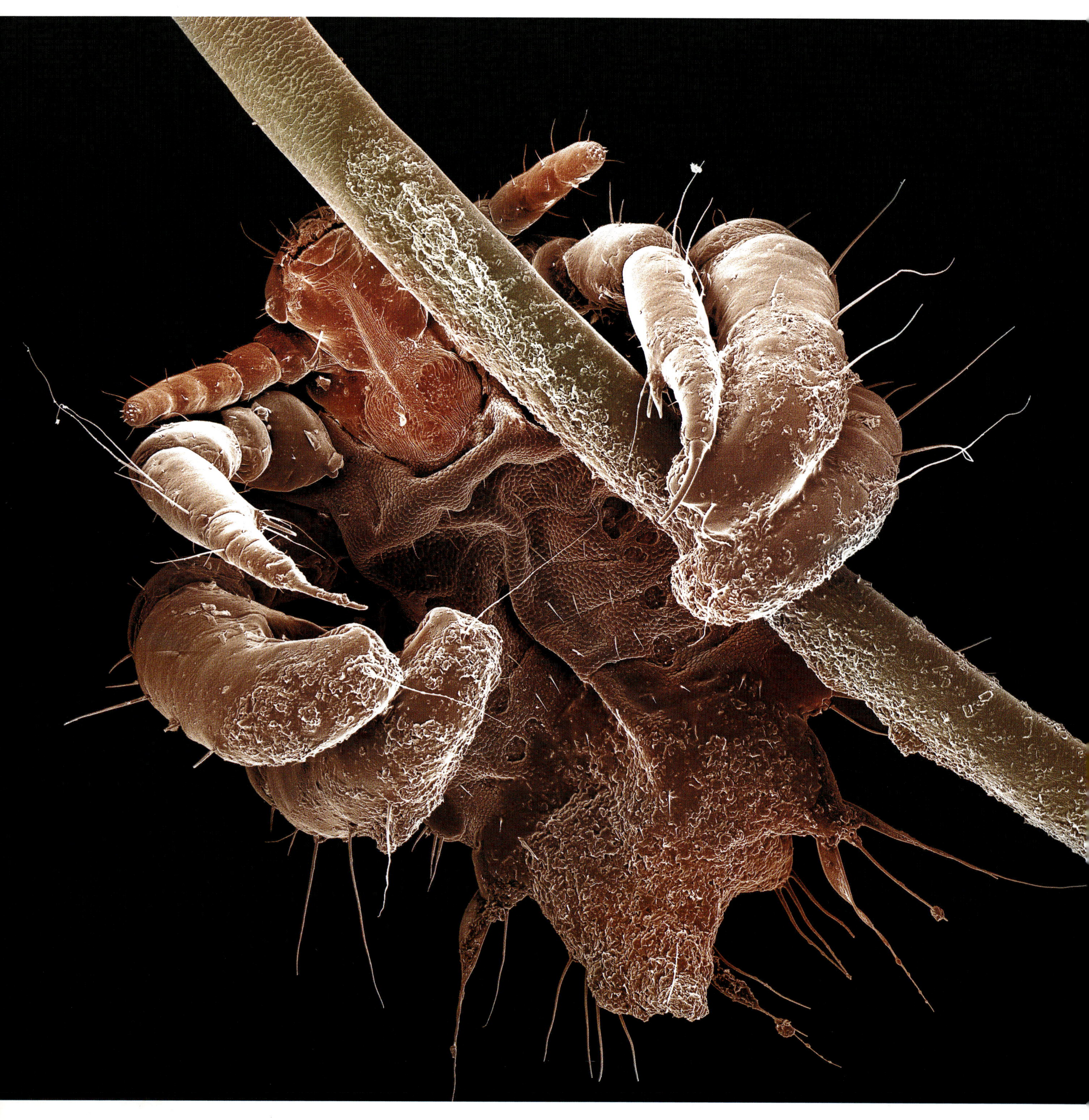

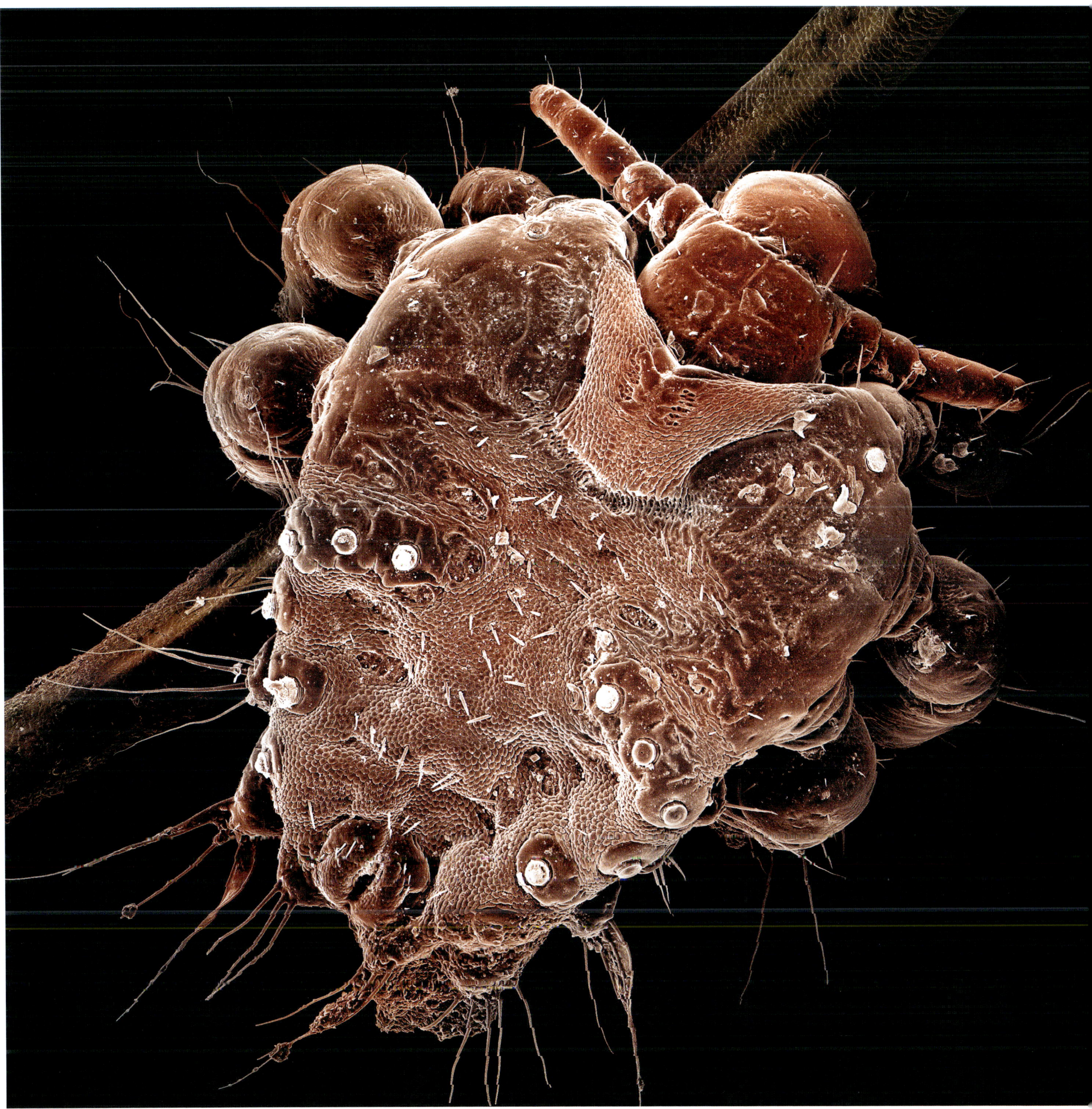

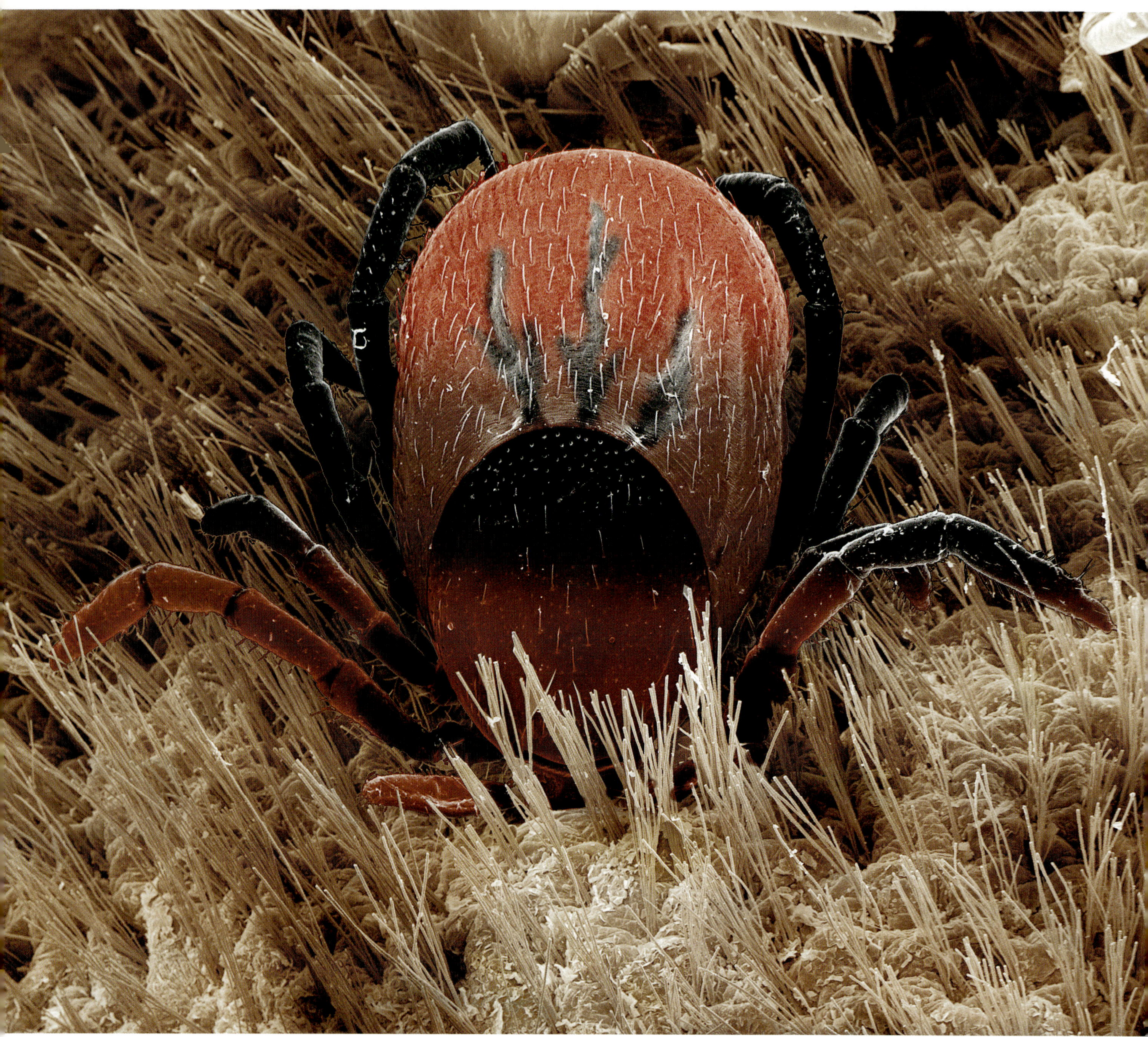

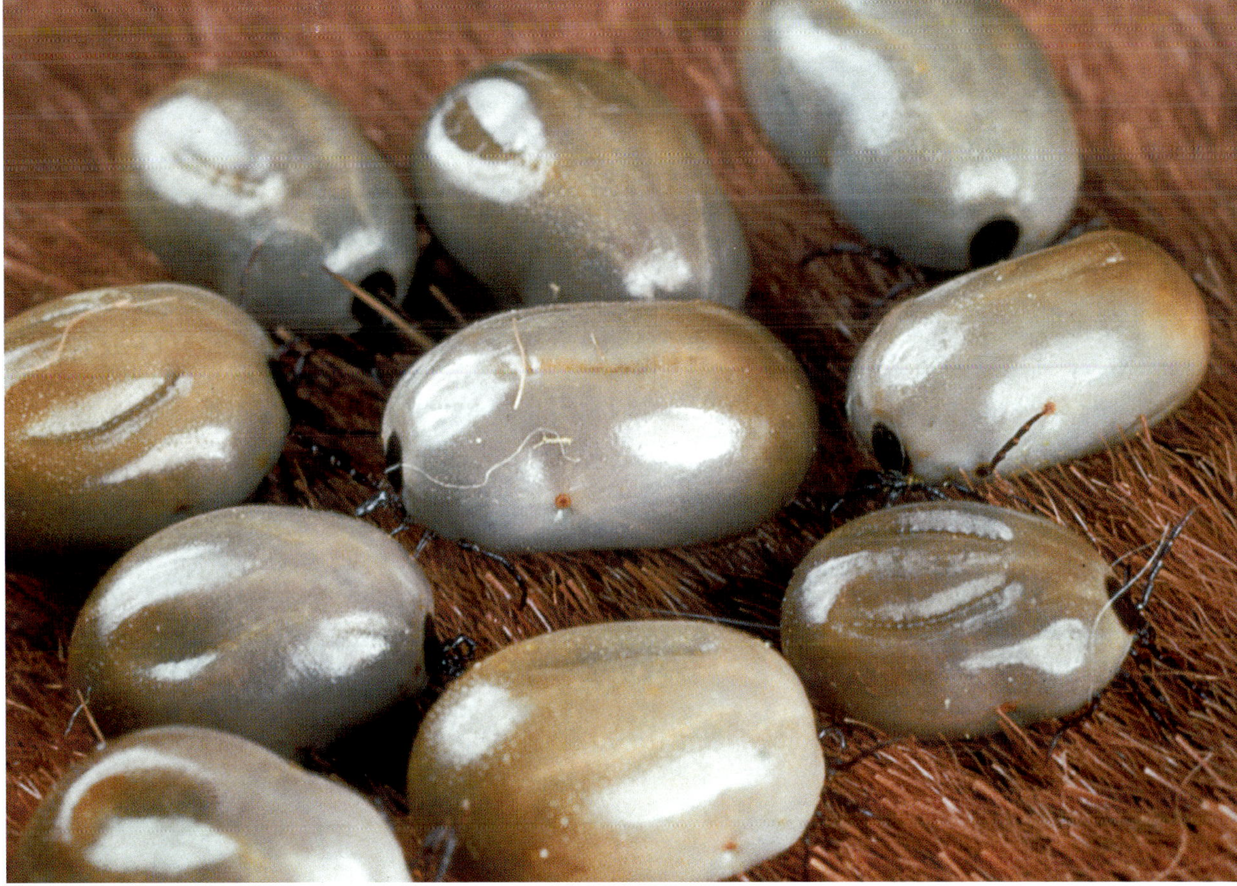

Black-legged or Deer Tick,

Ixodes scapularis (left)

Ticks transmit the widest variety of pathogens of any blood-sucking arthropod. *Ixodes scapularis* is a major carrier of Lyme disease. The central part of the mouthparts, the hypostome, is barbed, and embedded firmly into the skin while the tick is feeding.

Magnification: x70
Scanning Electron Micrograph

Sheep or Castor Bean Tick,

Ixodes ricinus (right)

A group of female ticks engorged after a blood meal. When full of blood, the outer cuticle of a tick can swell two hundred times its original size.

Magnification: x10
Macro-photograph

TICKS

Ticks are less mobile blood-feeders than lice or fleas, and in Britain only become a hazard during the warm months of summer. Ticks may occur anywhere but particularly in forests or farmland where large mammals, wild or domesticated, are found. Typically, they will wait on vegetation, with legs extended, for a potential host to brush past close enough for them to clamber on board. Having gained a foothold on shoes or clothing, a tick will crawl, seeking out a warm, dark, moist place such as the crotch or armpit. Once there, it will pierce the skin with barbed mouthparts to feed, engorging itself on blood to the extent that its body may swell to many times its previous size. *Ixodes* ticks can transmit an infection known as *Borrelia* or Lyme disease. First described in the town of Old Lyme in Connecticut, USA, it is the most common tick-borne infection in the United States and Europe. Caused by *Borrelia* bacteria, the disease manifests itself with skin rashes, cardiac, neurologic or flu-like symptoms, and painful joints. Without prompt treatment, the disease can cause chronic conditions that can be serious and difficult to treat. The spirochaete of the *Borrelia* bacteria is not transferred as soon as the tick bites, but takes some time to be passed on to its new host; this makes a tick an ideal vector since it may remain embedded in the skin for some considerable time. Lyme disease can even be passed from mother to child in the womb, and can be serious enough to cause stillbirth. *Borrelia* spirochaetes have also been detected in breast milk and in sperm, suggesting that breast-feeding or sexual activity may also provide another means of infection.

MITES

Closely related to ticks, mites tend to be much smaller, some of them microscopic. While most of them are free-living, some are parasitic. Most of us, at one time or another and without realising it, have played host to tiny worm-like follicle mites, *Demodex folliculorum*, in the hair follicles of our eyebrows and eyelashes. These mites, only 0.1-0.4 mm long, feed on dead skin cells and oils secreted by the skin, and are often one of the first gifts a baby receives from its mother, since they are transferred by direct contact. On the other hand, scabies is an unpleasant condition of the skin caused by another tiny mite, *Sarcoptes scabiei*. Only 0.3 mm long, it produces intense, itchy skin rashes when the female burrows under the surface, especially around the hands, feet, or male genitalia, and deposits eggs in the tunnel that she has excavated. The larvae hatch in a few days, move about on the skin, moult into a "nymph" stage, and then mature into adult mites. The presence of the scabies mites produces an allergic response, which may resemble eczema, and the itching tends to be more intense at night because the mites emerge from their burrows to wander about. The affected areas become red rashes with irregular shaped bumps called papules. There may be hundreds of them, but only a few will have associated burrows, which can be seen as thin wavy lines up to 1 cm long. The tiny mite lives deep down at the far end of the burrow. Scratching the rashes should be avoided as it may break the skin and make secondary infection more likely. The condition is highly

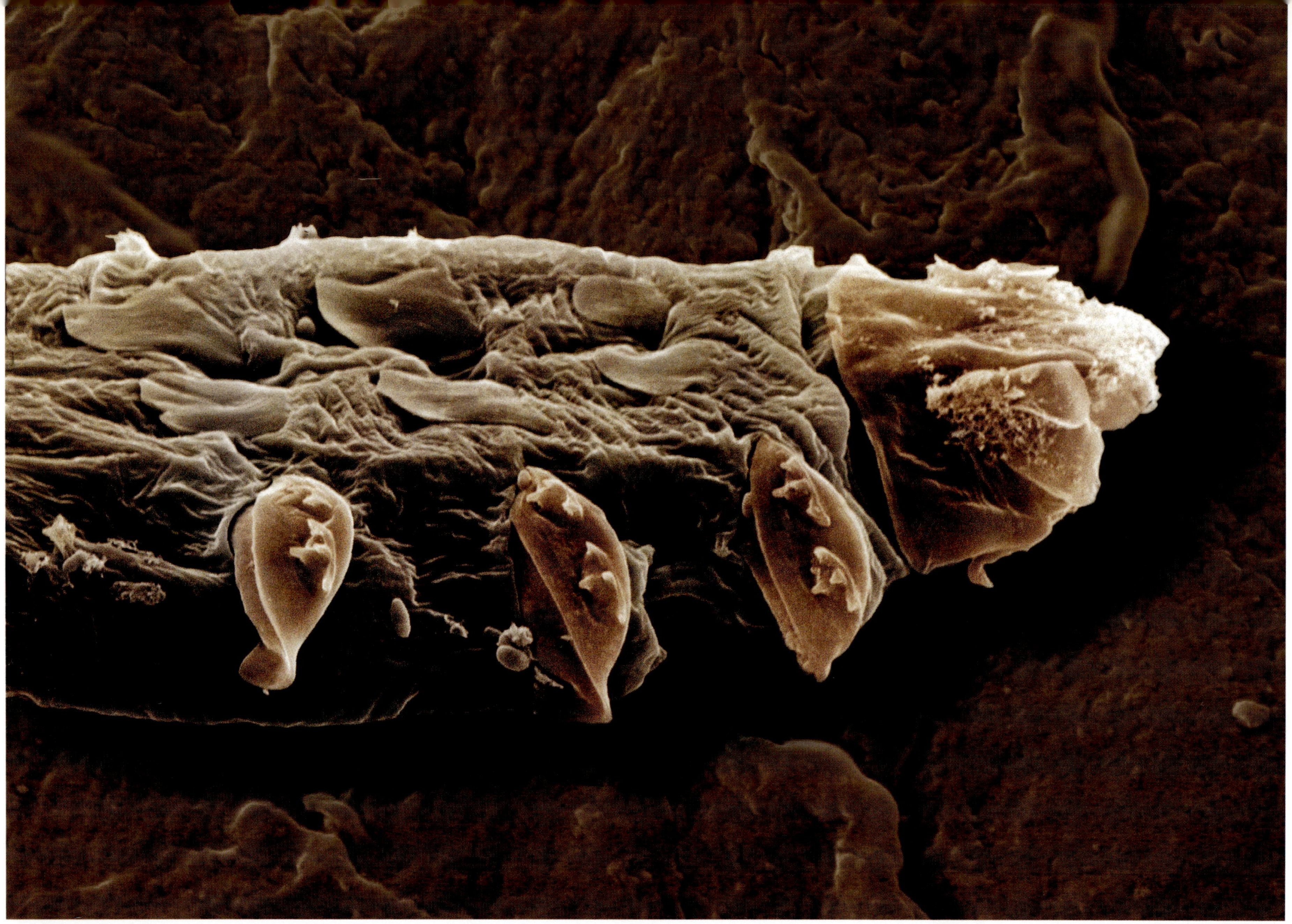

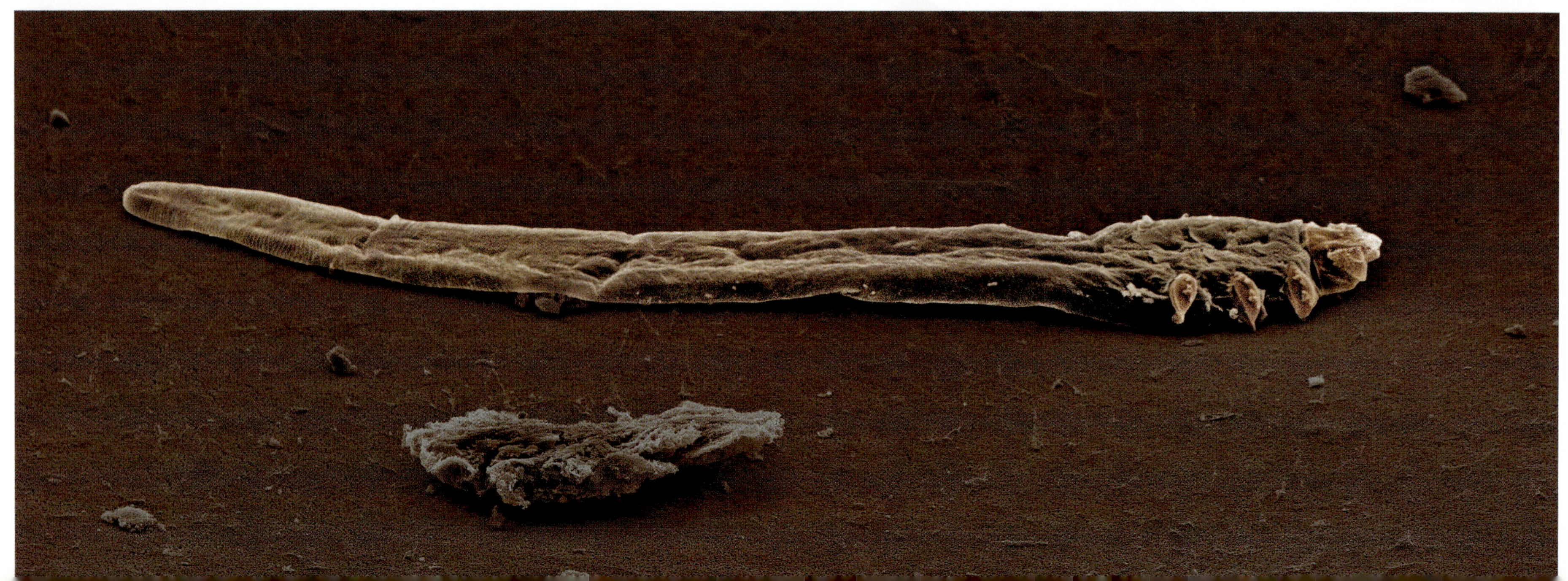

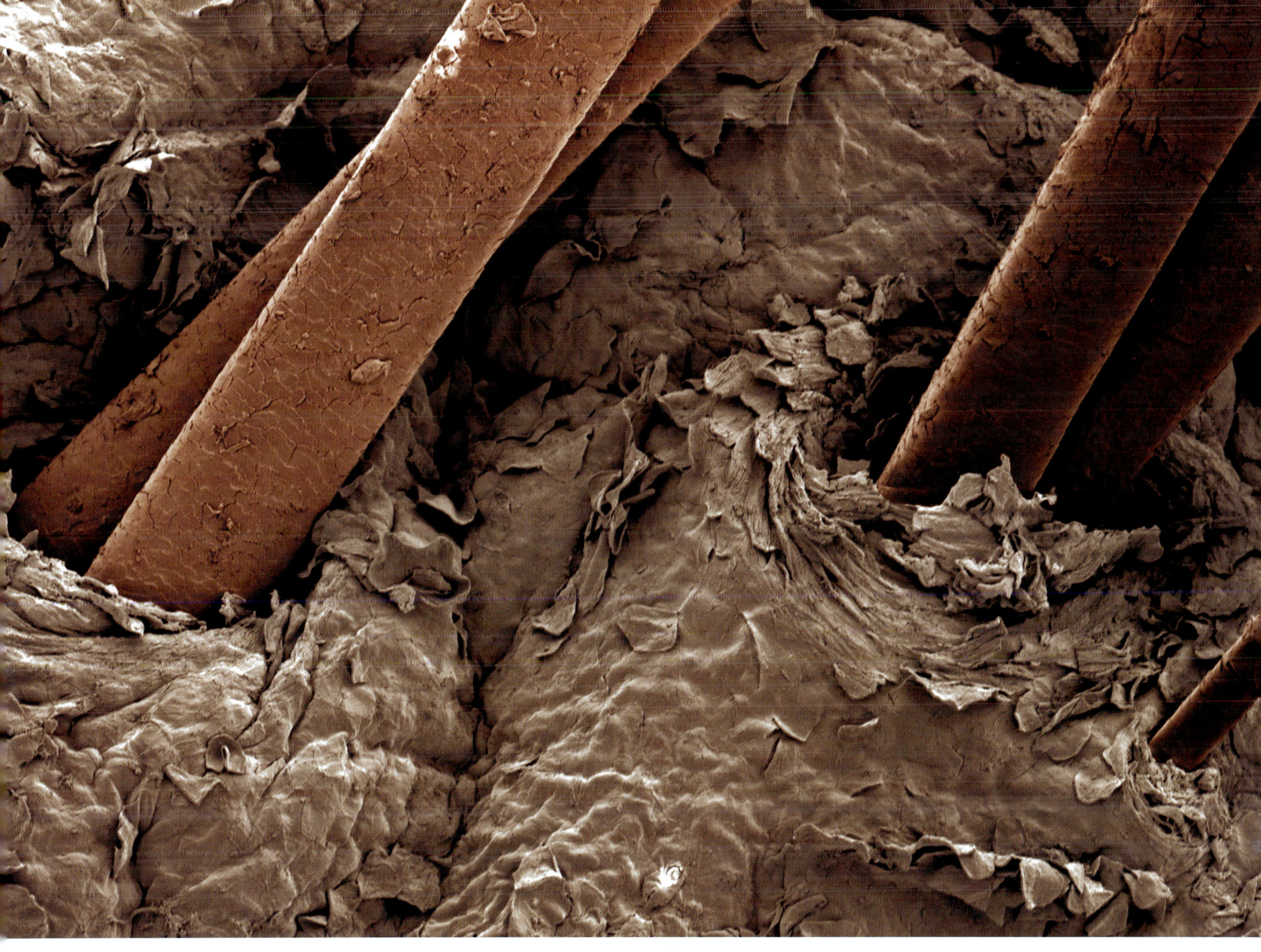

Follicle Mite, *Demodex folliculorum*

Microscopic with elongate bodies, follicle mites (see above left and below left) live on facial skin (see above) around the forehead, cheeks, eyelashes and ears, feeding on skin oils and debris. Many of us have a few, but are unaware of them. For most of the time, they live with their heads buried in hair follicles where they lay their eggs.

Magnification: x3500   *Scanning Electron Micrograph* (above left)
Magnification: x650   *Scanning Electron Micrograph* (below left)
Magnification: x500   *Scanning Electron Micrograph* (above)

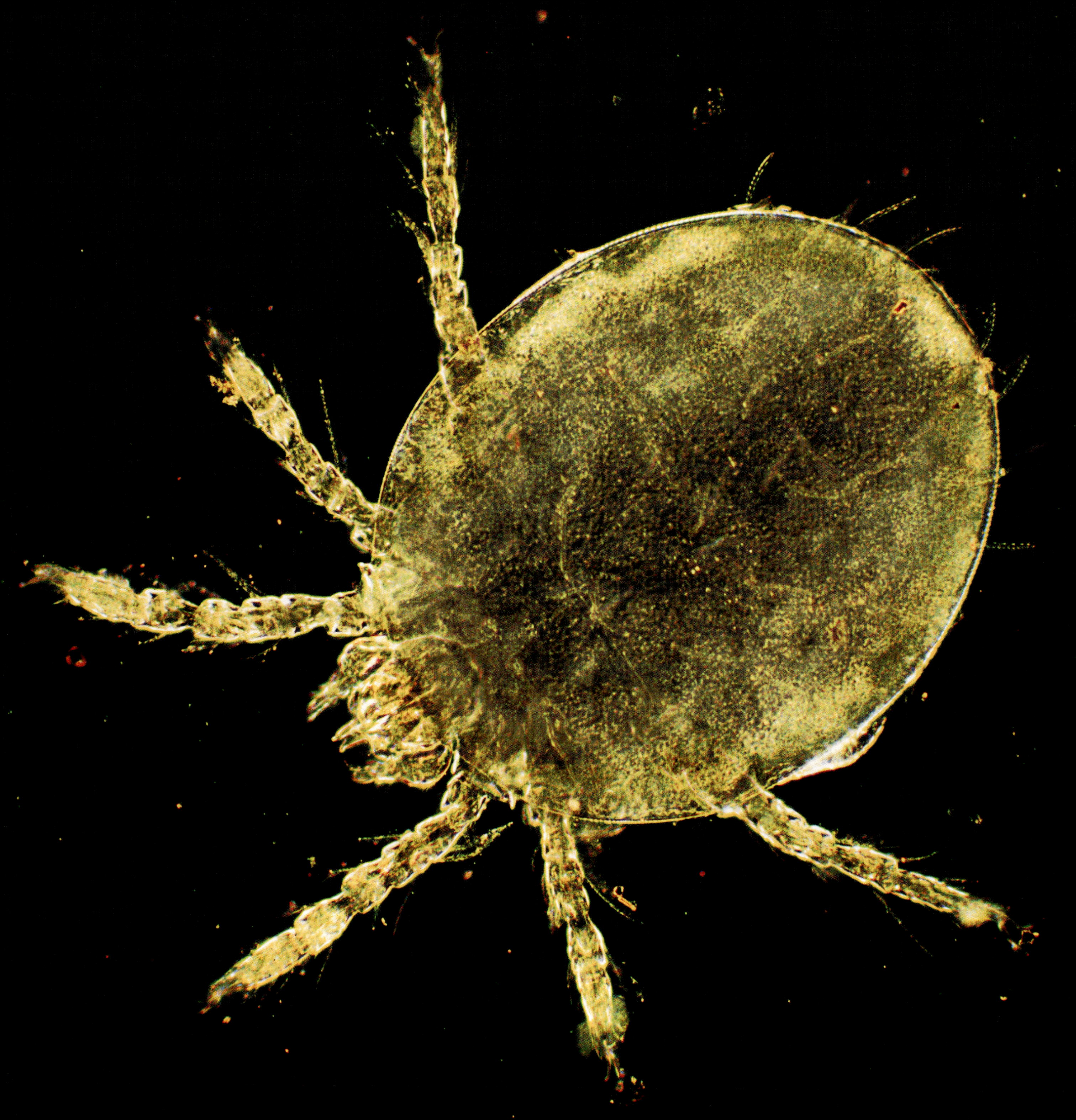

contagious, being transmitted from person to person through physical contact or personal articles such as clothing or towels.

Chiggers are the parasitic larvae of the harvest mite, *Neotrombicula autumnalis.* The adults are harmless, but the tiny, six-legged, red larval stage is parasitic on many animals, including us. Chiggers occur in several parts of the world; in parts of Asia, they carry "scrub typhus", but in Britain, they do not normally transmit disease. Here, the biting season is limited to the late summer and early autumn, hence the name harvest mite. Adult females lay hundreds of eggs at a time in scrubby, tall grassy areas picking damp but well drained sites often near riverbanks, under trees or bushes. The chigger larva moves to the tip of a blade of grass or a leaf to wait for a potential host to pass by. Once on our clothing, the larva crawls to seek a protected place such as on ankles underneath socks, or around the waist under the elastic of underwear. On the skin, they chew a hole into the surface, secreting digestive enzymes with the saliva that digest cellular contents to liquefy them for easy consumption. This process causes an allergic reaction: severe itching, accompanied by a red, swollen rash. After feeding, the larvae fall off, become nymphs, then mature into adults, which are not parasitic but feed on plant material.

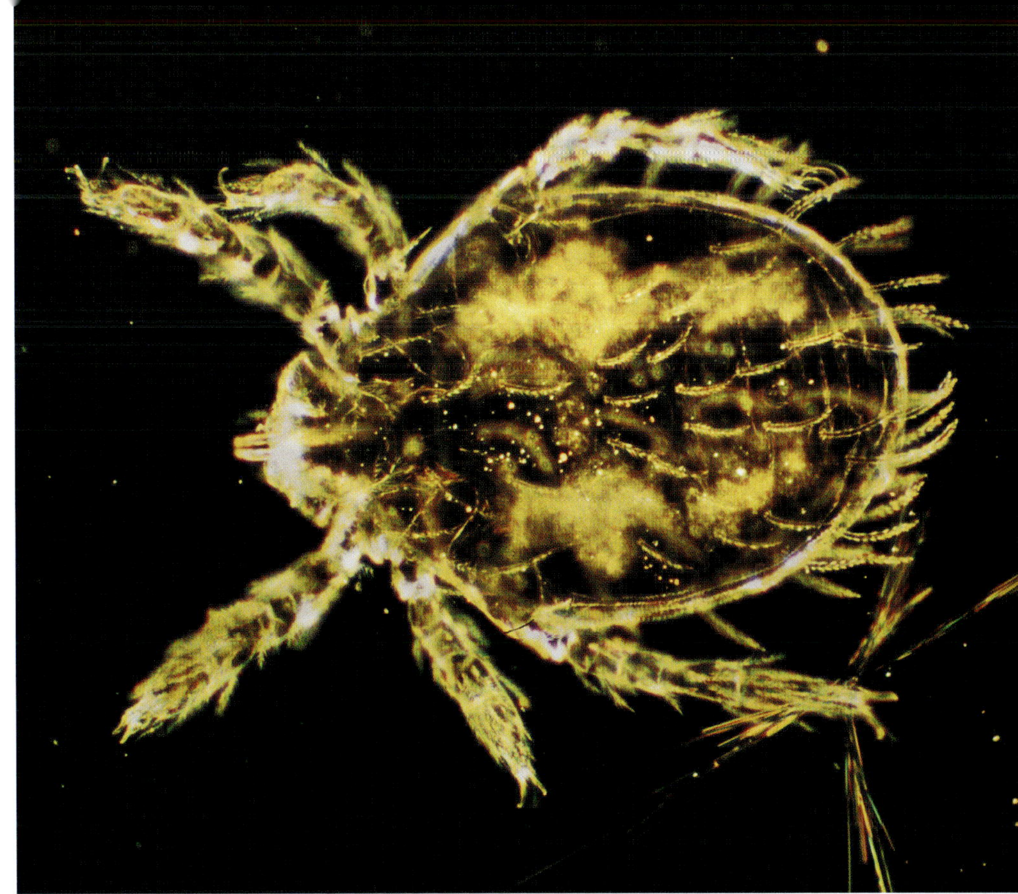

Harvest Mite Chigger, *Neotrombicula autumnalis* (left)

While adult harvest mites feed on plant material, their bright red larval nymphs, or chiggers, which have six legs, infest the skin of mammals including us. Chewing a hole in the skin with their sharp mouthparts, they inject enzyme-rich saliva and suck up the liquefied contents of skin cells, causing intense irritation to the victim.

Magnification: x100   *Dark Field Light Micrograph*

Scrub Typhus Mite Chigger,
Leptotrombidium akamushi (above right)

Leptotrombidium is closely related to the harvest mite. Only the larval nymph, or chigger, feeds on blood, infesting mammals and us. Their bites are irritating but, more seriously, they carry scrub typhus, a potentially fatal disease caused by the bacterium *Orientia tsutsugamushi,* endemic to parts of South-East Asia and the Pacific including Australia.

Magnification: x60   *Dark Field Light Micrograph*

Scabies Mite, *Sarcoptes scabiei* (below right)

The adult female scabies mite burrows into the surface layers of the skin and lays her eggs in the tunnel she has excavated, while feeding on skin cell contents and debris. The activity causes an itchy rash, an allergic reaction, and secondary bacterial infection can result from scratching.

Magnification: x370   *Dark Field Light Micrograph*

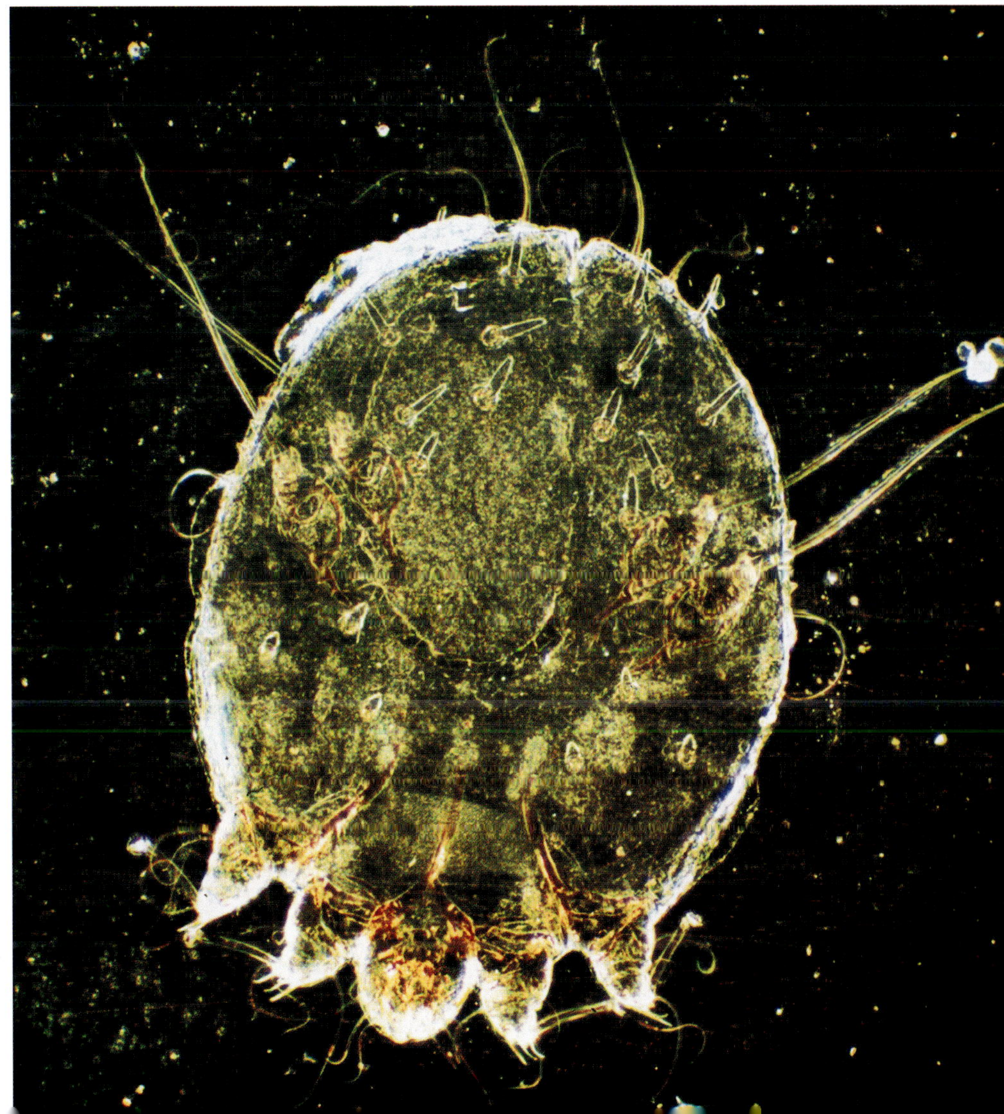

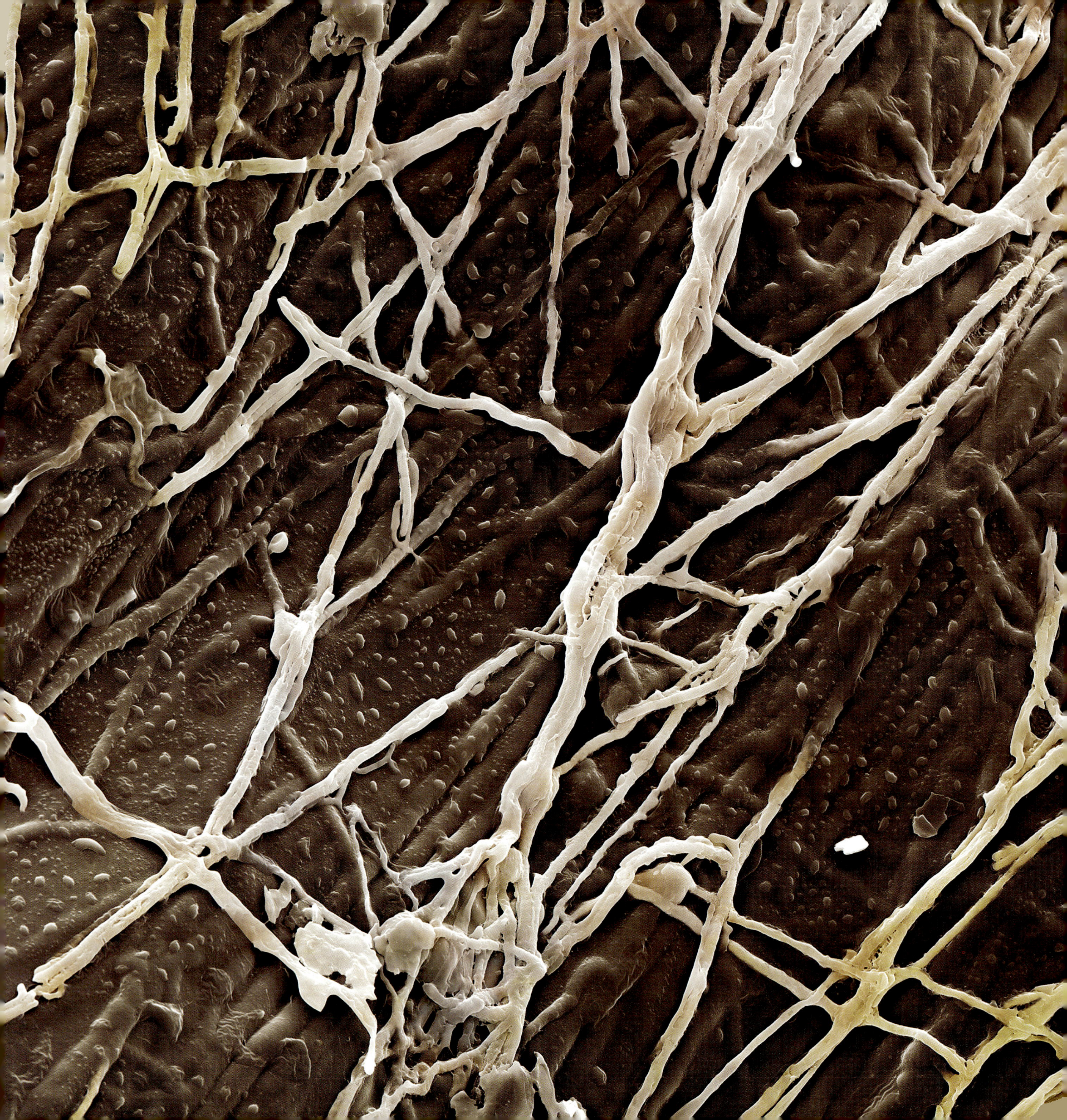

Athlete's Foot Fungus, *Tinea pedis*

Another itchy skin condition, athlete's foot, is a disease caused by the fungus *Tinea pedis*. The fungal spores are everywhere, and feet, especially the area between the toes, provide warmth and humidity encouraging growth. Strands of fungal hyphae grow rapidly across the skin surface, burrowing into the surface layers to extract nutrients. Fruiting bodies grow, producing spores that disperse through the air. Since it is contagious, the warmth and humidity around public swimming pools or sports changing rooms makes them common sources of infection, hence the name athlete's foot. In severe cases, blisters form and can lead to cracking of the skin, and if the toenails become infected, they become crumbly and hard to cut.

Magnification: x1800
Scanning Electron Micrograph (left)
Magnification: x1000
Scanning Electron Micrograph (right)

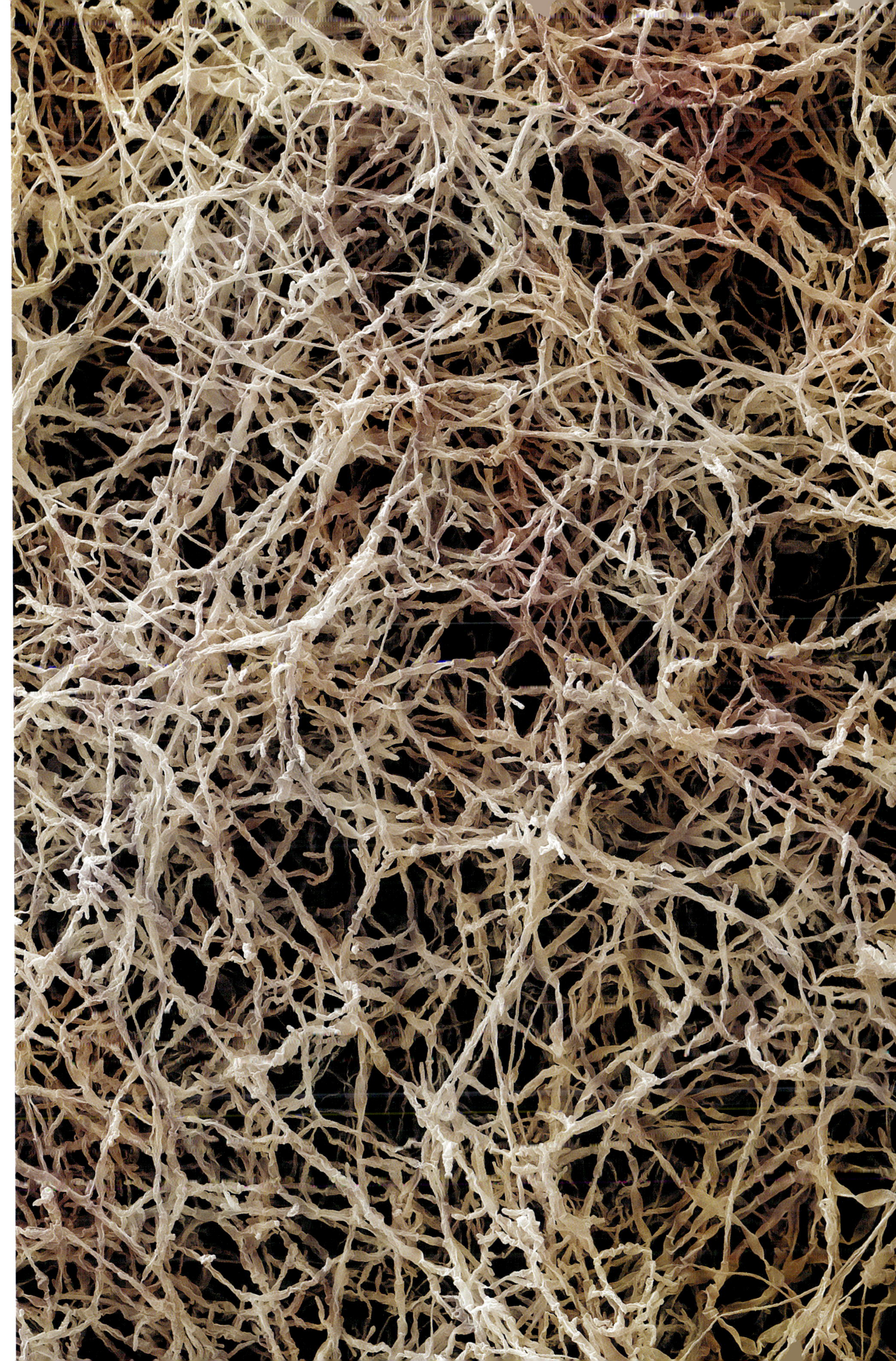

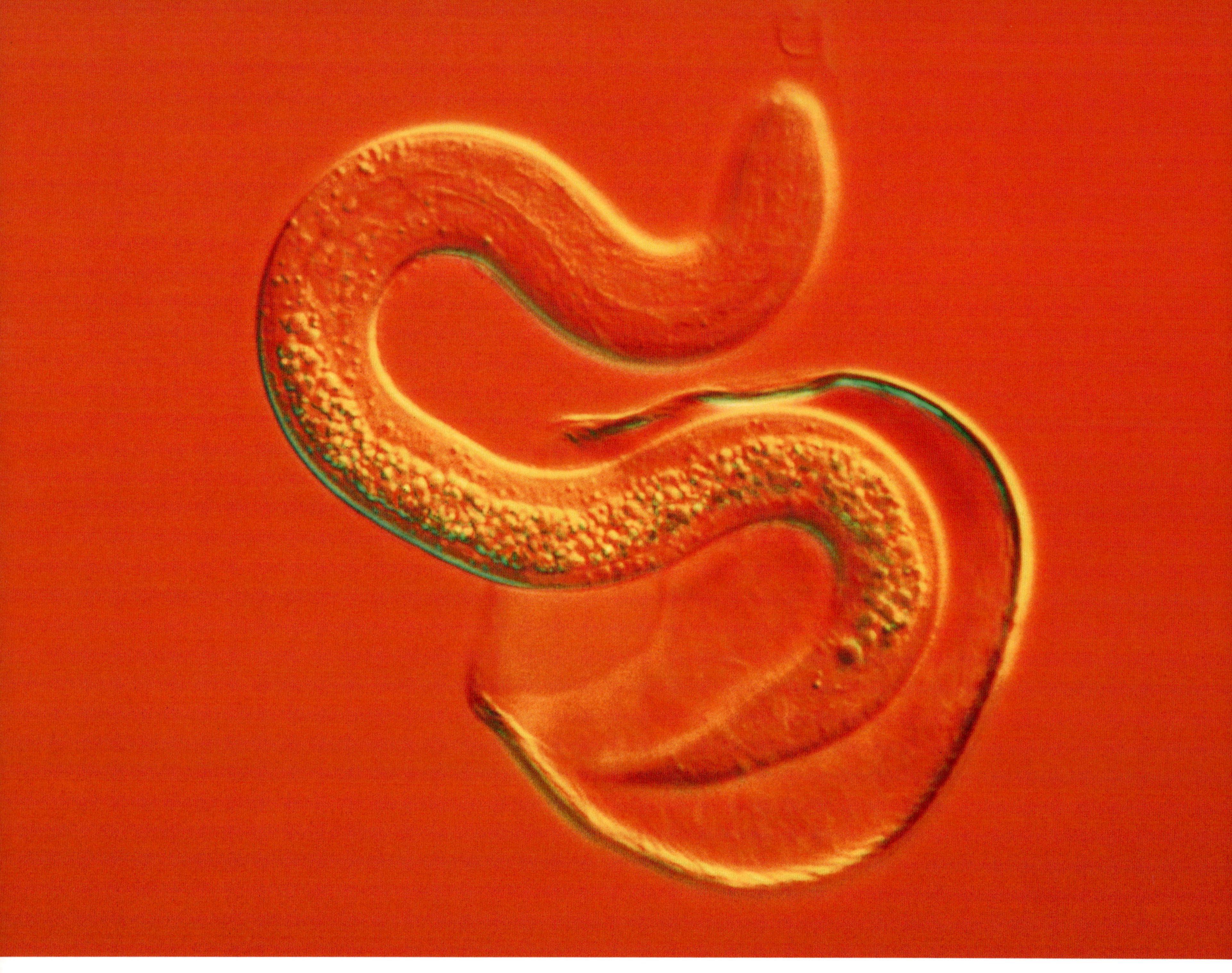

Intestinal Roundworm, *Ascaris lumbricoides* (above)

A larval roundworm hatches from an infective egg, a process that normally occurs in the human stomach. The larvae and eggs are microscopic, but as adults the worms can grow up to 30 cms long.

Magnification: x1600
Light Micrograph, using Nomarski Technique

Human Pinworm, *Enterobius vermicularis* (right)

In contrast to the roundworm, the pinworm is very small but is a very common infection of the human intestine, especially in children. The females carry thousands of eggs and make nocturnal excursions out of the anus to deposit some of them before crawling back inside.

Magnification: x30
Dark Field Light Micrograph

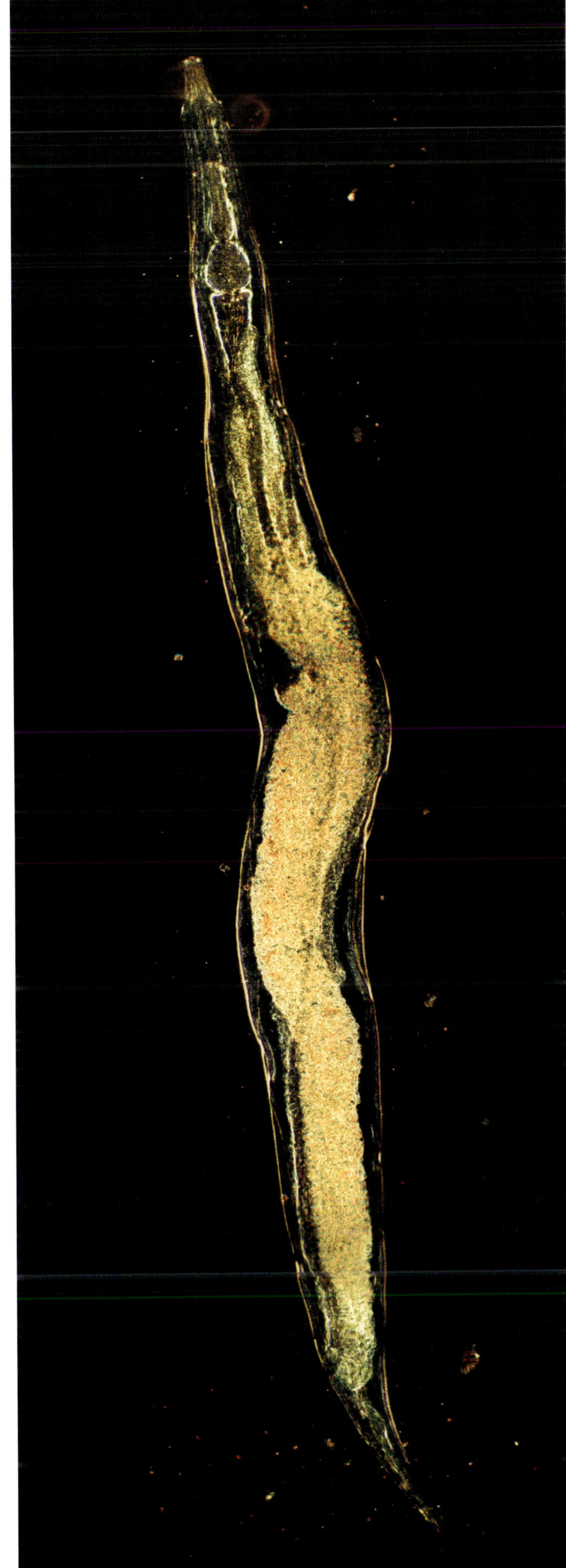

ROUNDWORM

There are many kinds of small worms that parasitise the human body, causing a variety of unpleasant diseases or conditions. Infections are quite common throughout the world, in fact the intestinal roundworm, *Ascaris lumbricoides*, the female of which grows up to 35 cm in length, infects around 1.5 billion people worldwide. The adult worms live in the small intestine and females can produce up to 200,000 eggs per day, which are passed out of the body with the faeces. There is no intermediate host, re-infection or transmission being achieved by touching the mouth with hands that have become contaminated or by ingestion of infective eggs in raw food such as fruit and vegetables. The larvae hatch inside the stomach, enter the blood stream and migrate to the lungs where they moult twice before making their way into the alveoli. From there, they are coughed up and swallowed, finally arriving in the small intestine where they mature and complete their life cycle.

PINWORM

The human pinworm, *Enterobius vermicularis*, infects the caecum and lower gut of over 200 million people each year. Unlike many other worm parasites, it is relatively rare in the tropics but quite common in the temperate regions of Western Europe and North America. The thin, white worms are small, the female only 1 cm long by 0.4 mm wide, and the male is even smaller. The female carries up to 15,000 eggs, and after migrating towards the anus, she makes regular nocturnal journeys outside, where contact with the air stimulates her to deposit eggs, before retreating back into the rectum. To infect a new host, the eggs must be ingested after which they hatch in the duodenum. Some of the eggs may hatch around the anus of the original host, in which case the larvae crawl back into the rectum and re-infect the caecum. Anywhere where there are large numbers of children gathered together, such as nurseries or playgroups, especially if conditions are unsanitary, are ready sources of infection, as one child may rapidly transmit the parasite to others. The emergence of egg-bearing females from the anus at night may cause itching, and scratching transfers eggs to the fingers and under the fingernails, aiding re-infection.

HOOKWORM

Hookworm, *Ancylostoma duodenale*, is another intestinal parasite that infects about one billion people worldwide. Heavy infections can be serious for young children, pregnant women, or the elderly. Hookworms have a life cycle that begins and ends in the small intestine where the adults, just over 1 cm in length, attach themselves to the intestinal wall to feed on blood. The female produces thousands of eggs, which are passed out of the body with the faeces. If the eggs successfully contaminate warm, moist soil, they will hatch, moult, and develop into infective larvae after about a week. The microscopic larvae enter the body of a new host by penetrating the skin, often through bare feet. From there, they are carried in the bloodstream to the lungs, where they are coughed up into the mouth and swallowed. Eventually they reach the small intestine, where the larvae grow into adult worms. The first sign of hookworm infection is an itchy rash at the site where the larva has penetrated the skin. Once the infection becomes embedded in the intestine, the symptoms are diarrhoea, loss of appetite, weight loss, and anaemia. Chronic hookworm infection can affect the growth and development of children due to the loss of iron and protein.

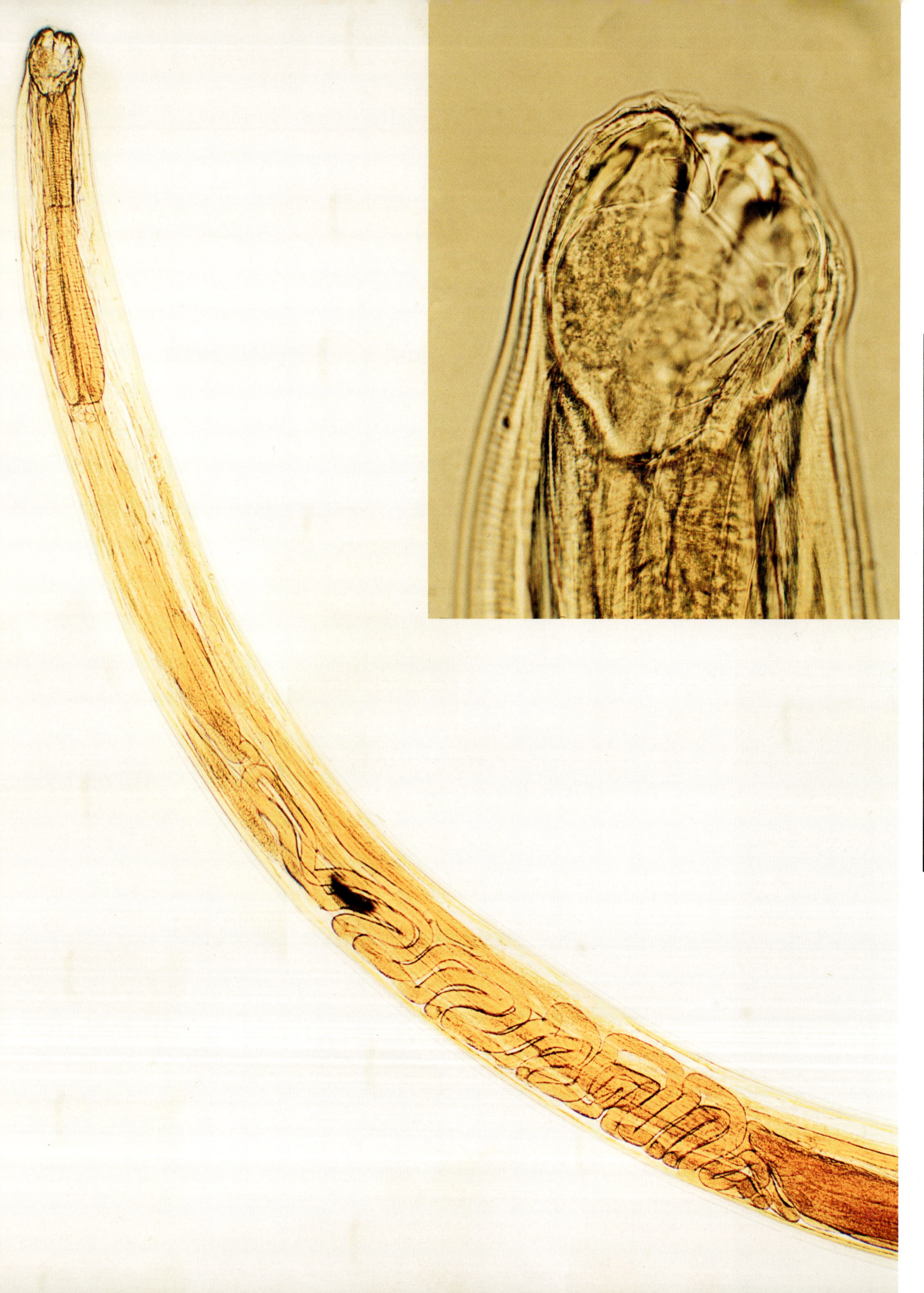

Hookworm, *Ancylostoma duodenale & Necator americanus*

One of the most common intestinal parasites, hookworms infect some one billion people worldwide. The close up photograph of the head of *Ancylostoma* (see inset, left) shows the curved jaws or 'hooks' that attach the worm to the intestinal wall. The tail of *Necator* (see right), photographed using the Nomarski technique, shows the reproductive organs.

Magnification: x300
Light Micrograph (inset, left)
Magnification: x50
Light Micrograph (left)
Magnification: x150
Light Micrograph (right)

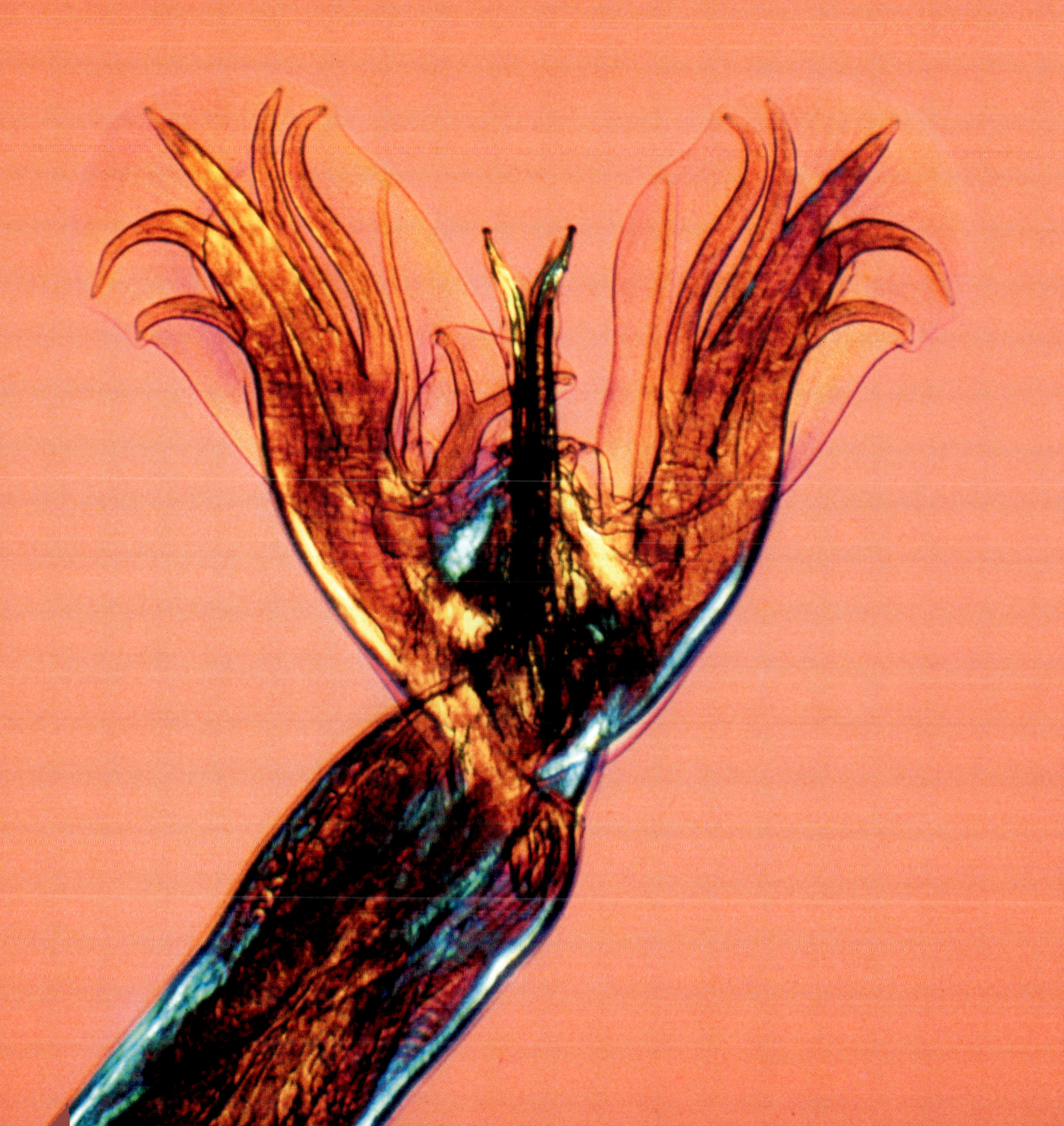

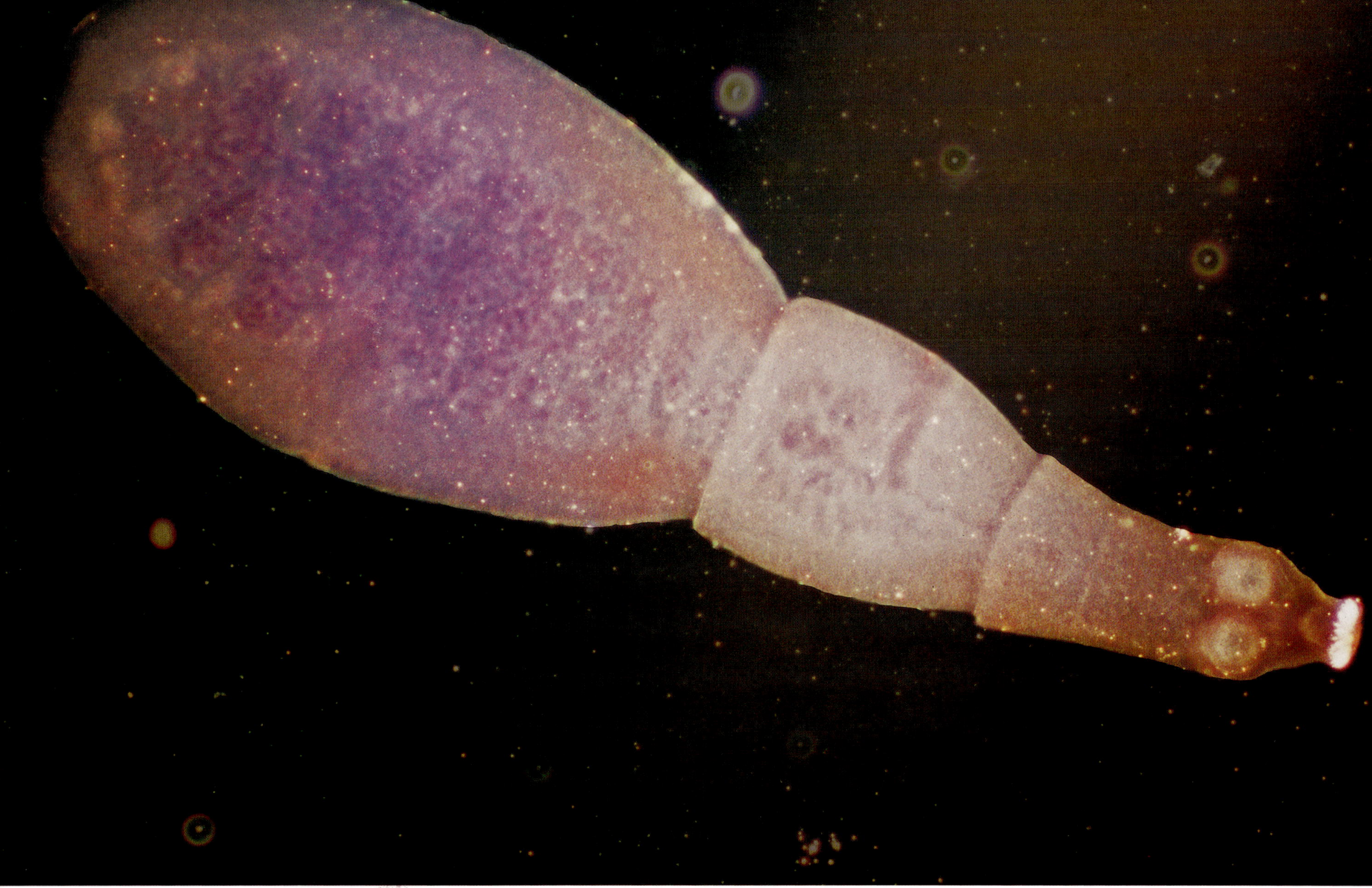

TAPEWORM

Tapeworms, related to parasitic flatworms, are long and segmented, living in the digestive tract of vertebrates where they absorb nutrients through their skin. *Taenia solium*, which infects humans worldwide and infects some 50 million people annually, can grow up to a length of 8 metres. It consists of a head, or scolex, which has hooked protrusions to attach to the wall of the intestine, and a body that is a series of up to 1,000 gradually ripening segments full of eggs, which are sequentially shed into the gut and migrate out of the anus to begin the next stage of their life cycle. The migrating segments move, causing itching and may appear on toilet paper or underclothing. The adult *Taenia* tapeworm is only found in humans, the definitive host, but the larvae, or cysticerci, are carried in the muscle and subcutaneous tissue of pigs. We become infected if we eat undercooked pork that is contaminated with larvae, and, occasionally, we can also act as the intermediate host if we ingest the eggs or auto-contaminate ourselves when we already have a tapeworm. The condition can be serious, the symptoms of an adult tapeworm in our gut causing belly discomfort, nausea, hunger, and weight loss, and if the larvae encyst, they can spread to the central nervous system, causing seizure, even death.

Tapeworm,

Echinococcus granulosus (left)

A small tapeworm with a ring of hooks and some suckers on its head, or scolex, to cling onto the wall of the intestine. The normal definitive host for *Echinococcus* is a carnivore such as a dog, with herbivores such as cattle as intermediate hosts. It can also infect humans, sometimes causing serious complications. The eggs hatch into tiny hooked larvae, which migrate through the bloodstream to various parts of the body, even the brain, forming hydatid cysts. These cysts can grow to the size of a football and become life threatening.

Magnification: x50
Light Micrograph

Tapeworm,

Taenia solium (right)

A close-up view of the head, or scolex, of *Taenia solium*, reveals the double row of hooks, and four suckers, that enable it to attach to the intestinal wall.

Magnification: x100
Light Micrograph

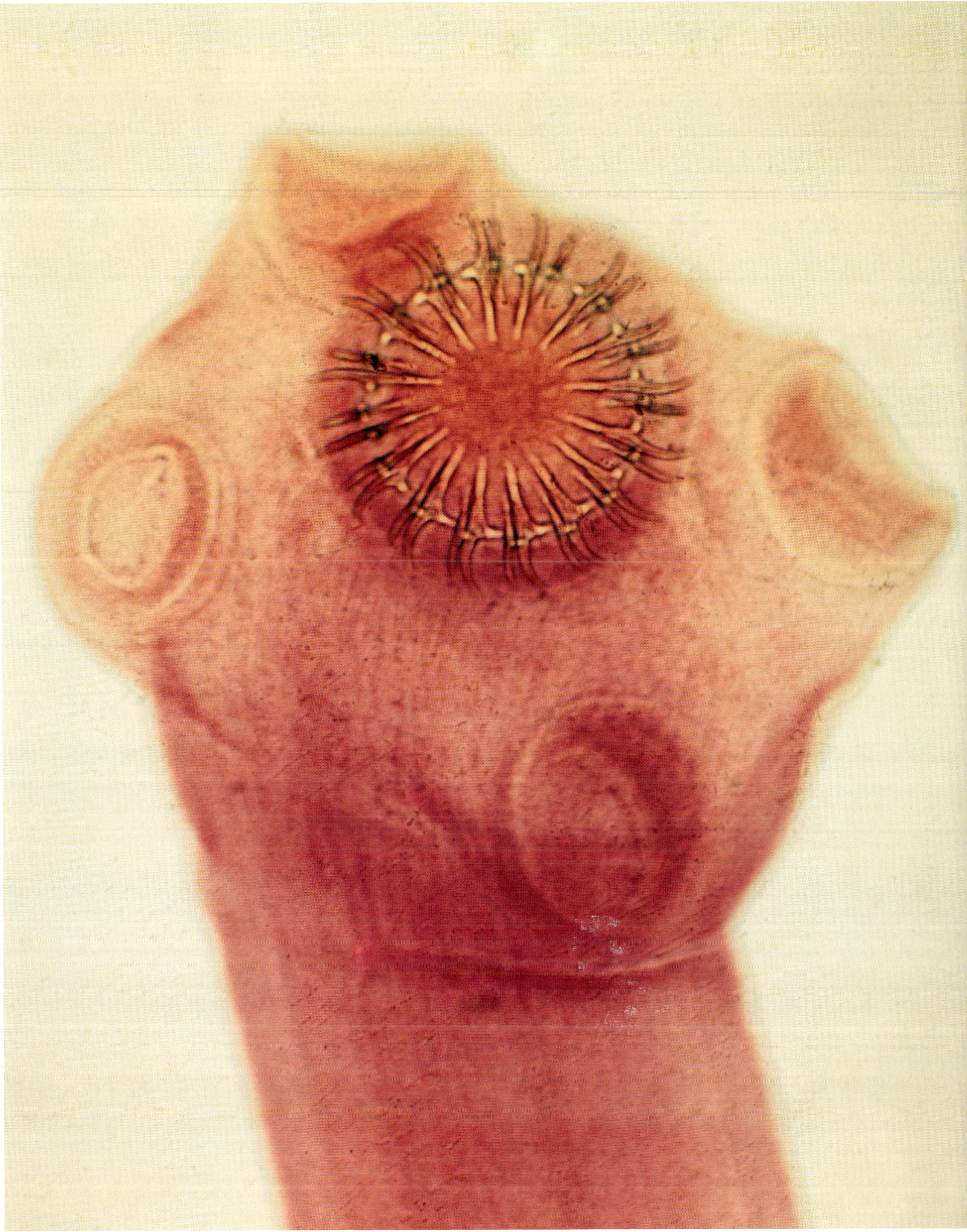

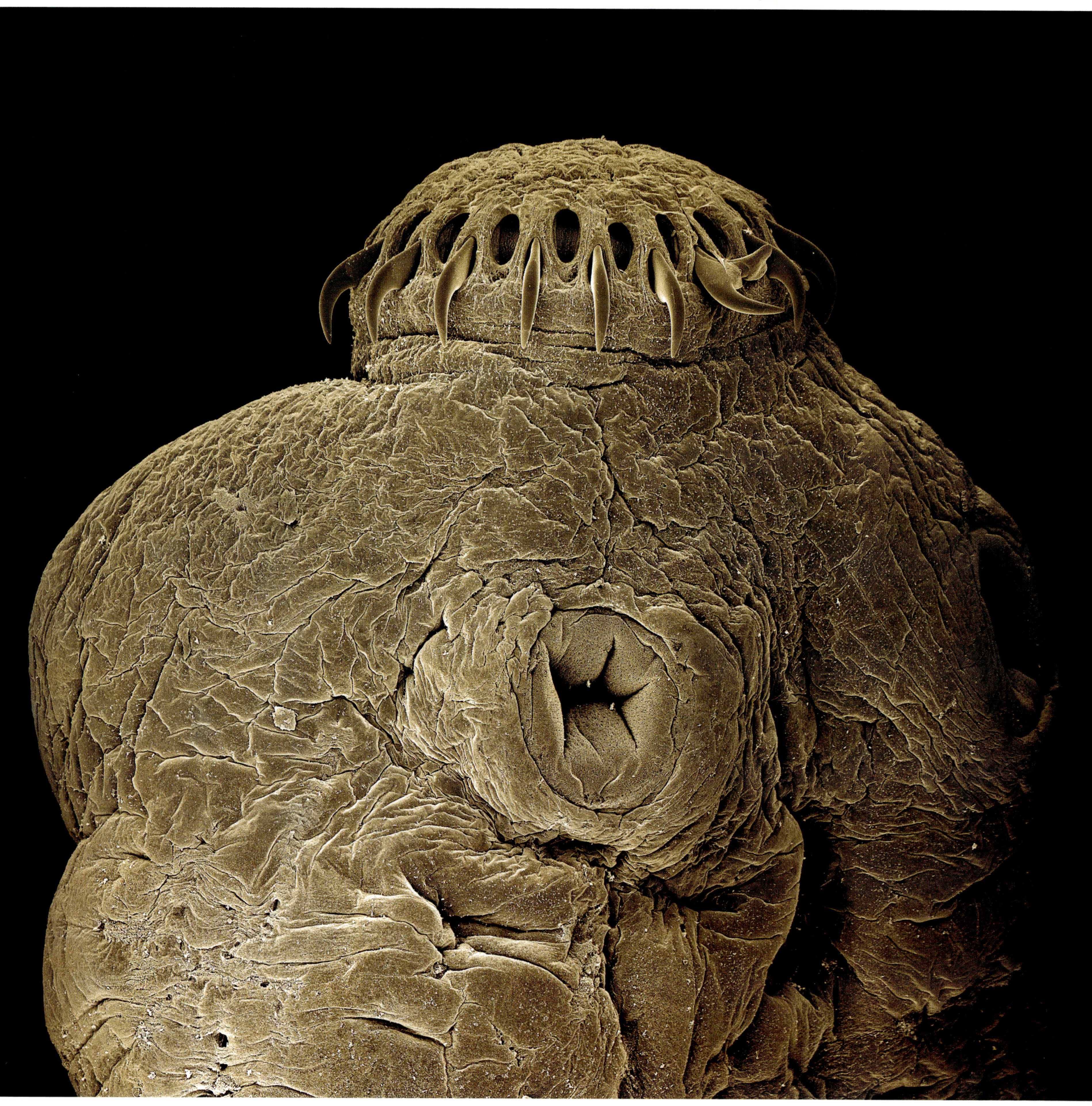

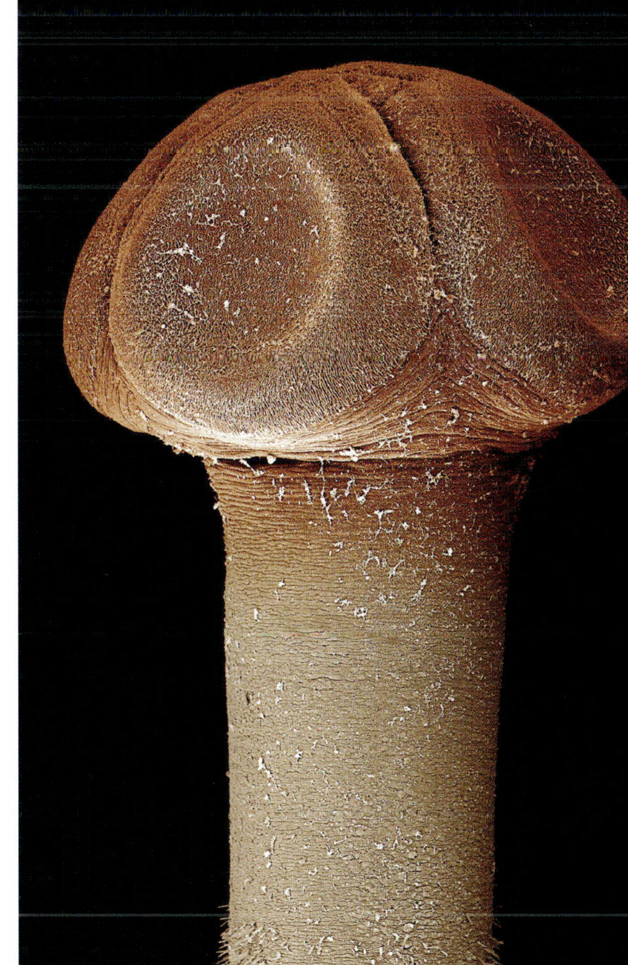

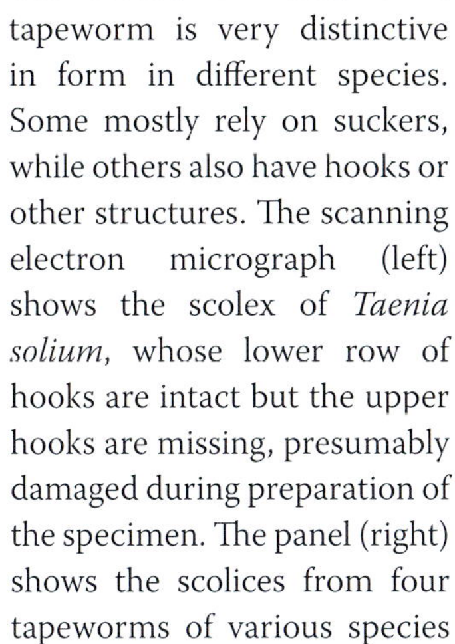

Tapeworm Scolex

The scolex, or head, of a tapeworm is very distinctive in form in different species. Some mostly rely on suckers, while others also have hooks or other structures. The scanning electron micrograph (left) shows the scolex of *Taenia solium*, whose lower row of hooks are intact but the upper hooks are missing, presumably damaged during preparation of the specimen. The panel (right) shows the scolices from four tapeworms of various species found in sharks.

Magnification: x300
Scanning Electron Micrograph (left)
Magnification: unknown
Scanning Electron Micrographs (right)

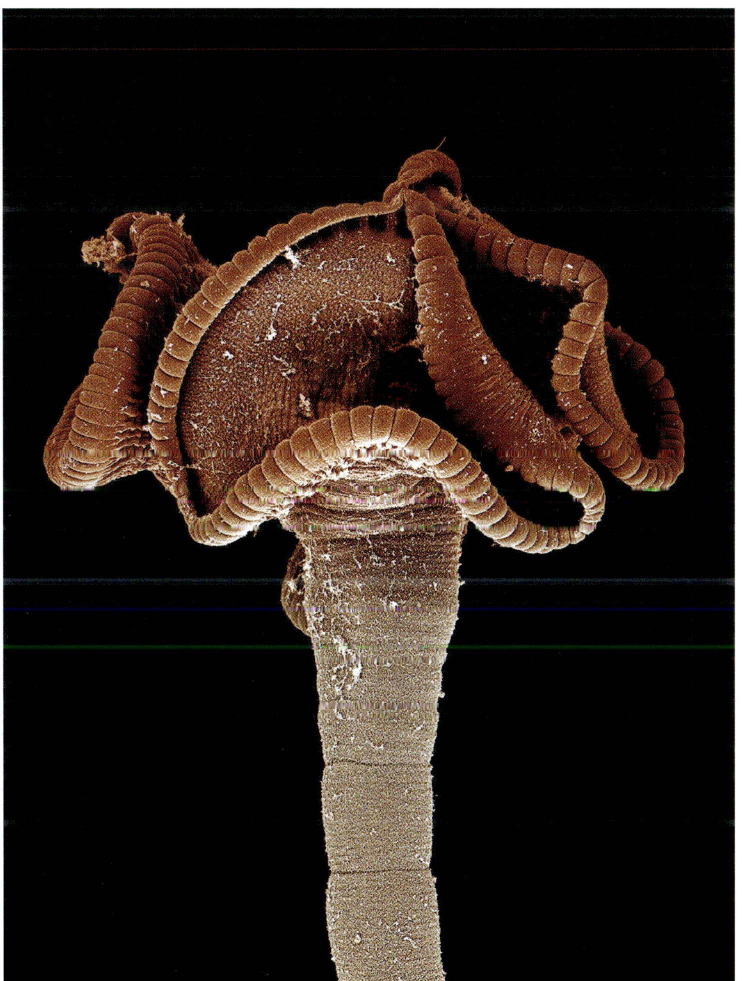

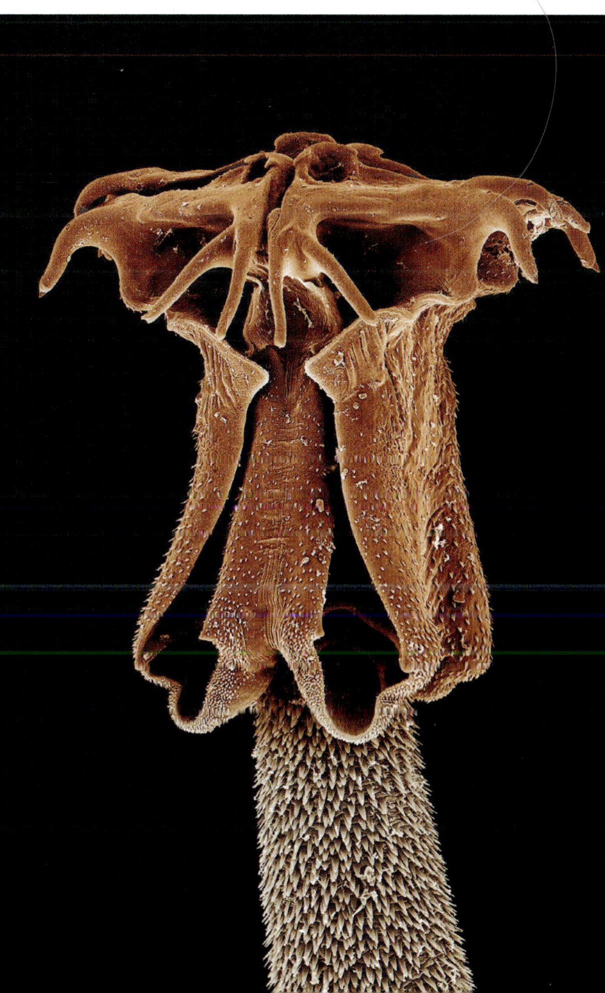

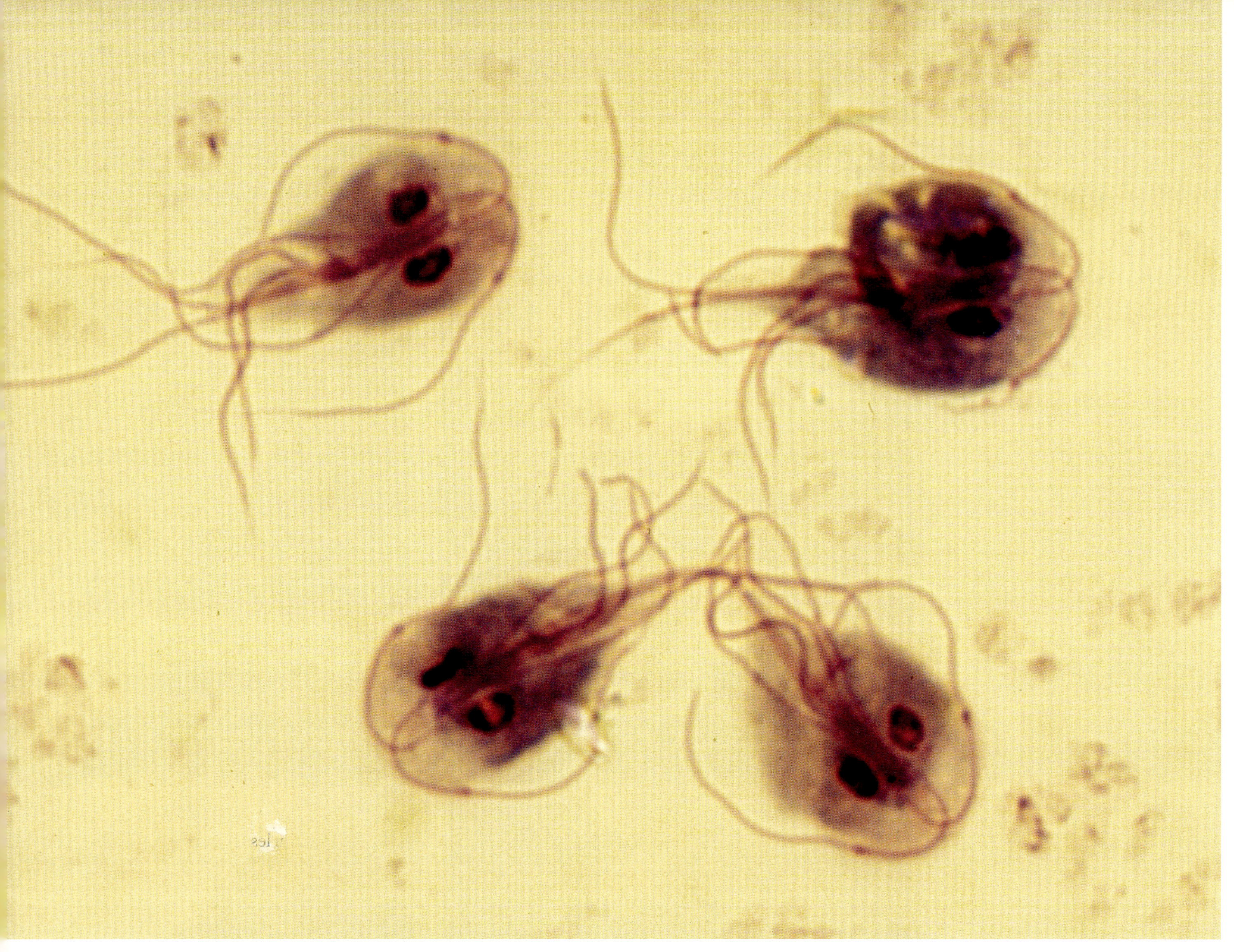

Flagellate Protozoan Parasite, *Giardia lamblia* (above)

A major cause of intestinal disease worldwide, *Giardia* infects human through the consumption of contaminated water and food. Once ingested, the trophozoite, or motile stage, which is distinguished by having two nuclei, emerges to feed on mucus in the gut, causing abdominal pain, gas and diarrhoea.

Magnification: x5000   *Light Micrograph*

Lancet Liver Fluke, *Dicrocoelium dendriticum* (right)

Flukes are flat leaf-shaped parasites that infect the bile ducts of the liver, causing fever or anaemia, and sometimes the rupture of the liver capsule, a potentially fatal condition. Flukes have multi-host life cycles, for example, *Dicrocoelium* requires a mammalian herbivore such as a sheep or rabbit, as well as a snail, and an ant.

Magnification: x25   *Dark Field Light Micrograph*

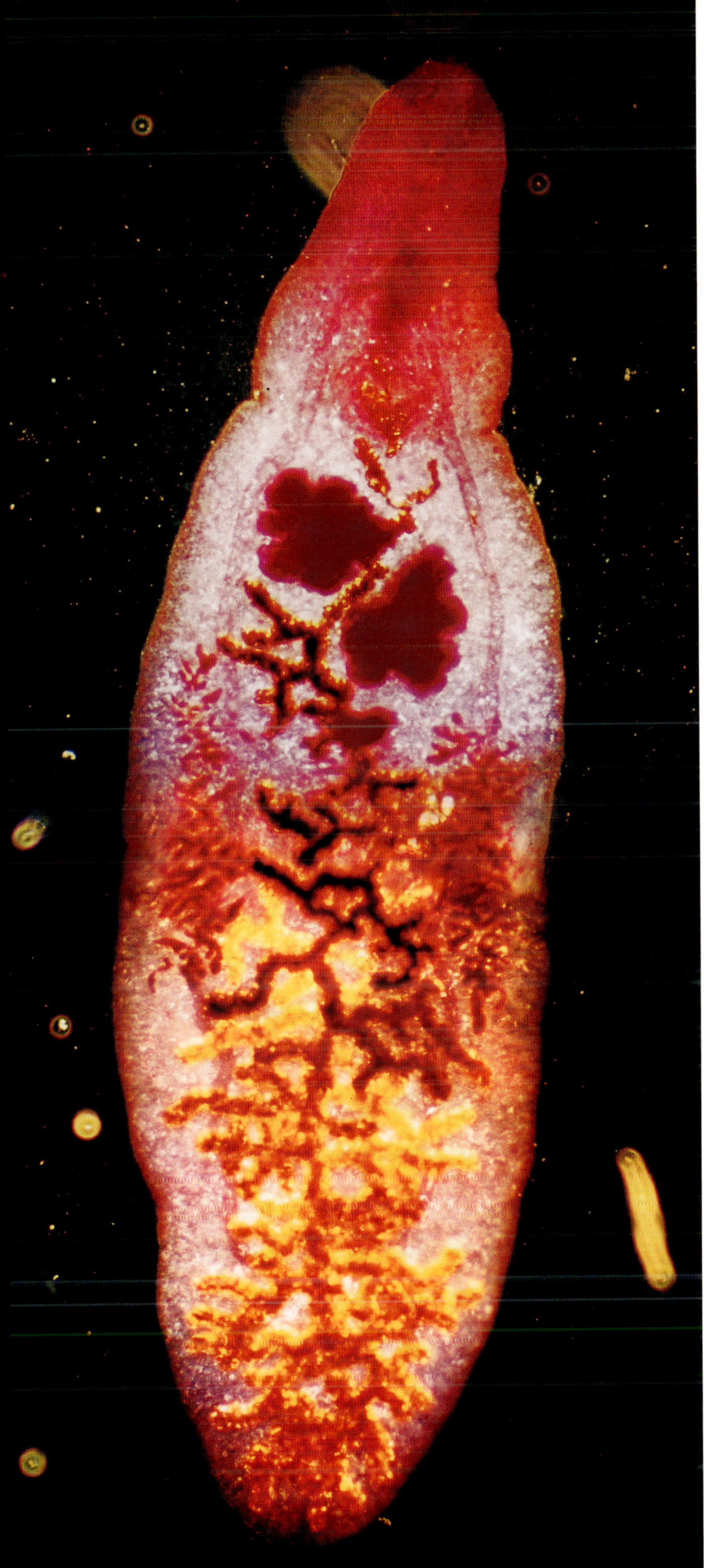

FLUKE

Found throughout the world, flukes are leaf-shaped parasitic flatworms that live in the bile ducts of the liver of large mammals, and some of them can infect humans. The liver fluke, *Fasciola hepatica*, is about 3 cm long; its hosts are normally sheep and cattle, but it can also infect us if we eat plant foods such as watercress that are contaminated. The life cycle of *Fasciola*, like many flukes, is complicated. While in the bile ducts, the parasite produces eggs, which pass into the duodenum, then exit the body with faeces onto grazing pasture. These hatch into mobile larvae called *miracidia*, which infect the snail, *Lymnaea truncatula*, by penetrating its body wall. There, the larvae develop into another mobile stage called *cercaria*, which leave the snail, when conditions are warm and moist, to attach to blades of grass or other vegetation. The *cercaria*, an infective stage, encyst and, providing conditions remain moist, can survive for many months, but if ingested by a cow, a sheep or us, the cyst releases young flukes, which migrate to the liver to feed on blood and begin producing eggs.

Another parasitic flatworm, the lancet fluke, *Dicrocoelium dendriticum*, requires three different hosts, a mammalian herbivore such as a cow, sheep or rabbit, a snail and an ant. The adult fluke, living in the liver of the mammal, produces eggs, which are passed out of the gut with faeces. Snails ingest the eggs, which hatch into larvae that are expelled in the snail's slime trail. Ants are attracted to the slime trail as a source of moisture, and unwittingly ingest the larvae, which proceed to control the ant's behaviour, forcing it to climb tall stems of grass where it remains motionless, waiting to be eaten by a grazing mammal to complete the life cycle. There have been cases where humans have been infected. Usually, it causes fever and anaemia, but if large numbers of cysts are ingested they can cause a traumatic hepatitis, and in some cases, the liver capsule may rupture into the peritoneal cavity causing death due to peritonitis. In some cases, a potentially fatal infection by the bacterium *Chlostridium oedematiens* can also take hold in lesions produced by the young larvae as they migrate into the liver.

GIARDIA

Once diagnosed, the presence of worms in the digestive system is relatively easy to treat. There is one condition, however, that can be more difficult to eliminate. Usually associated with recent foreign travel, it is normally contracted by the consumption of water or food contaminated by the faeces of infected humans or animals. The parasite is a tiny flagellate protozoan, *Giardia lamblia*, and a typical infestation of *Giardia* can number in the millions, causing extreme discomfort, with trapped gas, flatulence, diarrhoea, and greasy stools that tend to float. The difficulty with digestive problems caused by protozoa or bacteria is that they can be notoriously difficult to diagnose since the symptoms tend to be similar. Some of the most violent conditions, causing vomiting and diarrhoea, are bacterial in origin.

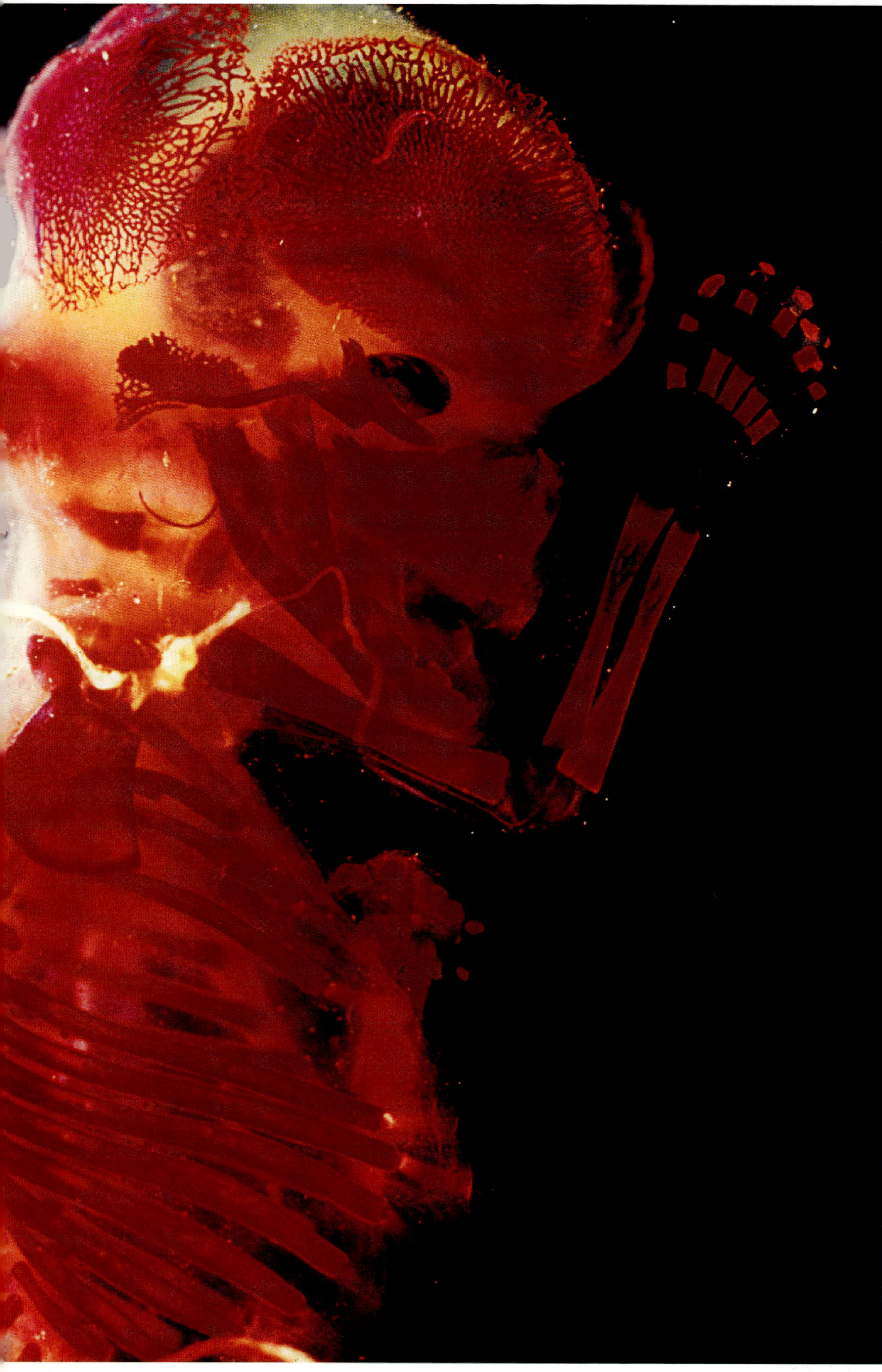

THE PERFECT PARASITE?

Humans are fragile creatures, easily destroyed by predators or disease. Considering the variety of hungry organisms that invade or feed on the human body, it is surprising that we have survived until the 21st Century, but we owe our immortality as a species to the most perfect and efficient 'parasite' of all, the human embryo. Classing our own offspring as parasites may seem unfair and open to debate. A parasite, after all, is defined as an organism that invades and benefits from an individual of a different species. But the human embryo still exhibits a number of traits characteristic of a parasite. During coitus, male spermatozoa invade the female body. With a tail that thrashes about to provide movement, a spermatozoid might almost be considered as an organism in its own right. It vigorously swims, carrying its package of human DNA, together with millions of other spermatozoa, to meet the female ovum or egg as it travels towards the uterus. Just one sperm will penetrate the ovum and fertilise it, resulting in a zygote, a genetically unique pre-embryo. After just 72 hours, the new entity organises itself, forming an inner group of cells, which will develop into the embryo itself, and an outer group, or trophoblast, which will become the placenta. At this point, the mother is all set to reject this foreign body, for on day 5 or 6, as it secretes an enzyme to prepare the uterus lining as an implantation site, the mother reacts by producing antibodies. The pre-embryo, however, suppresses the mother's immune system, thus beginning a lifelong relationship, involving years of dependence.

Human Embryo, *Homo sapiens*

The human embryo controls the immune system of the human female to prevent its rejection as it implants itself in the uterus lining. For the next nine months, it draws nutrients from its host, the mother, via a structure called a placenta.

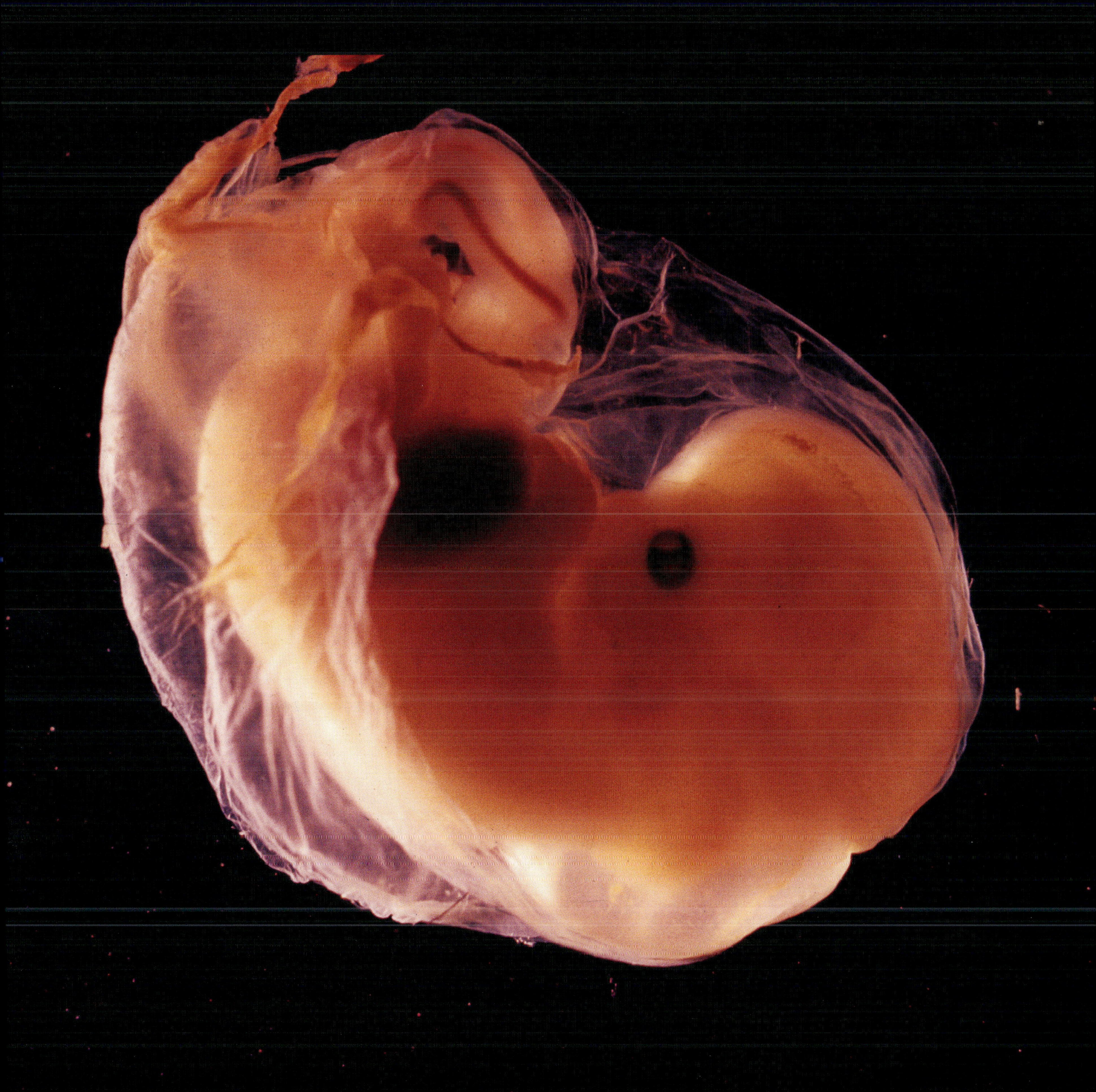

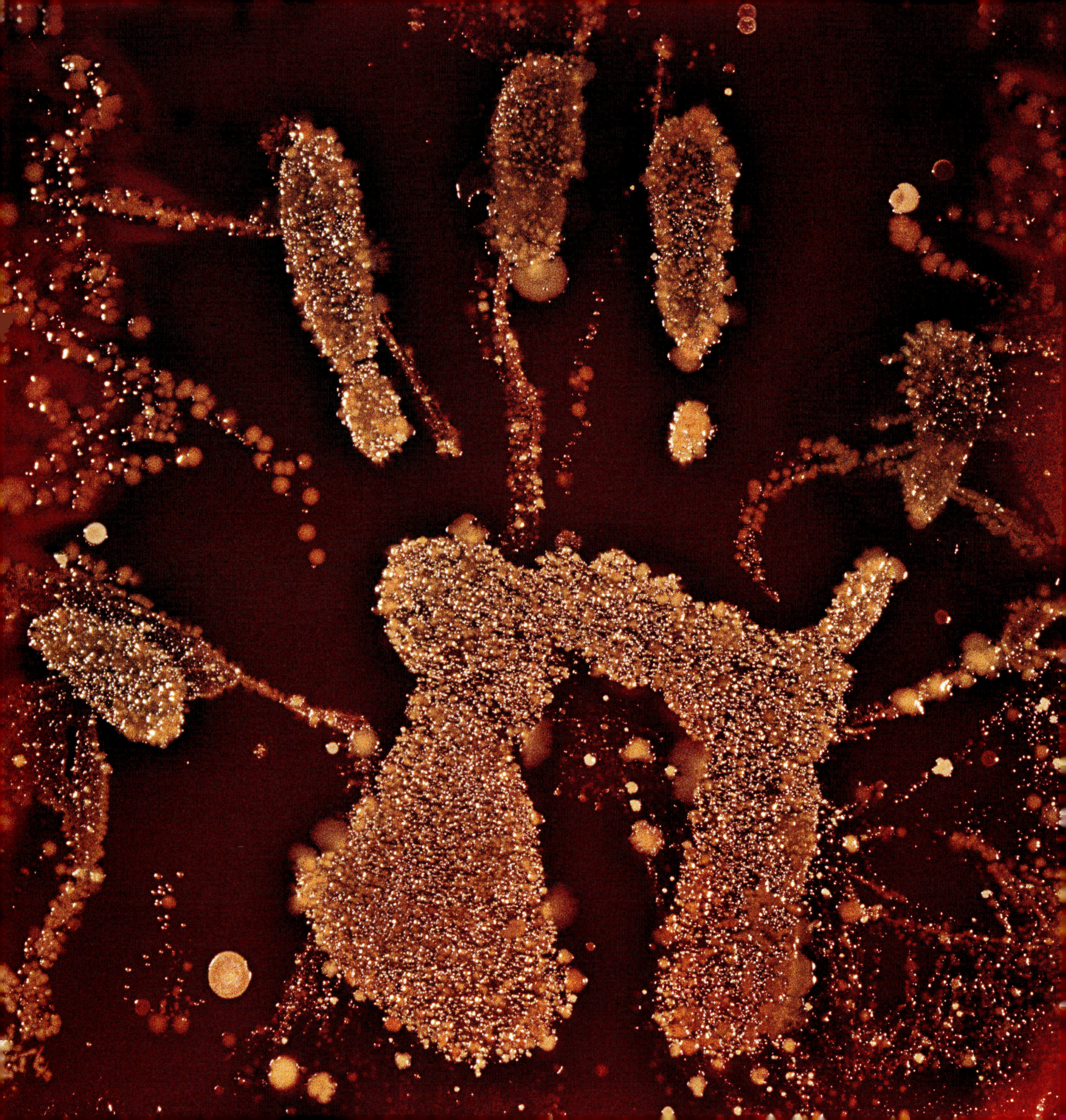

BACTERIA

Bacteria are tiny, diverse and incredibly prolific, playing a vital role in fixing nitrogen from the atmosphere and in the recycling of nutrients. Unicellular organisms, they exist in a variety of forms, mostly shaped as rods, spheres or spirals, and occur in every conceivable habitat, from the highest mountains to deep in the Earth's crust. Bacterial cells, unlike those of plants and animals, do not possess nuclei or membrane-bound organelles, and multiply by simply dividing in two, a process that under ideal conditions can repeat every few minutes. They exist in huge numbers. It is estimated that there are some two hundred different species of bacteria just living on the surface of our skin, and hundreds more inside us. Most of these are beneficial, helping to keep us healthy, and protect us against infections. There are a few pathogenic species or strains however producing toxins that are excreted, endotoxins that are liberated when the bacteria die, or enzymes that attack healthy cells.

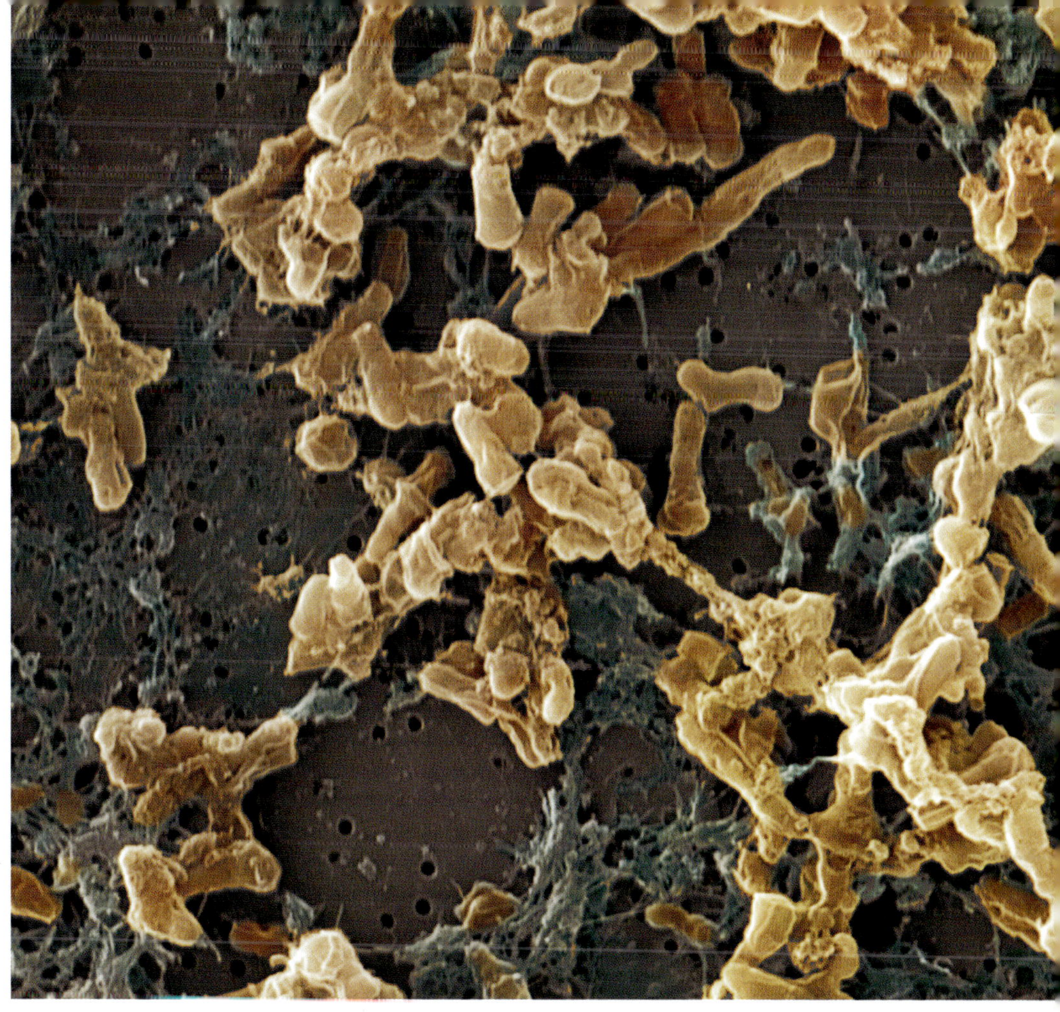

Human handprint with associated bacteria (left)

A washed hand was stroked over 'clean' hair, then pressed down onto a nutrient agar gel. After incubation for 48 hours, many different kinds of bacterial colonies had grown.

Plague Bacterium, *Yersinia pestis* (above right)

The cells of *Yersinia* are rod-shaped and non-motile. The cause of bubonic plague and the 'Black Death', it is primarily carried by the fleas of rats, and transmission to humans occurs when infected fleas feed on our blood. It kills by cutting off immune system pathways needed to fight off the bacterial invasion, leaving the host without normal natural defences.

Magnification: x15000   *Scanning Electron Micrograph*

Bacterium, *Chlamydia trachomatis* (below right)

The bacterium *Chlamydia trachomatis* causes one of the most common sexually transmitted infections worldwide, with over 50 million known new cases worldwide each year. *Chlamydia* has a unique life cycle, invading host cells and replicating inside them before bursting out to re-infect other cells or be dispersed to a new host. The condition can be asymptomatic but can lead to infertility in both women and men, as well as other complications.

Magnification: x60000   *Scanning Electron Micrograph*

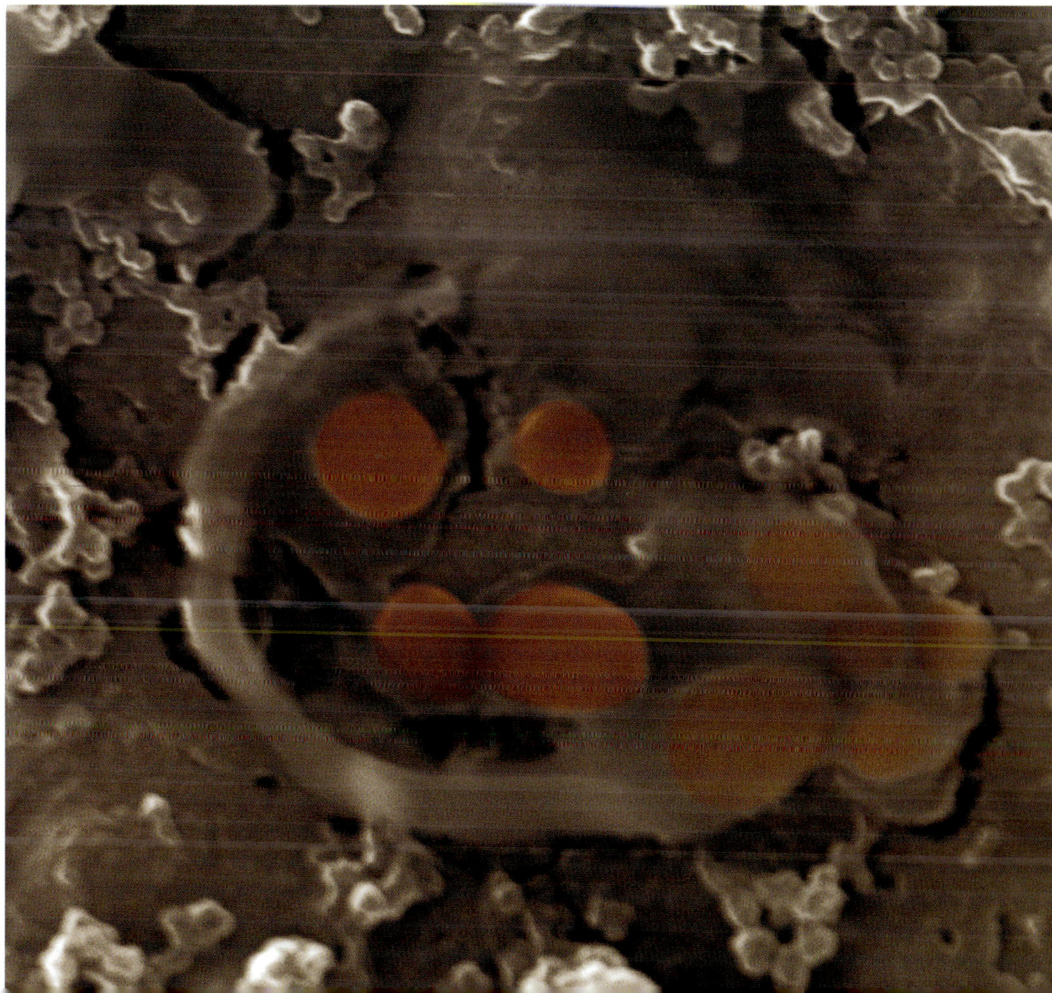

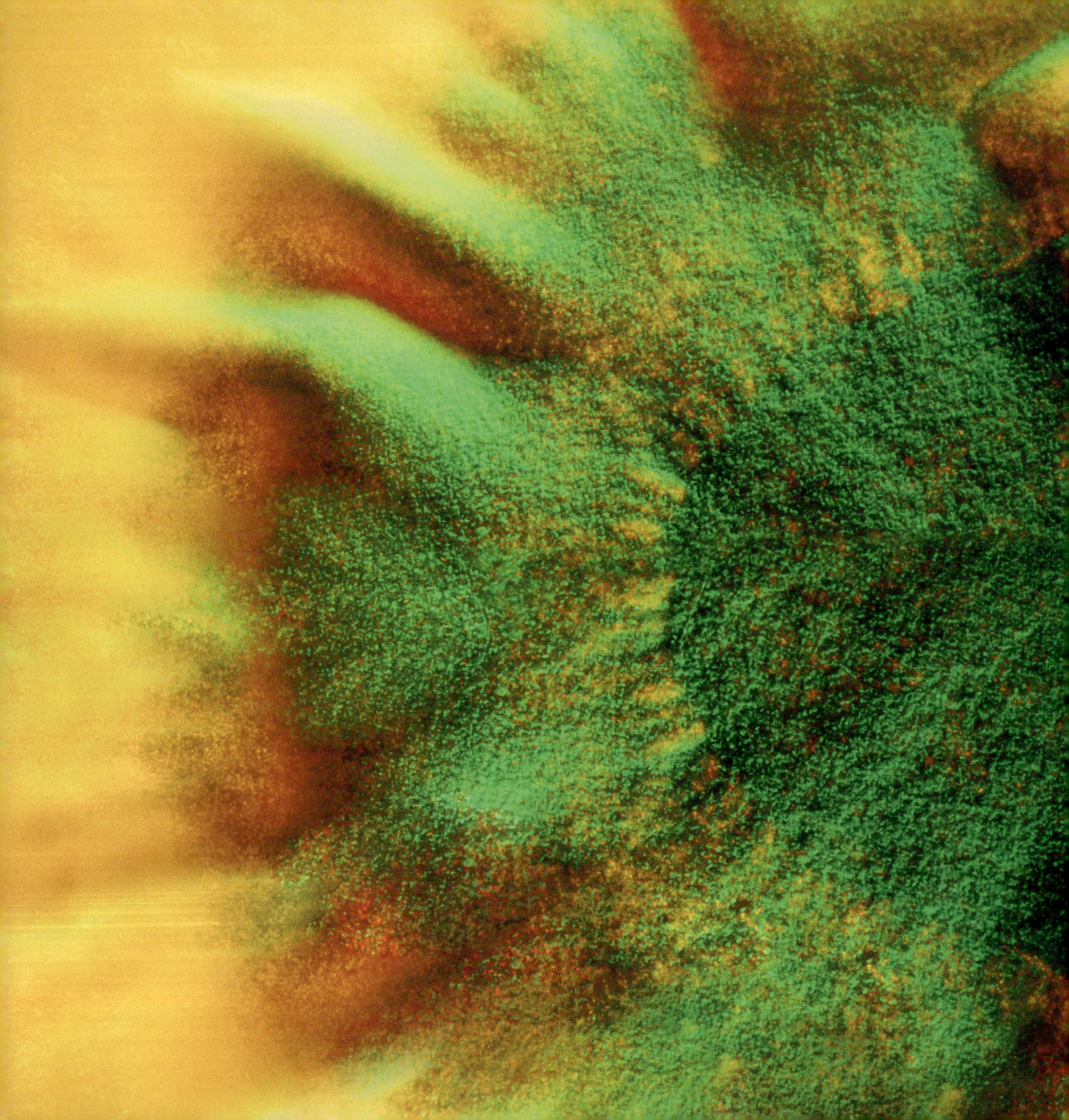

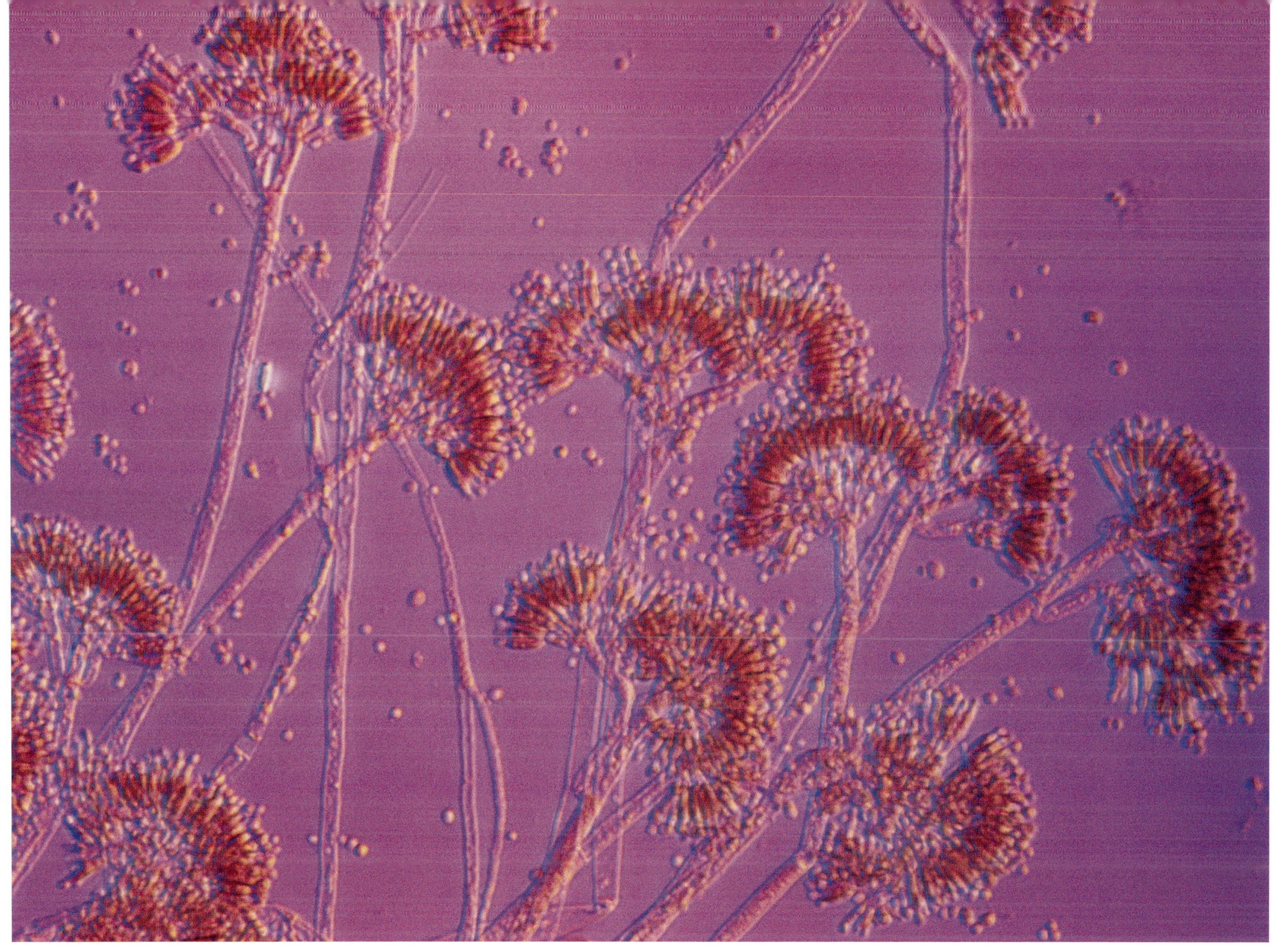

ANTIBIOTICS

Antibiotics, for example penicillin and tetracycline, are chemical substances, produced by organisms such as moulds and other bacteria that either destroy or inhibit the growth of harmful bacteria. Some damage their cell membranes, preventing the synthesis of bacterial cell wall components, and others interfere with their protein or nucleic acid synthesis. The bacteria under attack, however, can produce enzymes that may neutralise antimicrobial compounds, and evolve resistance to weak, short-term or repeated exposure to antibiotics, which is why it is always important to complete a course of antibiotics when treating a bacterial infection.

The Mould Penicillium

A penicillin culture (see left) shows the production of green spores. The spores, produced on fruiting bodies called conidiophores (see above) are carried away on air currents. Moulds of many different kinds grow everywhere in nature and are familiar as furry growths on old bread or fruit. They are important in the production of certain foods, such as cheese, and antibiotics.

Magnification: x6 — *Macro-photograph* (left)
Magnification: x2750 — *Light Micrograph* (above)

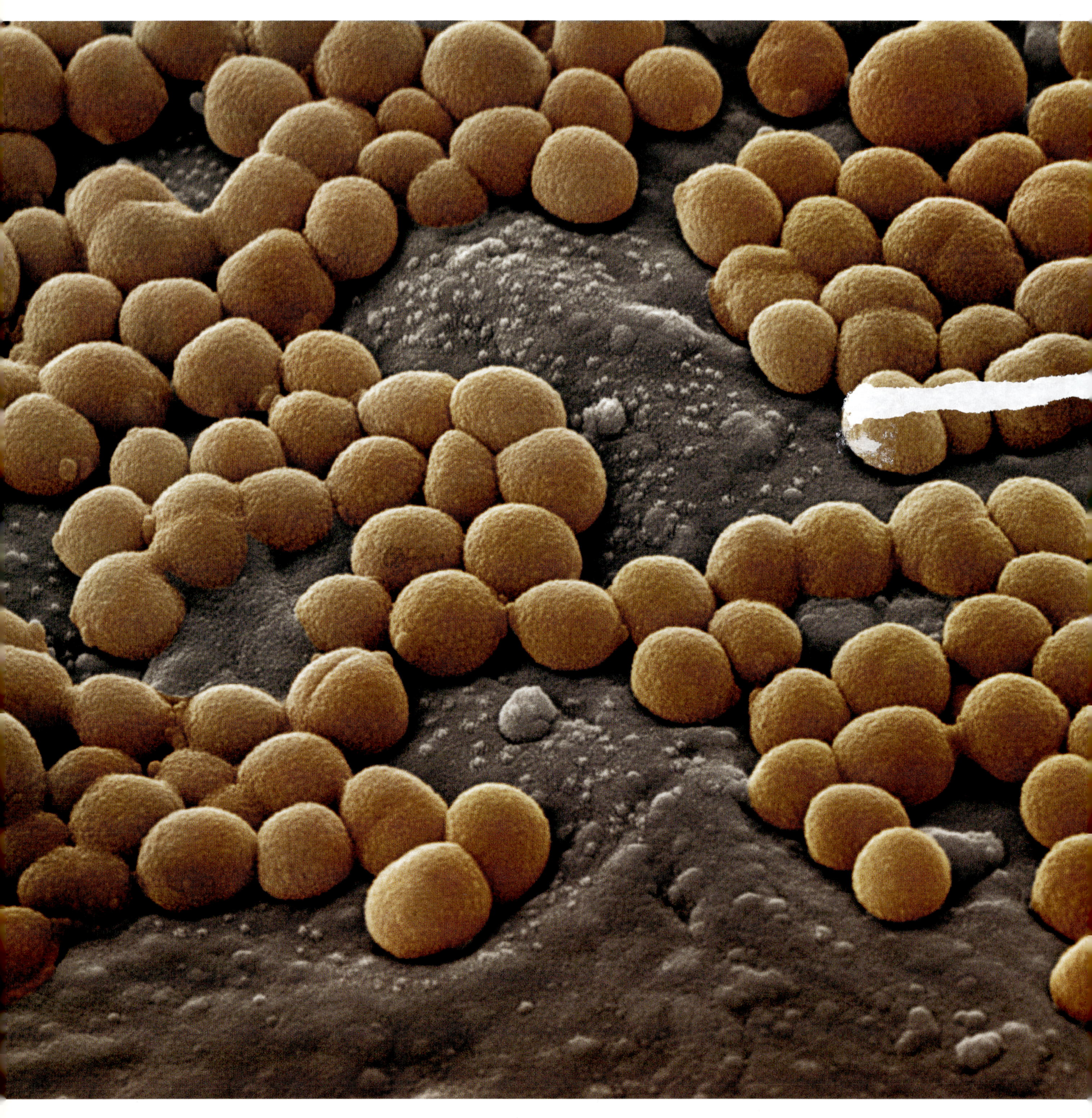

BACTERIUM: Staphylococcus

Staphylococcus aureus is an extremely common bacterium, one of many different kinds of bacteria that are found on the surface of human skin, helping to keep it healthy. Certain strains however can cause food poisoning, and contribute to the development of wound infections, urinary tract or blood infections, and pneumonia. *Staphylococcus*, and occasionally *Streptococcus*, bacteria can cause impetigo, a contagious, but not dangerous, skin condition that frequently affects young children since it quickly spreads in schools and playgroups. It first appears as an itchy area on seemingly healthy skin, which quickly develops into a blister containing a yellow substance. Later, the top of the blister becomes crusty and weeps, while new blisters develop in the same place or on other parts of the body. Impetigo usually begins on the face, especially around the corners of the mouth, the nose and back of the ears. Children and adolescents suffering from eczema are especially likely to develop this condition. Eczema itself is not caused by a parasite but is an allergic reaction caused by contact with an irritant material or substance. It results in dry, flaky skin, accompanied by intense itching.

BACTERIUM: MRSA

Most strains of *Staphylococcus* are sensitive to treatment by antibiotics, however a few strains have evolved resistance, even to the last resort antibiotic methicillin causing a condition called methicillin-resistant *Staphylococcus aureus*, more commonly known as MRSA. MRSA infections cause a range of symptoms depending on the part of our body affected, from skin wounds to blood. Many hospitals are hotbeds for MRSA because many different strains of bacteria are being treated simultaneously with a range of antibiotics, and therefore develop resistance. *Staphylococcus*, for example, exists in a variety of strains, each possessing subtle natural genetic mutations that make it different. Some of these will inevitably be more resistant to an antibiotic, and over time, strains will further mutate, increasing their survival ability.

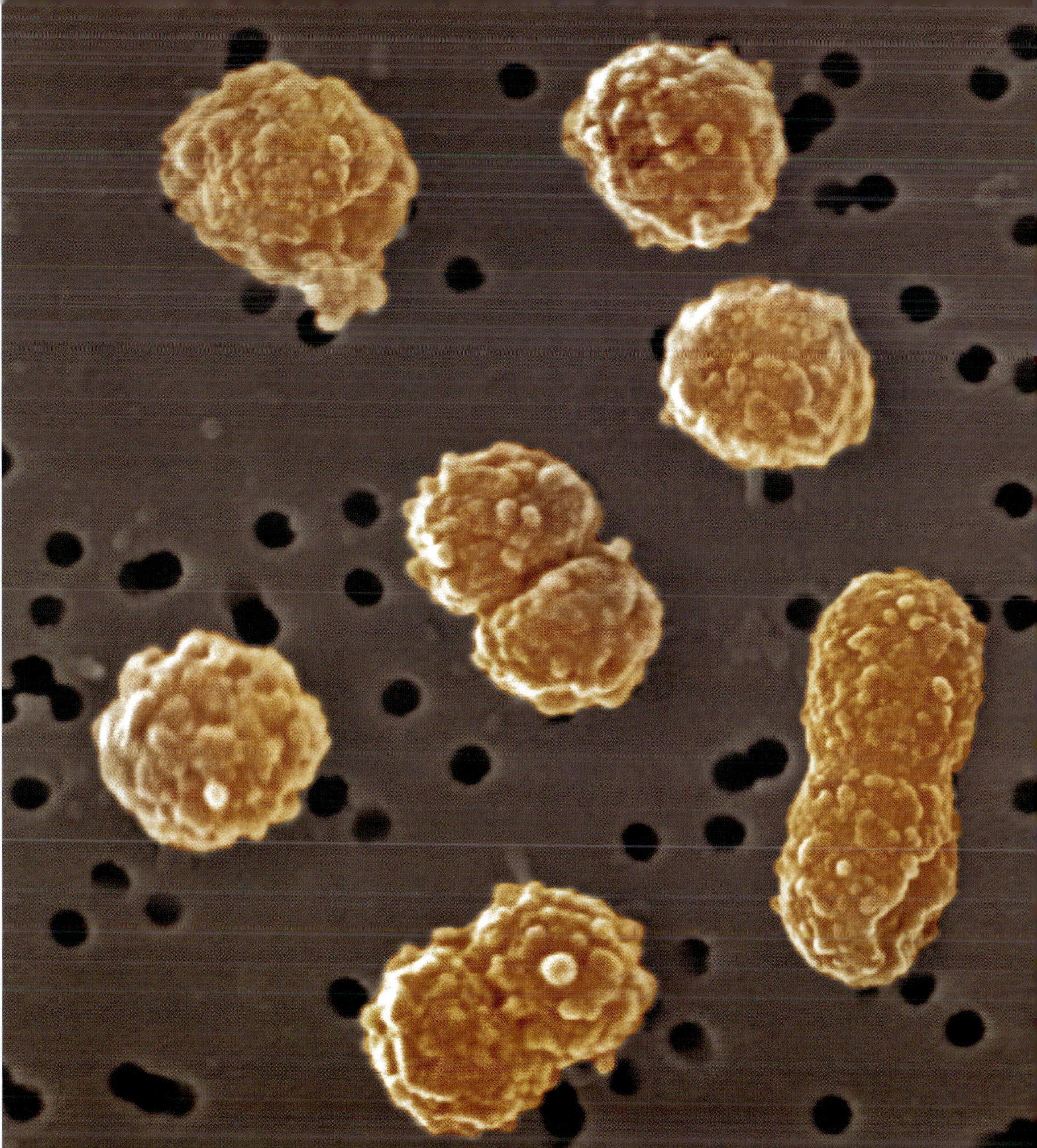

Bacterium, *Staphylococcus aureus,* and **MRSA**

Staphylococcus (see left) is an extremely common bacterium, one of many different kinds of bacteria that are found on the surface of human skin, playing a vital role in keeping it healthy. Some strains of *Staphylococcus*, however, can cause skin problems such as impetigo, as well as food poisoning and blood infections. Most strains of *Staphylococcus* are sensitive to treatment by antibiotics, but a few have become resistant, causing a condition called methicillin-resistant *Staphylococcus aureus*, more commonly known as MRSA. The sample MRSA (see above) has been killed by the action of a new experimental biocide, by causing damage to the cell membrane.

Magnification: x24000	*Scanning Electron Micrograph* (left)
Magnification: x33000	*Scanning Electron Micrograph* (above)

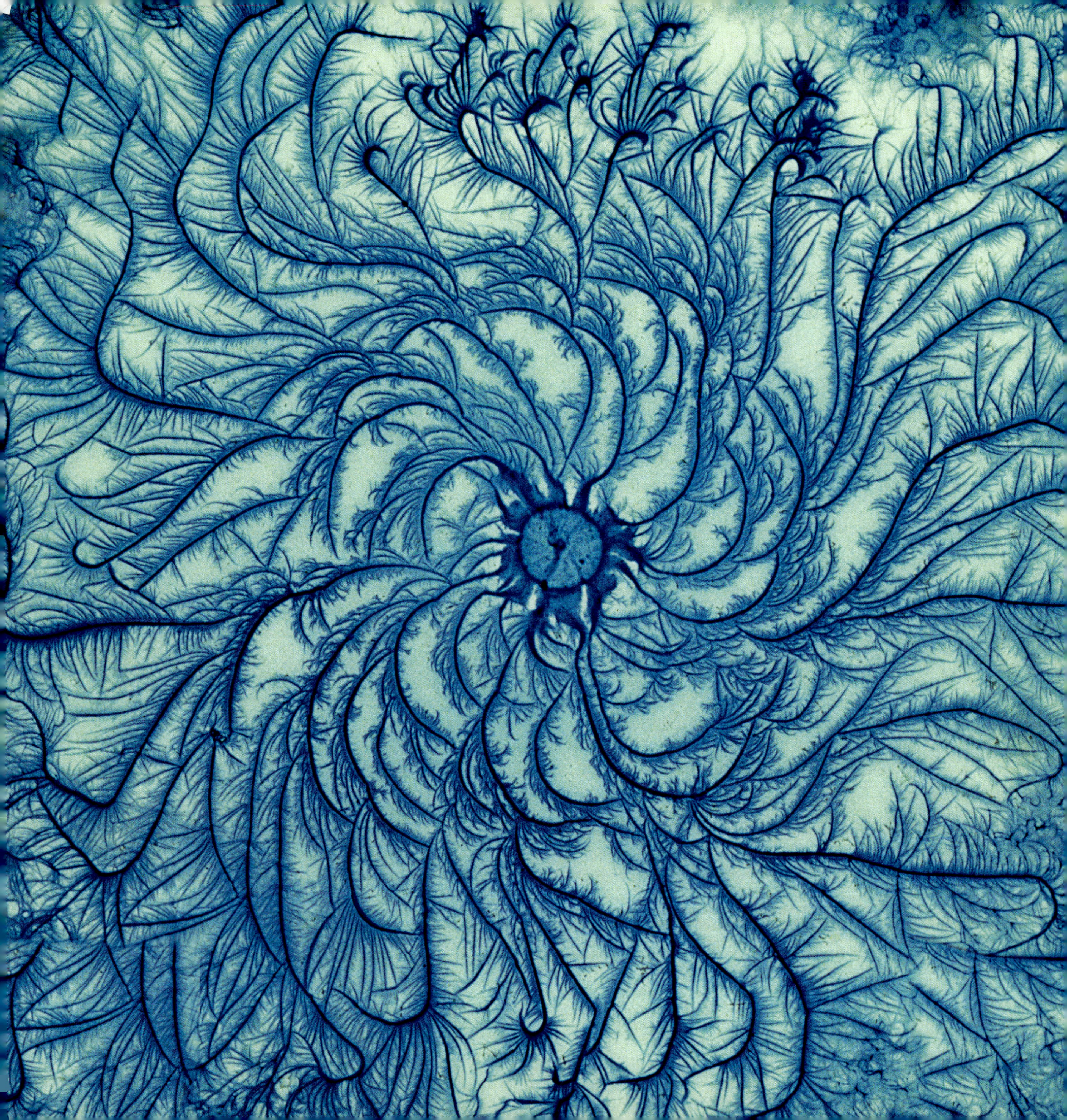

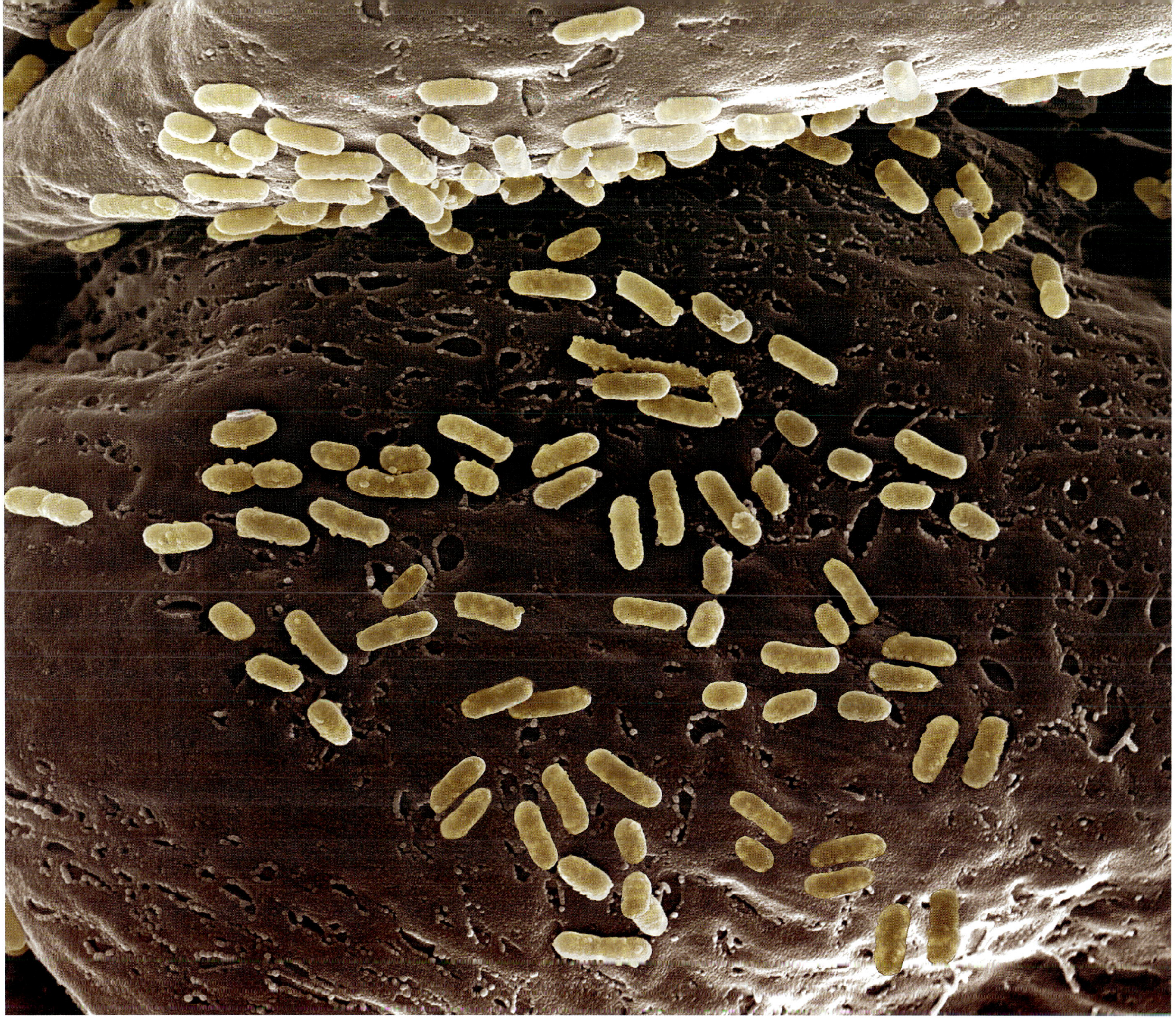

Bacterium, *Proteus sp.* (left)

A bacterium, *Proteus sp.*, whose cells are mobile, form intricate feathery patterns as the colony grows on the surface of agar. Proteus species normally inhabit the soil, decomposing organic matter, but can cause urinary infections.

Magnification: x3   *Macro-photograph*

Bacterium, *Escherichia coli* (above)

Escherichia coli is a rod-shaped bacterium common in nature and normally found in our intestines, serving a useful function by suppressing the growth of harmful bacterial species, synthesising vitamins, and fermenting lactose with the production of gas. Some strains of *E. coli* are harmful, however.

Magnification: x7600   *Scanning Electron Micrograph*

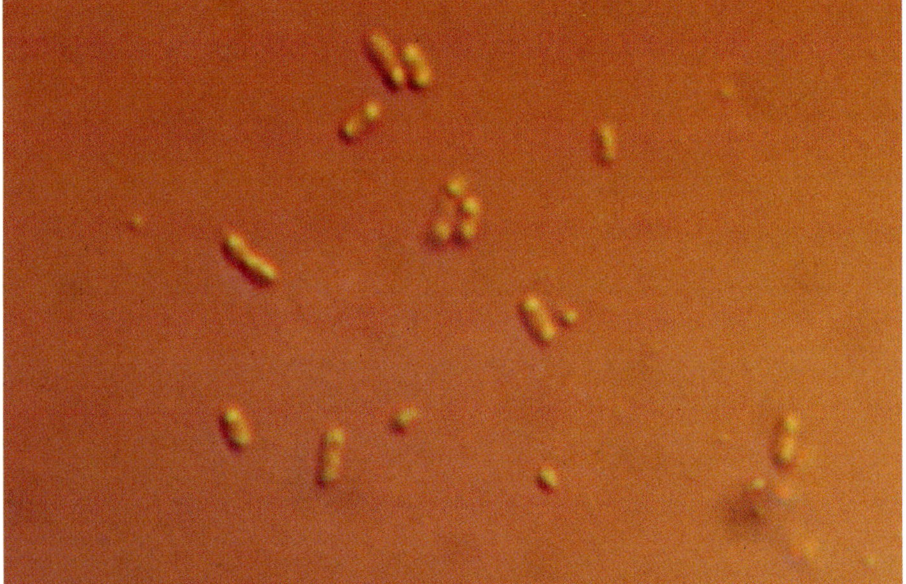

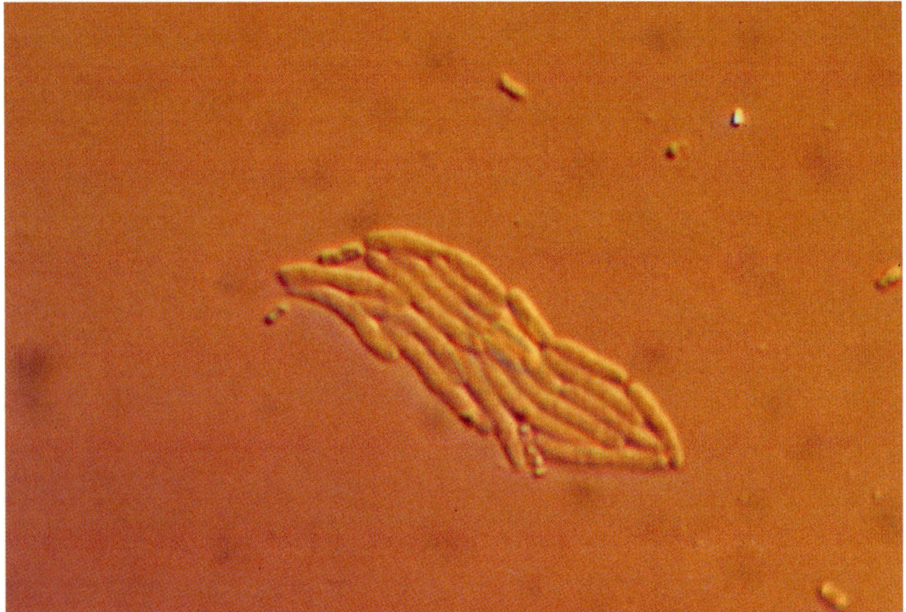

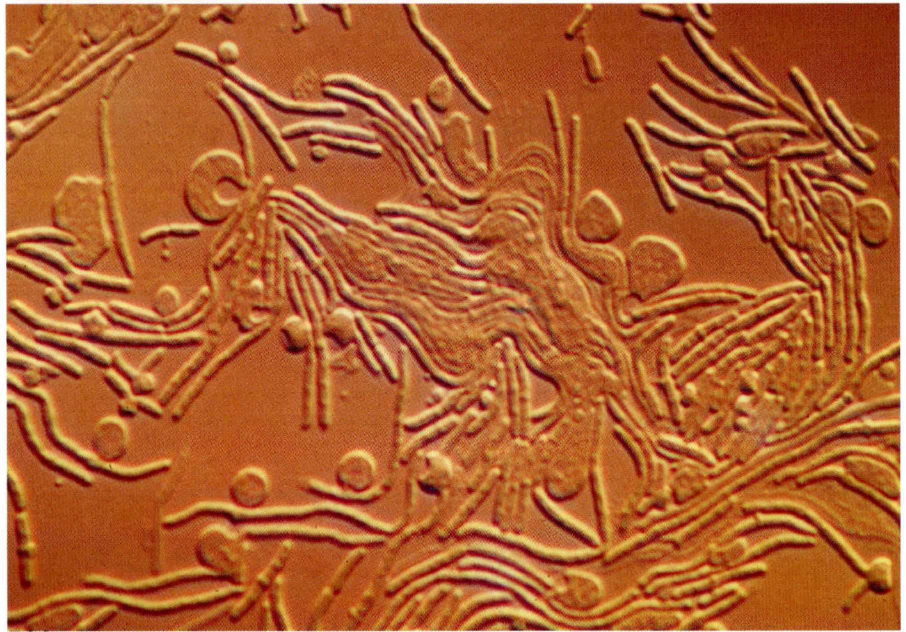

BACTERIUM: E. coli

Escherichia coli is a normal inhabitant of the intestines of all animals, including humans, serving a useful function to the body by suppressing the growth of harmful bacterial species and by synthesising vitamins. It is one of the world's most studied organisms, however, since certain strains of *E. coli*, for example 'strain 0157', cause violent food poisoning and have been responsible for large-scale outbreaks of the condition from contaminated foodstuffs. Symptoms can range from mild diarrhoea and dehydration to severe abdominal cramps and blood in the stools. Occasionally, the condition can lead to haemolytic uremic syndrome (HUS), which kills red blood cells, and can cause kidney failure, sometimes with fatal consequences. For example, in 1996, 20 people died in Scotland after attending a church lunch in Strathclyde, and in the same year, there were nearly 9,000 victims and at least eight deaths, from an outbreak in Sakai city, Japan.

CHLAMYDIA

Chlamydia is one of the most common sexually transmitted infections worldwide. It is caused by the bacterium *Chlamydia trachomatis* and infects the genitals of both men and women causing inflammation, pain and discharge. If detected in time it is easily cured with antibiotics but not all infected people exhibit symptoms of infection and the condition may linger for months, even years, before being discovered. Without treatment, it can spread to other parts of the body such as the eyes causing long-term damage; it can also lead to infertility and many other problems, including the transmission of the disease to an embryo during pregnancy.

Bacterium, *Escherichia coli* **affected by antibiotic**

Certain strains of *E. coli*, for example 'strain 0157', cause violent food poisoning and have been responsible for large-scale outbreaks from contaminated foodstuffs. In the scanning electron micrograph (see right), the rod-shaped cells of *E. coli* are under attack by the antibiotic cephalosporin. The cells elongate and blebs form on their surfaces as they disintegrate. The series of Light Micrographs (see left) show what happens to *E.coli* in contact with cephalosporin; growth looks normal until, in the bottom picture, the cells elongate and burst.

Magnification: x25200   *Scanning Electron Micrograph* (right)
Magnification: x3400   *Light Micrographs* (left)

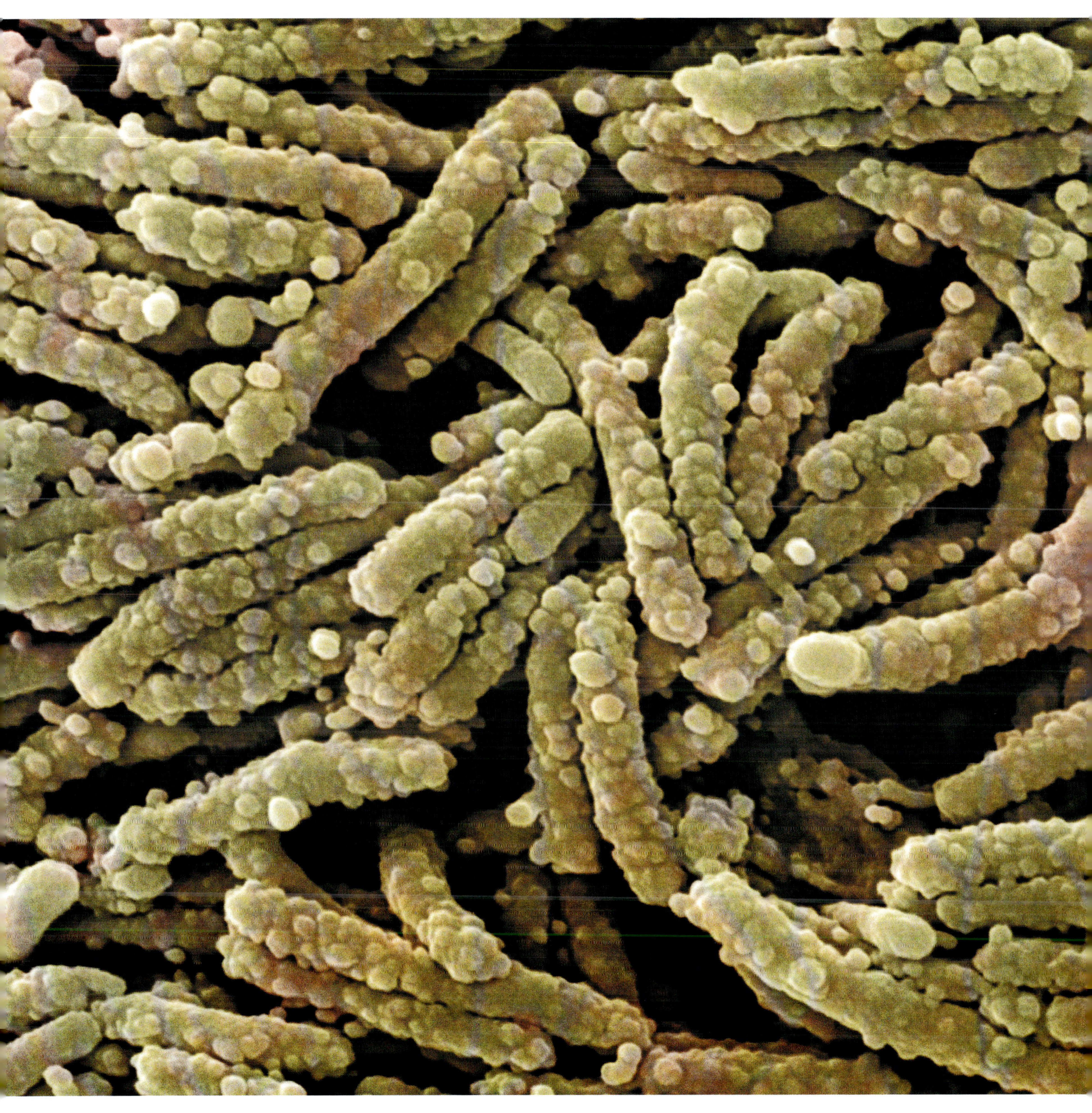

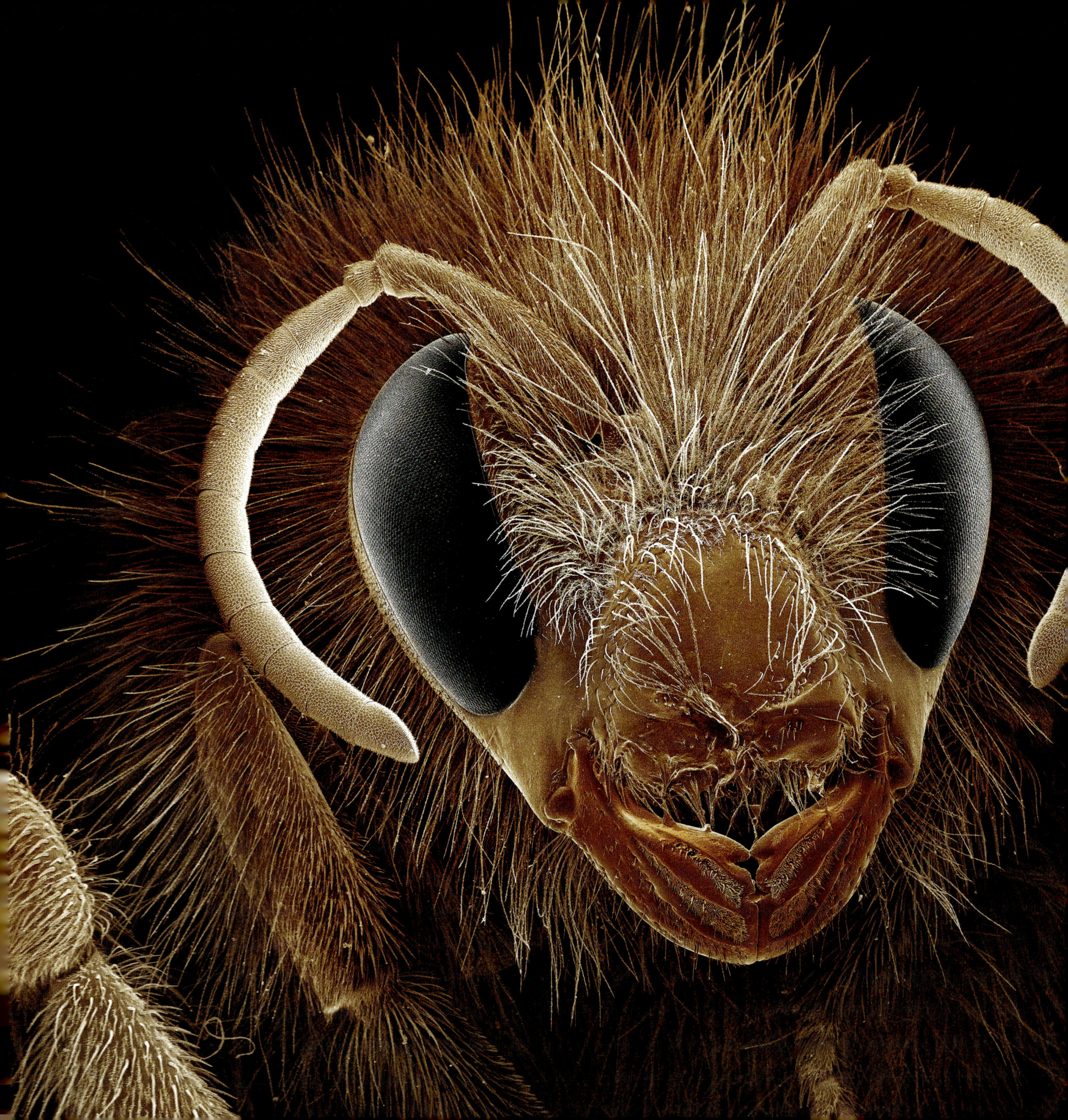

GARDEN JUNGLE

A garden is an extension of the home, an ordered wilderness, planned and manicured to squeeze a great variety of habitats into a small area, and even the smallest contains a rich variety of wildlife. The larger the garden, the more the variety, including perhaps a vegetable patch, a greenhouse, shrubs, trees, an orchard for traditional fruits, a boundary hedge, and a pond. We may provide the control and enforce the design, but each of these habitats attracts its own succession of creatures, to feed, reproduce, and take their place in nature's chain. The activity of gardening dates back at least to the time when people exchanged a nomadic, hunter-gatherer lifestyle for one of semi-permanent settlement and started to grow plants for food. Cultivation of the land and the rearing of domestic animals brought advantages, and gave rise to the growth of permanent communities. The creation of ornamental gardens came much later, with some of the earliest evidence seen in Egyptian tomb paintings from around 1500 BC, and of course in Mesopotamia, where the hanging gardens of Babylon, supposedly created by the Assyrian King Nebuchadnezzar II (604-562 BC), became renowned as one of the wonders of the world. In England, landscape gardening became fashionable for manor houses and stately homes in the 18th Century, with the more romantic style of cottage flower gardens developing later. Cottage gardens are typically random and carefree in form, with plants growing in dense clumps. Originally, they were created by workers that lived in cottages in country villages, to provide them with food and herbs, and with flowering plants for decoration.

The method of planting everything close together reduced the amount of weeding and watering required. Clear any patch of soil in the garden and it will be quickly colonised by weeds. The soil contains so many dormant seeds that when the ground is disturbed, some are brought to the surface, germinating rapidly as soon as they come into contact with the light. Weeds are hardy plants, tenacious, aggressive, natural survivors. They grow fast, spread rapidly, are able to live in hostile conditions and are usually very resistant to disease. Their seeds, often produced in vast quantities, can survive long periods before germinating. If left, they will overwhelm cultivated plants for light and nutrients, reducing the yield of any crops. If neglected, lawns become havens for daisies, clover and dandelions, but the summer flowers of these and other weed species provide pollen for bees. Many weeds are valuable wildlife plants; nettle and groundsel, for example, provide food for caterpillars; brambles, ivy and thistles attract butterflies. In addition, weeds contain important nutrients and minerals brought up from the subsoil and therefore improve soil fertility, and many can be eaten or have uses in herbal remedies. The quintessential English garden, if somewhat untidy, has charm and character, with herbaceous borders informally planted to mimic nature with colourful flowering plants such as the rose, lavender, clematis, buddleia and honeysuckle. In the warmth of summer, a heady cocktail of perfumes attracts butterflies and bees that are so much a part of the garden ambience, working busily from flower to flower to tap nectar, and helping to pollinate the plants.

Bumblebee, *Bombus terrestris*

The bumblebee, *Bombus terrestris,* is large, furry, and flies rather slowly. Feeding on nectar and pollen, it has immense value as a pollinator of many plants. Only the queen and worker bees can sting; unlike honeybees, when bumblebees sting they do not die.

Magnification: x60   *Scanning Electron Micrograph*

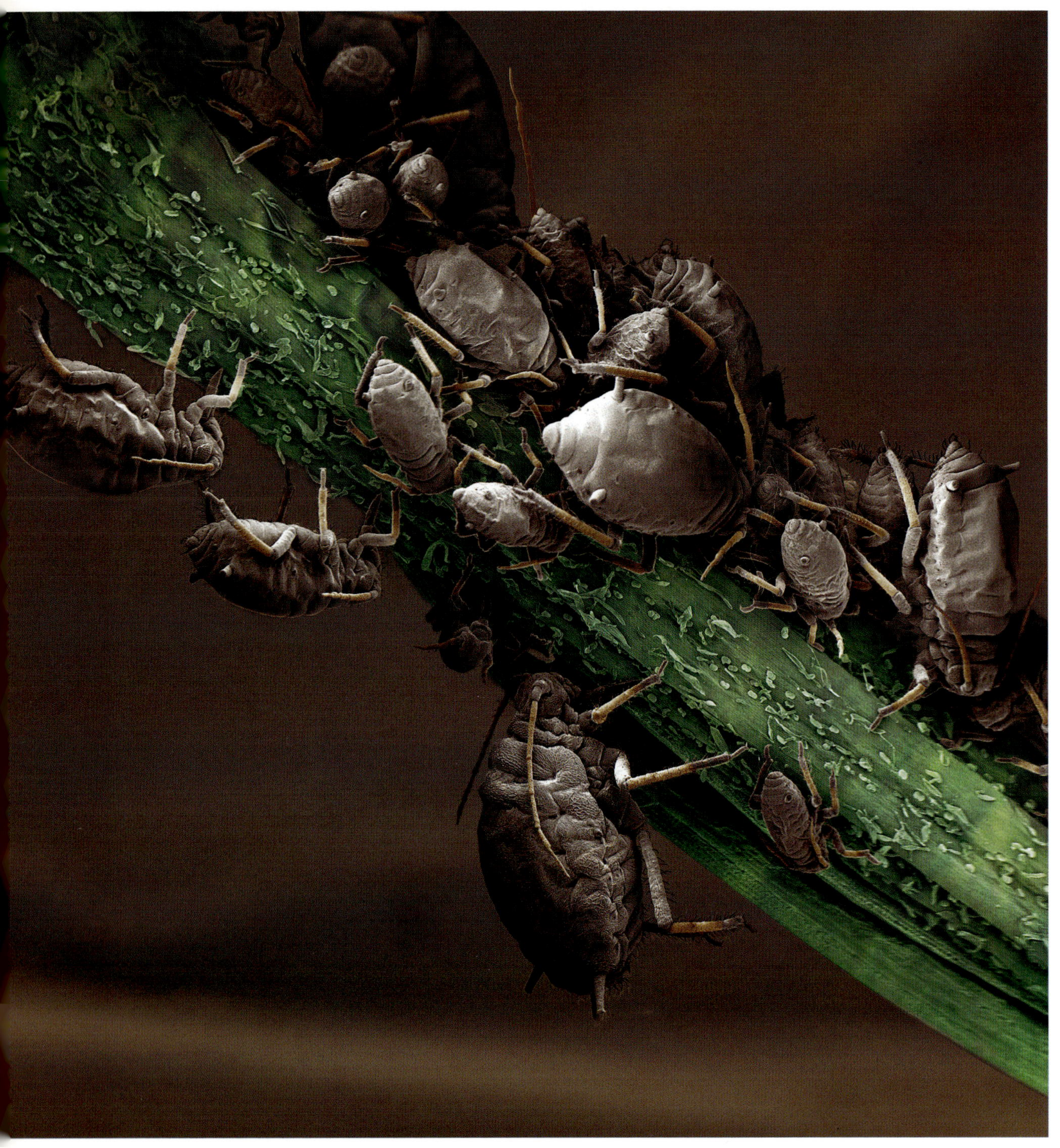

By growing for our pleasure a flowing succession of flowering plants and vegetables through the seasons, we produce an unnatural situation, a rich concentration of potential food sources for tiny creatures all in one place. Consequently, huge numbers of opportunists, not all of them welcome, are attracted to our gardens and their populations can sometimes swell unchecked. Many of them are sap-feeders, the equivalent to plants of blood feeders, which use sharp mouthparts to pierce the stem of a host plant to tap the flow of nutrient sap. Take aphids, for example. There are over 500 species in Britain, of which a few are familiar garden pests. Otherwise known as greenfly or blackfly, they are small, flask shaped bugs, just a few millimetres long, that infest certain plants, especially young shoot tips, flower buds and the soft underside of young leaves. When serious infestations occur, aphids huddle together in their thousands, causing distorted, weak growth of leaves and shoots. They can also transfer viruses when they move from plant to plant.

For much of the year, aphid colonies consist of wingless females that give birth to live young by a process called parthenogenesis that does not require any fertilisation by males. The females just keep producing perfect replicas of themselves, a relentless conveyor belt that helps their population swell to epidemic proportions. Under perfect conditions, a single aphid could give rise to over one billion descendants in just three months, happening so quickly that predators often cannot consume the aphids fast enough to curb the tide. Winged forms develop whenever overcrowding or deterioration in a host plant induces the need to move to another plant, so throughout the warm summer months, an aphid infestation can spread quickly. In the autumn, males are produced and sexual reproduction takes place to produce eggs,

Black Aphid, *Aphis sp.*

Aphids are small, flask shaped bugs that infest flower buds, tender shoot tips, and the soft underside of young leaves. In serious infestations, aphids huddle together in their thousands, using their sharp mouthparts to pierce the outer cell layers of a host plant to feed on nutrient sap. They are often protected by ants, in return for the sweet 'honeydew' excreted by the aphids from their anuses.

Magnification: x40 *Scanning Electron Micrograph* (above)
Magnification: x150 *Scanning Electron Micrograph* (right)

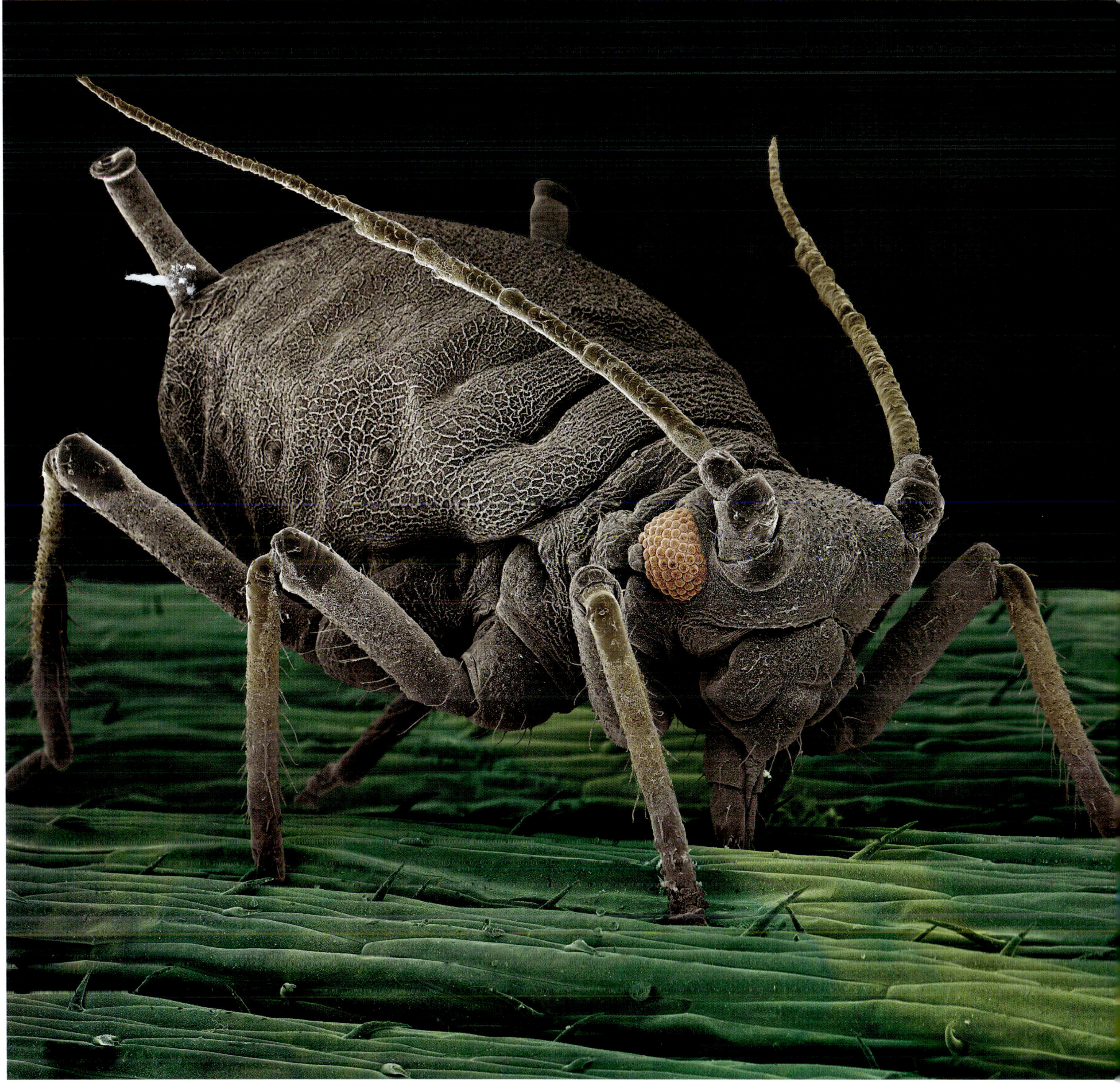

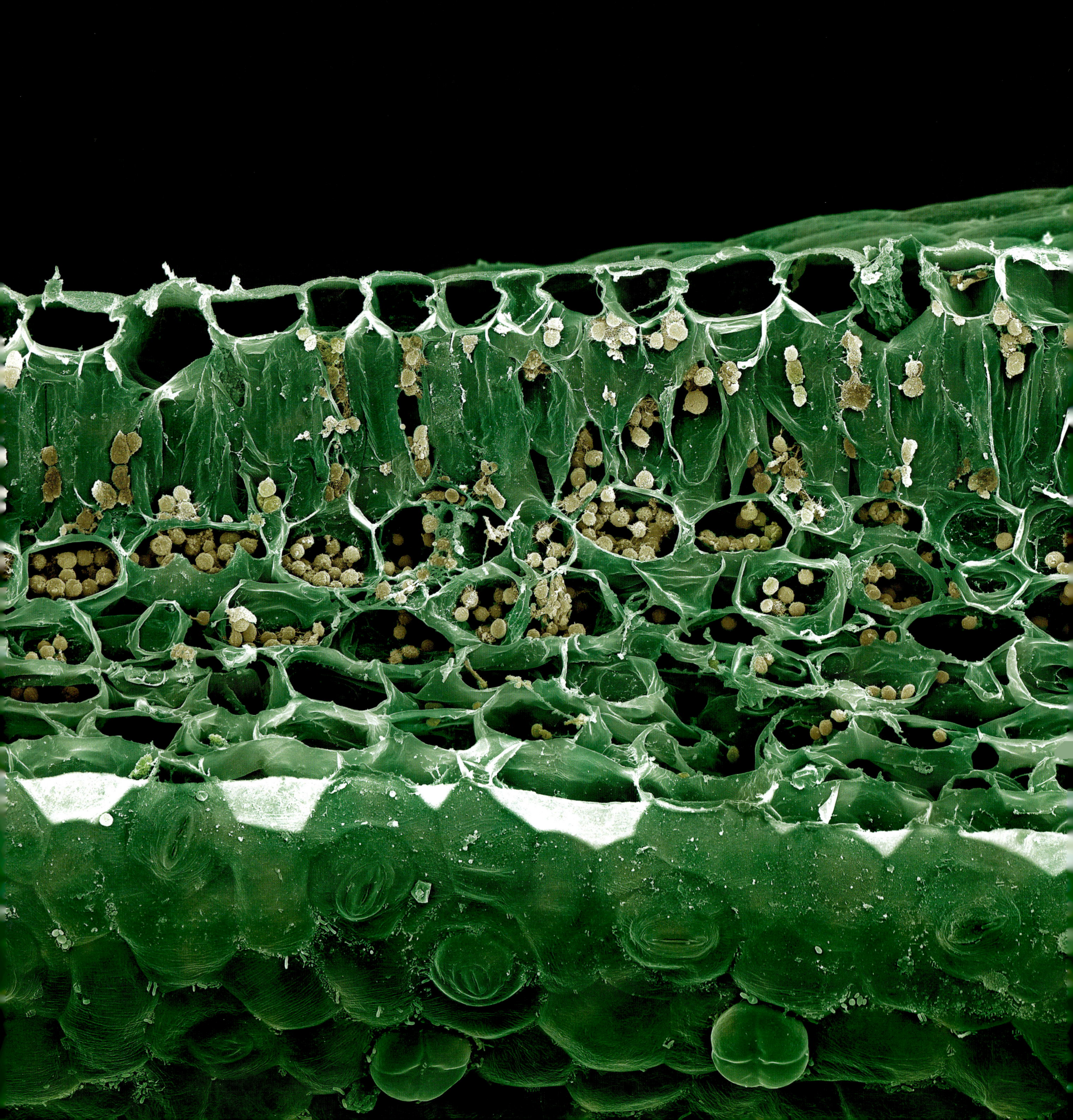

which are laid on a woody tree or shrub to survive over winter. If the weather stays mild, some adult aphids may remain active, especially in greenhouses. In the spring, the eggs hatch and the young aphids feed on tender new leaves close by. By early summer, as the foliage grows older and tougher, winged forms migrate to suitable summer host plants.

Forethought in garden planning can help control aphid populations, by encouraging the predators that feed on them, such as ladybirds, hoverflies and lacewings, together with their respective larvae. Planting marigolds near plants often attacked by aphids reduces infestations because marigolds have strongly aromatic leaves, which repel most invertebrates. Hoverflies are one exception. They like to visit marigold flowers and, after feeding, the adults deposit eggs on colonies of greenfly or blackfly nearby. When the eggs hatch, the larvae feed on the aphids. Aphids, however, have an unexpected means of defence – ants. Many kinds of aphid excrete a sweet substance called honeydew in copious amounts from the anus. It consists of partially digested plant sap and other wastes, and is a prized food item for ants, flies, and bees. Certain aphid species have a symbiotic, or mutually supportive, relationship with ants; the aphids producing honeydew when stimulated by the ant's antennae, while the ants tend the aphids, protecting them by fighting off predators, transporting them to food plants and sheltering aphid eggs in their nests during the winter.

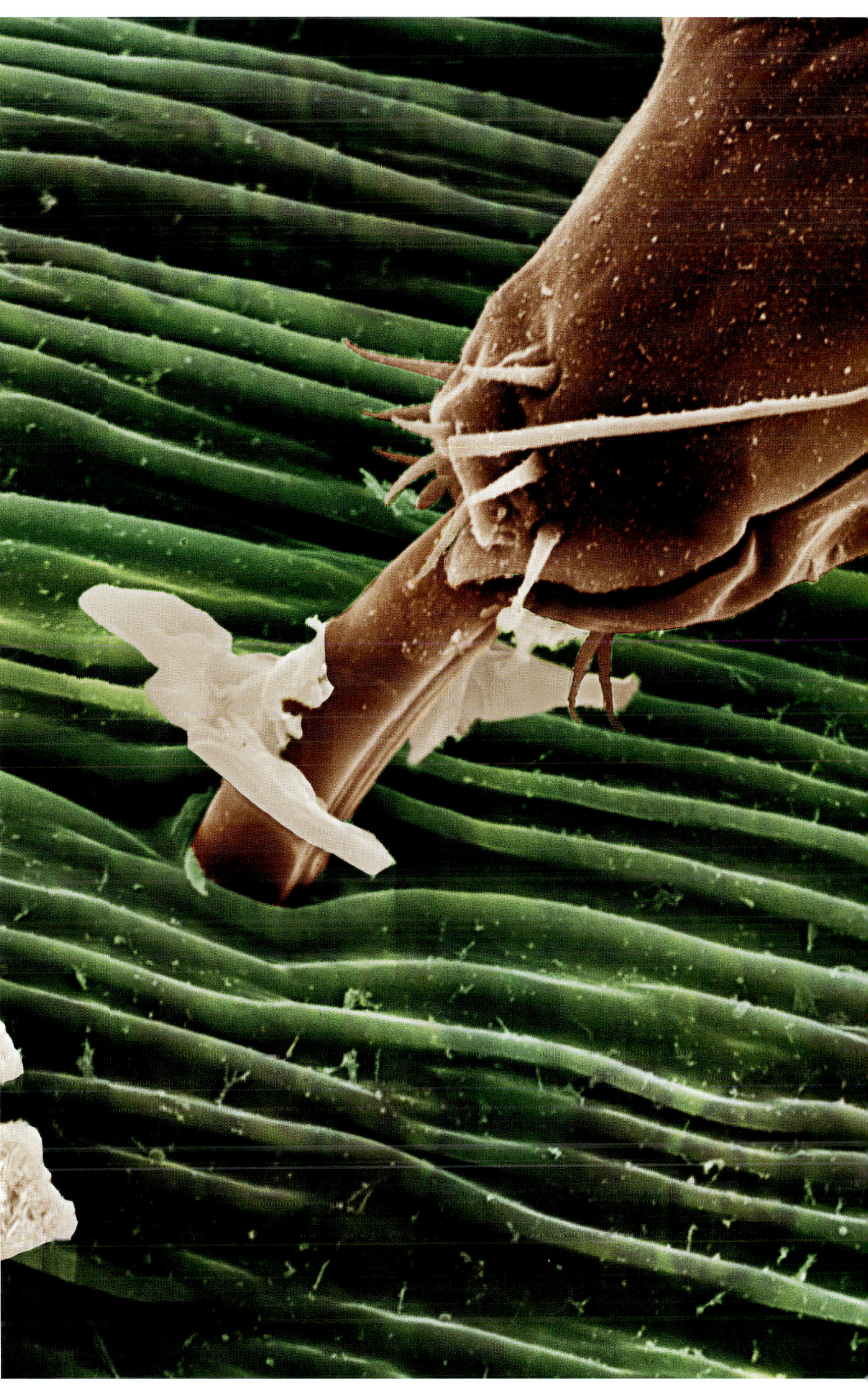

Black Aphid, *Aphis sp.*

A section of leaf epidermis (see left) shows the cells and sap spaces inside, tapped by aphids when feeding. The aphid pierces the leaf surface using its sharp mouthparts, or stylet, (see right) then sucks up the juices.

Magnification: x500
Scanning Electron Micrograph (left)
Magnification: x3500
Scanning Electron Micrograph (right)

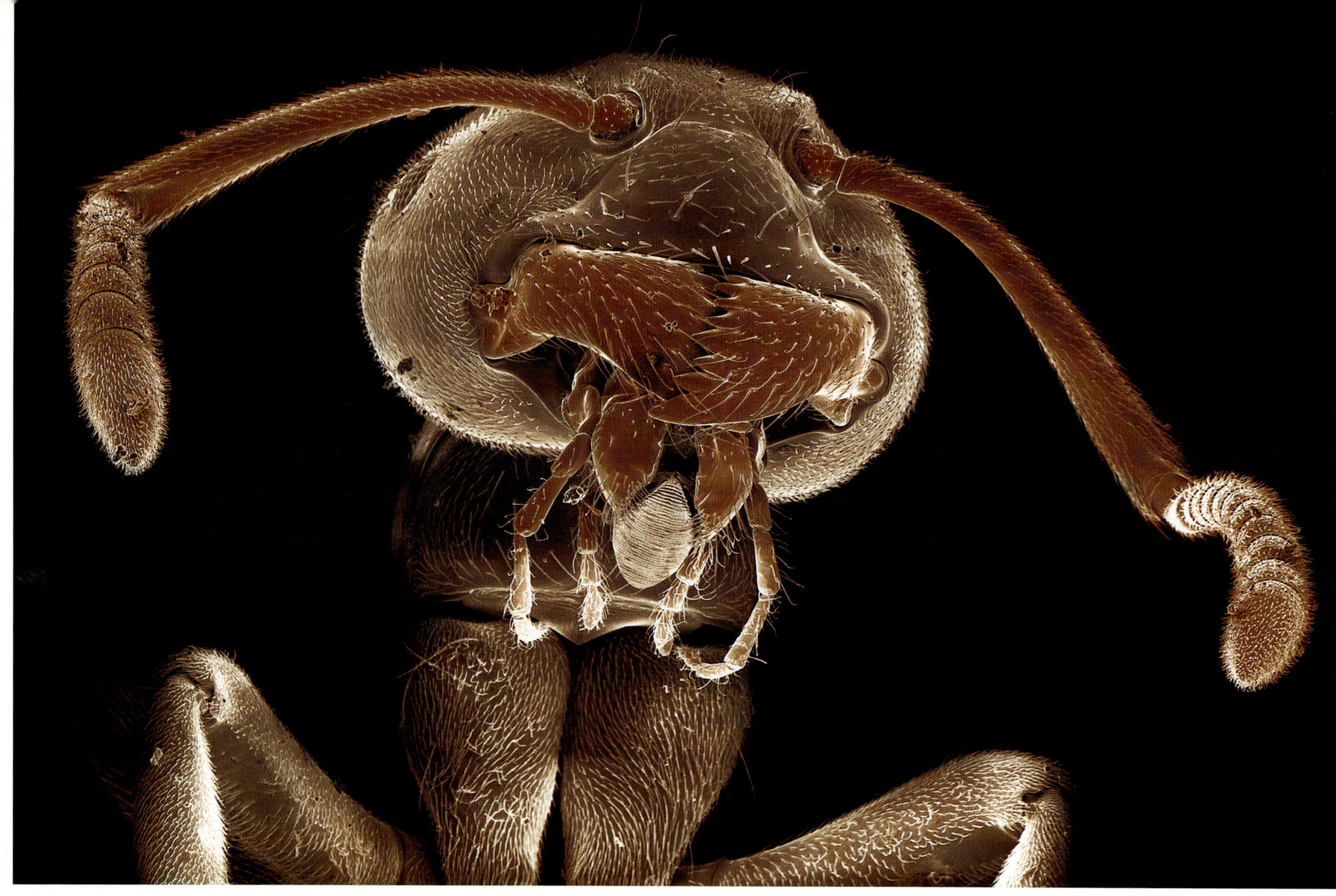

ANTS

Ants are social insects, related to bees and wasps. Their colonies, which can consist of millions of individuals, are highly organised with individual ants of different types: workers, soldiers, drones and queens, each with special tasks to perform that contribute to the colony as a whole. Ant colonies have been described as 'super-organisms' because they appear to operate as single entities. By their nature predatory and aggressive, some species are known for attacking and taking over the colonies of other ant species. Some attack colonies to steal eggs or larvae, which they eat or keep alive to work as slaves. Battles between colonies sometimes leave thousands of ants dead.

There are some 36 species of ant found in Britain. Some have stings, others such as wood ants can squirt formic acid if threatened; all have well developed biting jaws. Some species have a caste of workers with large heads and jaws called soldiers, whose job it is to defend the colony from predators or other ants. At certain times of the year, usually in late summer, winged males and females are produced and these take flight in swarms to mate in the air. This behaviour only happens in certain weather conditions, so it is usual to see many swarms in the air at one time. The males die soon after mating, the queens drop their wings, find new nest sites and hibernate over winter, producing eggs in the spring.

Black Garden Ant, *Lasius niger*

Garden ants have a symbiotic relationship with aphids. In exchange for the sweet 'honeydew' excreted by the aphids, the ants not only protect them by fighting off predators but actively farm them, transporting them to new food plants and sheltering their eggs over the winter months in their nest. A worker ant (see above) gently carries an aphid egg in its powerful jaws. Garden ants sometimes enter houses in search of sweet foods, setting up pheromone trails for others to follow to the food source. Here (see right), a garden ant is seen with grains of brown sugar.

Magnification: x90
Scanning Electron Micrograph (above)
Magnification: x75
Scanning Electron Micrograph (right)

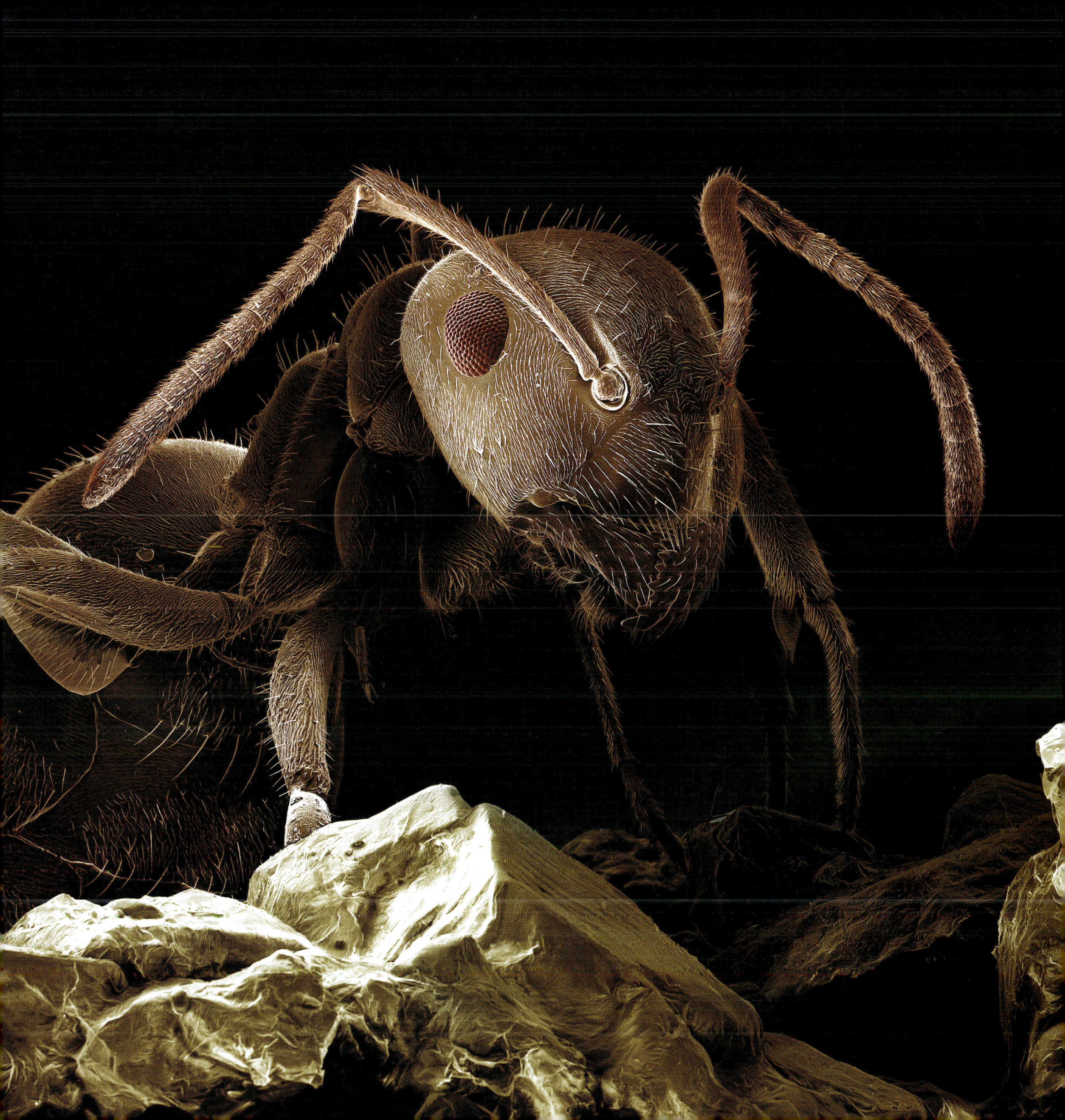

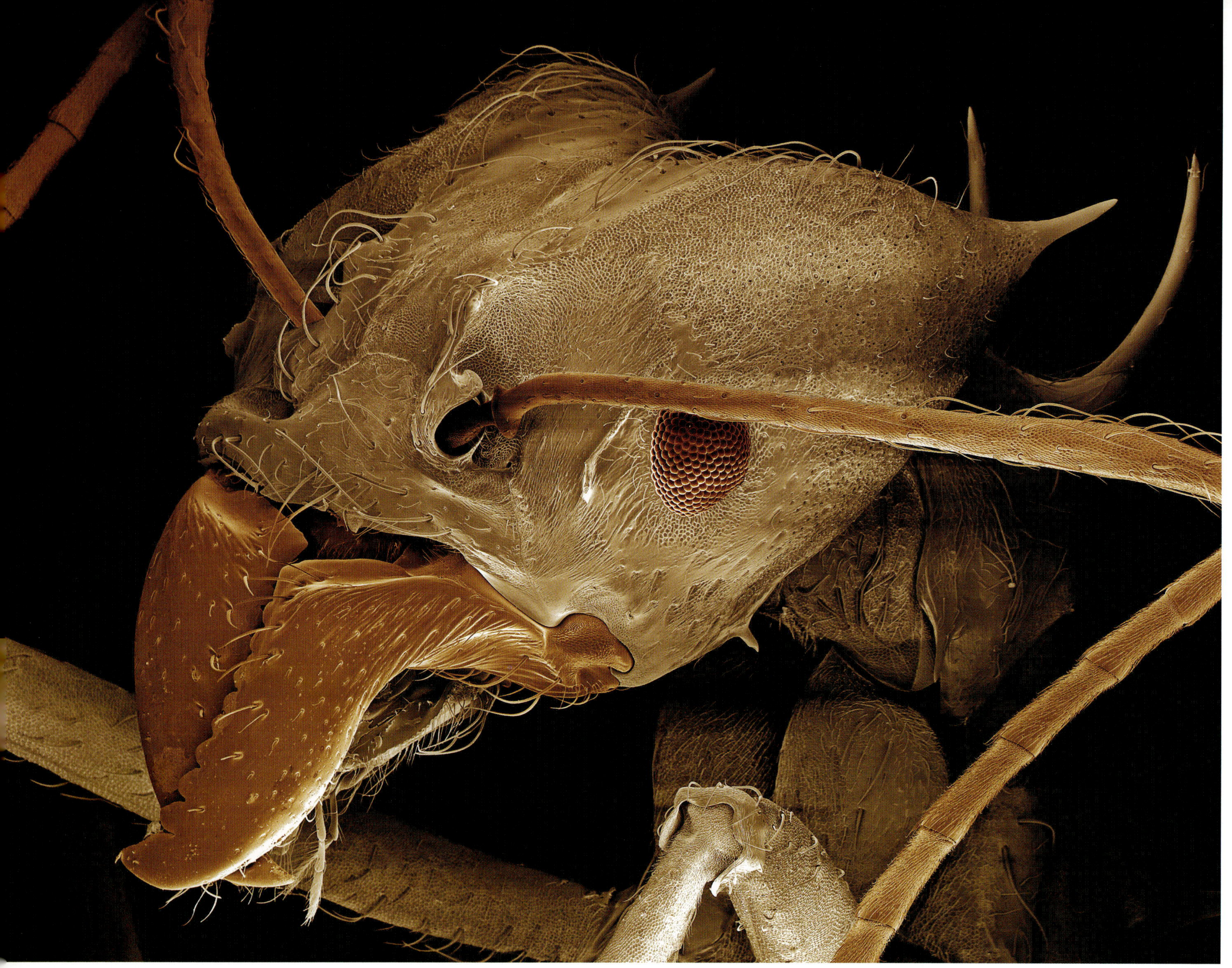

Leaf Cutter Ant, *Atta sp.*

The powerful jaws of this tropical American ant are used to cut out and carry sections of leaves, which are carried back to underground gardens in their nests. These highly organized, social ants have evolved a system of agriculture to supply them with their food, based on the careful cultivation of a fungus using the cut leaves as compost. The fungus only grows in their nests, and the symbiotic relationship is a fine example of mutualism.

Magnification: x200   *Scanning Electron Micrograph*

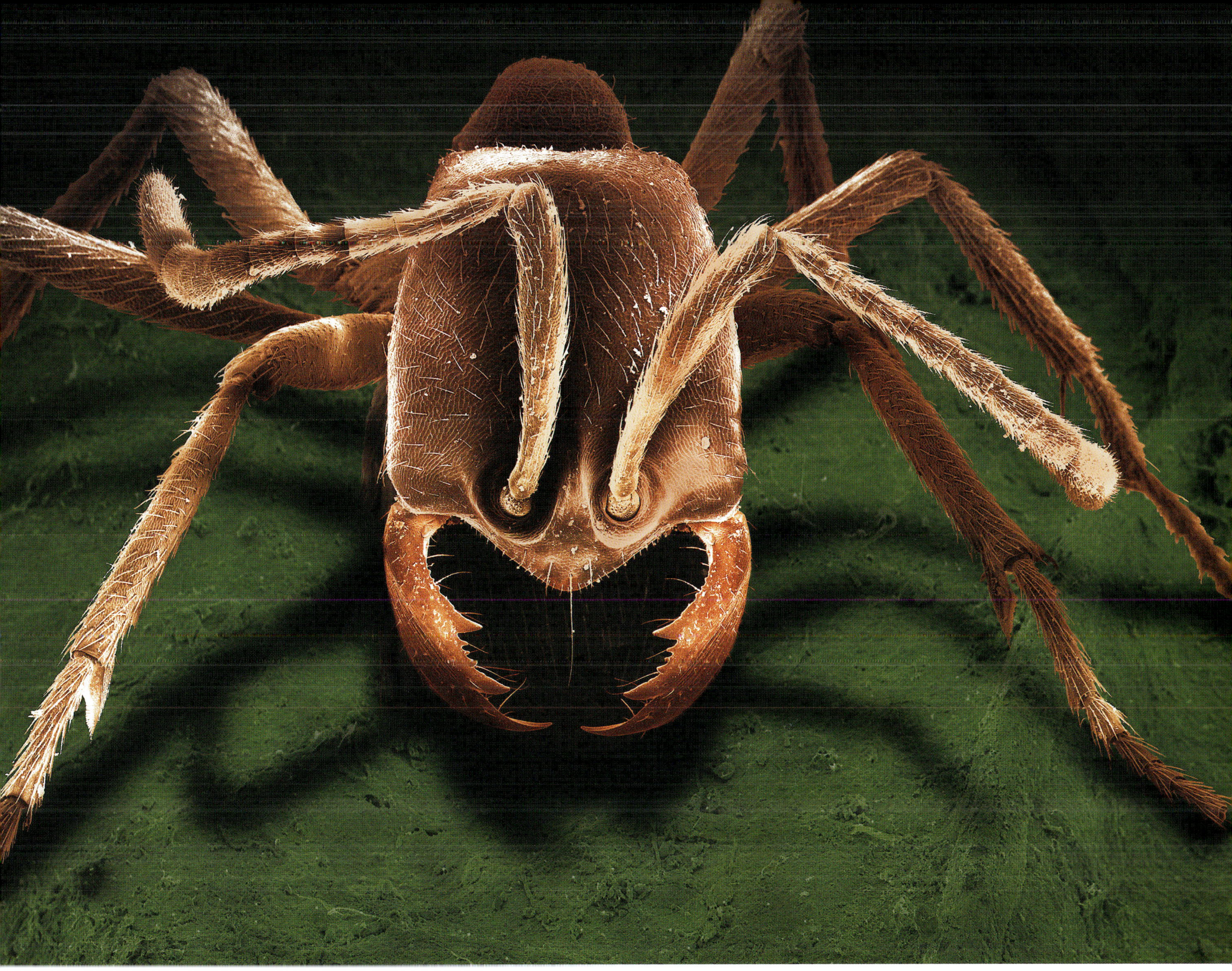

Driver Ant, *Dorylus sp.*

Driver ants, often called safari ants, from East and Central Africa, live in enormous colonies. When food is in short supply, they form broad marching columns killing and consuming any living creature that cannot escape. Their jaws are extremely strong, their bite painful and swarms of them can overwhelm quite large animals.

Magnification: x90   *Scanning Electron Micrograph*

BEETLES and WEEVILS

Beetles form the largest order of insects, with some 4000 species found in Britain. They have evolved to exploit many habitats and lifestyles. Among the familiar beetles found in our gardens are the small colourful ladybirds and the larger ground beetles that busily scurry around looking for prey. Beetles have armour-plated bodies and two pairs of wings, the foremost of which is modified to form a hard protective covering for the delicate hind wings. Although heavy for their size, most beetles can fly, opening the wing cases before deploying their thin hind wings for flight. The larvae of beetles are soft-bodied and maggot-like with a distinctive head, and impressive jaws.

LADYBIRDS

Brightly coloured ladybirds, red or yellow with black spots, are perhaps the best known of our native beetles, with 45 species known in Britain. The colours are a defensive strategy to deter predators; if threatened, ladybirds discharge a pungent, yellow fluid from leg joints and other parts of the body, a process called reflex bleeding. The two-spot ladybird, *Adalia bipunctata*, and the larger seven-spot ladybird, *Coccinella septempunctata*, are the most common species, both of them valuable friends of the gardener, for the adults and their larvae, as well as many other species of ladybirds, are voracious predators of aphids, scale insects and other pests. Indeed, ladybirds are bred commercially for use as biological control agents in greenhouses. Their wild population can fluctuate enormously, from vast carpets of them to being hard to find, being dependent on weather conditions and food availability in the spring. A warm damp spring season encourages aphids and other prey species, enabling ladybirds to swell their numbers. The long, warm summer and mild winter of 1975 set the scene for unusually high numbers of aphids and ladybirds in the spring of 1976. In the summer drought that followed, aphids initially thrived and ladybirds reached plague proportions. As the hot summer continued and plants browned and wilted in the heat, the aphids died out rapidly, and the growing shortage of food caused ladybirds to migrate in huge numbers, eventually resulting in a population crash, and leaving the ground littered with thick carpets of small red corpses.

Ladybirds spend the winter sheltering in garden sheds, window frames, in hollow plant stems, or log piles. Sometimes ladybirds will huddle together, a few at a time, although the 16-spot ladybird has been found in large clusters of over a thousand. In a good summer, ladybirds can breed more than once, the females depositing their eggs in small batches, usually on the underside of leaves of plants already infested by aphids or scale-insects, so there will be plenty of food available when the eggs hatch. The larvae of the 7-spot ladybird are steely-blue with yellow or white spots. They forage actively, consuming several hundred aphids during the three-week larval period, and the adults are even more voracious, so even just a few of these friends in the garden can be beneficial. A few British ladybird species are not carnivorous, for example the small, yellow and black 22-spot ladybird, *Psyllobora vigintiduopunctata*, is non-predatory, eating mildews and other microscopic fungi.

LESSER STAG BEETLE

The lesser stag beetle, *Dorcus parallelopipedus*, is similar in appearance to the female of Britain's largest beetle the rare greater stag, but has a grey-black body with a large head and thorax, and a single spur on the tibia of its middle legs. It grows to 3 cm in length, and is quite common, especially in the south of England, coming into gardens where there are orchards, big trees or old hedges. The adult beetles may feed on the sap of deciduous trees but largely rely on body fat to keep them going since they only live for about two months during late spring and summer. The larvae feed on rotting wood from broad-leaved trees. After pupating, the adults emerge and disperse by flying.

Seven-Spot Ladybird, *Coccinella septempunctata*

The familiar red and black ladybirds are perhaps the best known of Britain's beetles, and are valuable allies of the gardener, since both larvae and adults feed voraciously on aphids.

Magnification: x50   *Scanning Electron Micrograph*

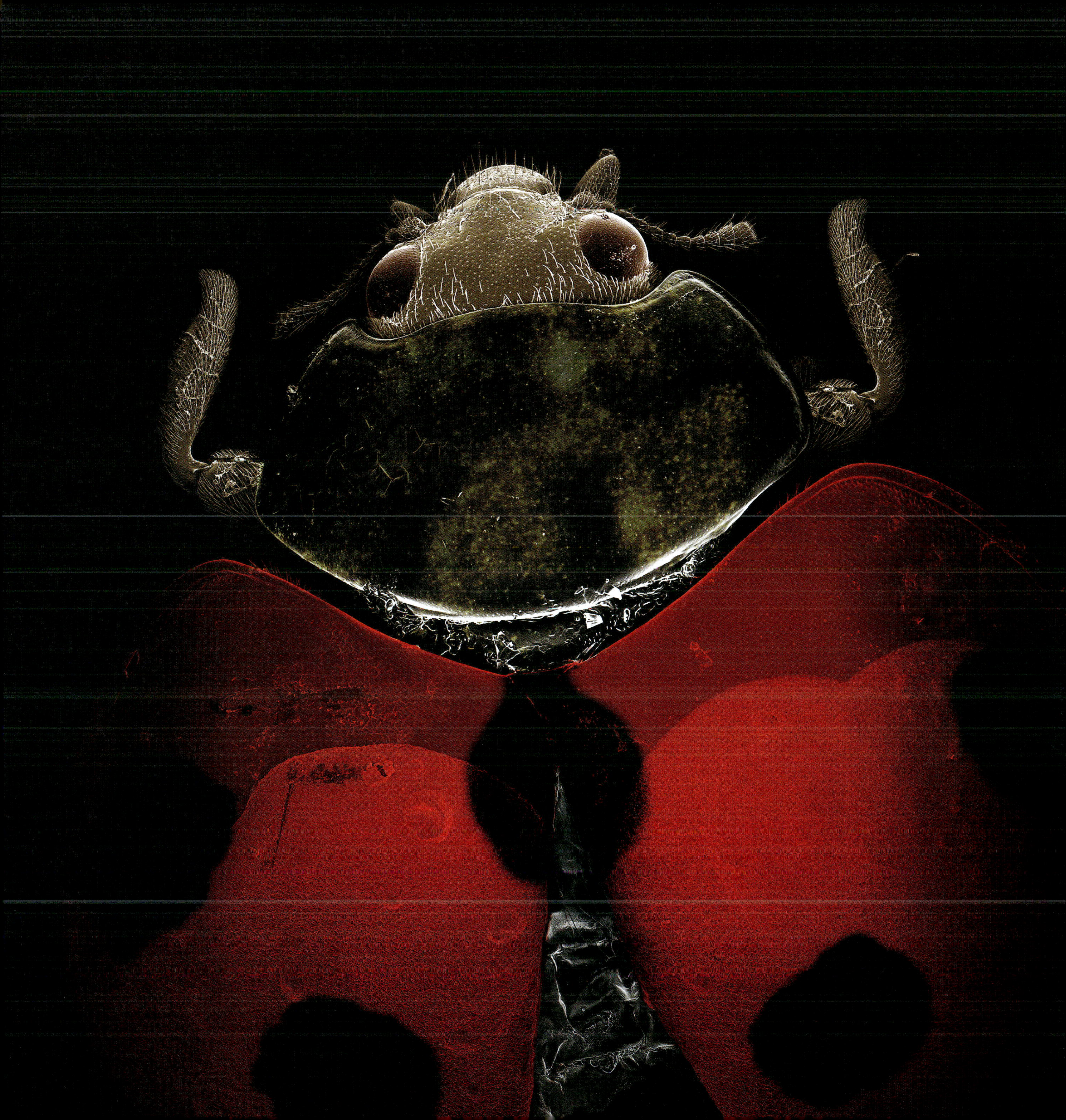

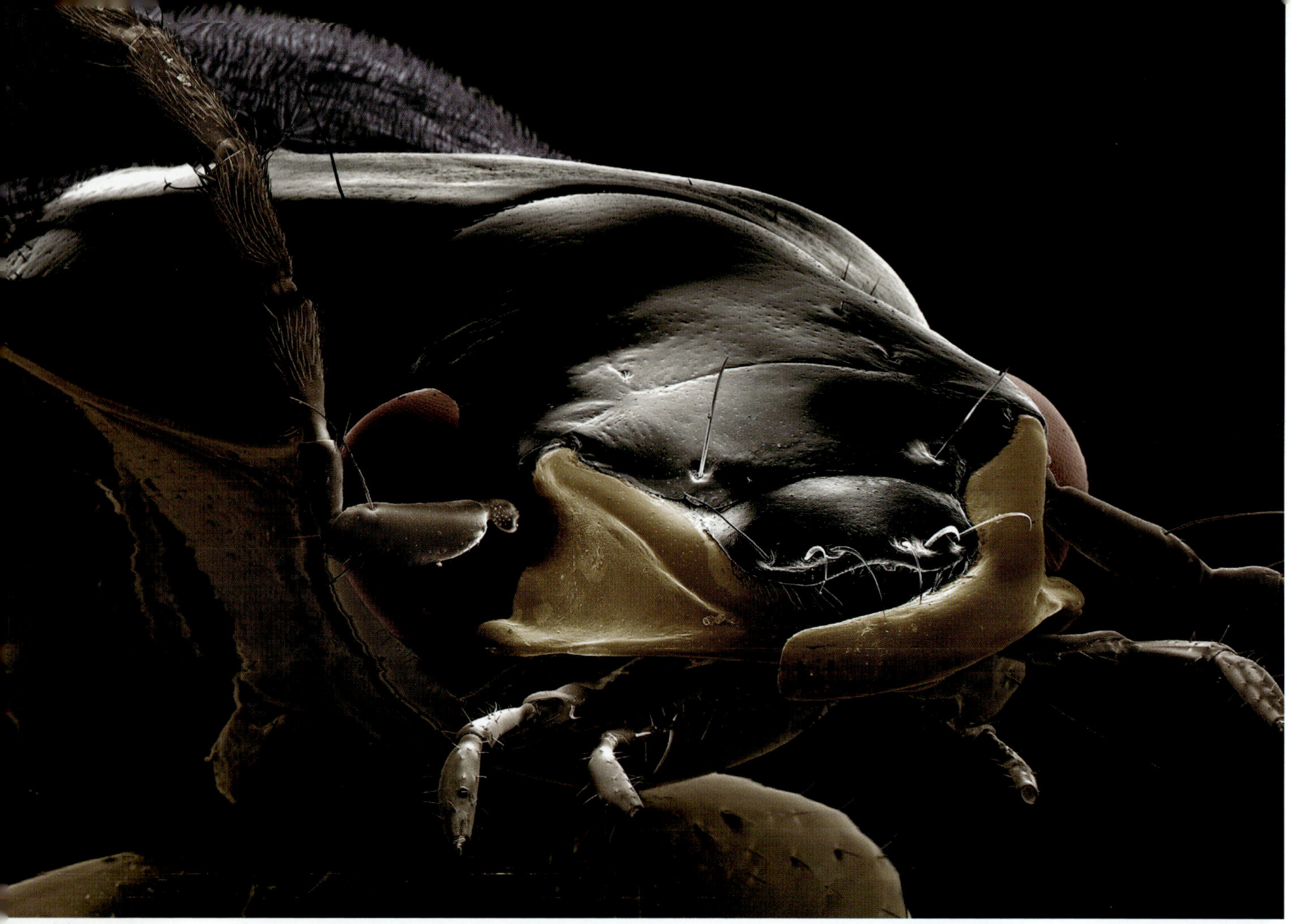

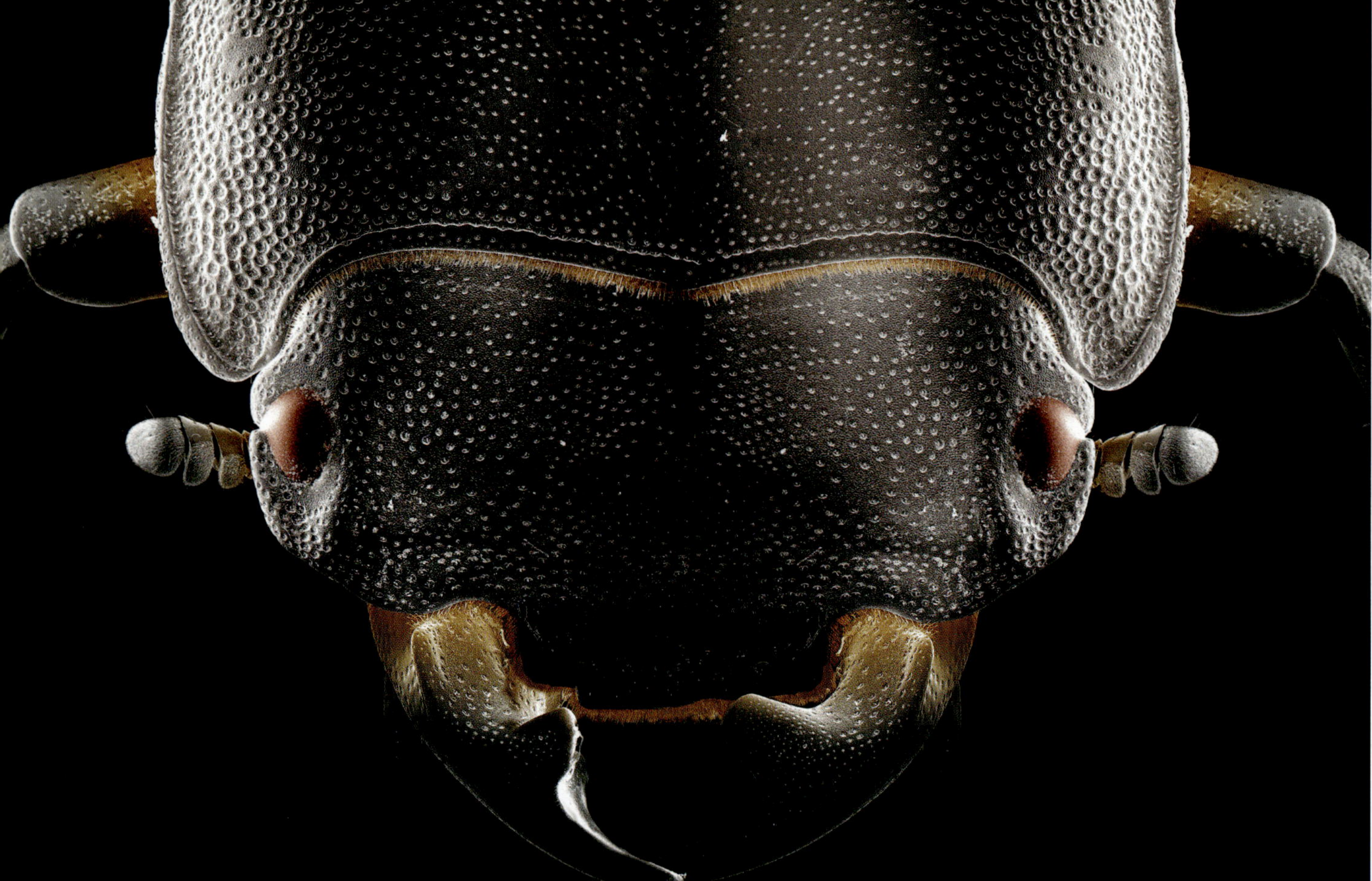

Ground Beetle, *Agonum sp.*
(above left)

Fast runners, ground beetles are predators, hunting all kinds of small creatures, including worms, slugs and small insects. There are about 350 species of ground beetles in Britain, some of which lack wings and their wing cases, or elytra, are fused together, giving them extra protective armour as they scramble about.

Magnification: x50
Scanning Electron Micrograph

Lesser Stag Beetle,
Dorcus parallelopipedus
(below left)

A diminutive version of its larger, rarer cousin, adult lesser stag beetles, and their larvae, frequent soft, rotting tree trunks and can sometimes be found in compost heaps.

Magnification: x20
Scanning Electron Micrograph

Green Tiger Beetle,
Cicindela campestris (right)

Closely related to *Agonum*, the green tiger beetle is a long-legged fast-running beetle and can fly well over short distances. They are fierce hunters.

Magnification: x50
Scanning Electron Micrograph

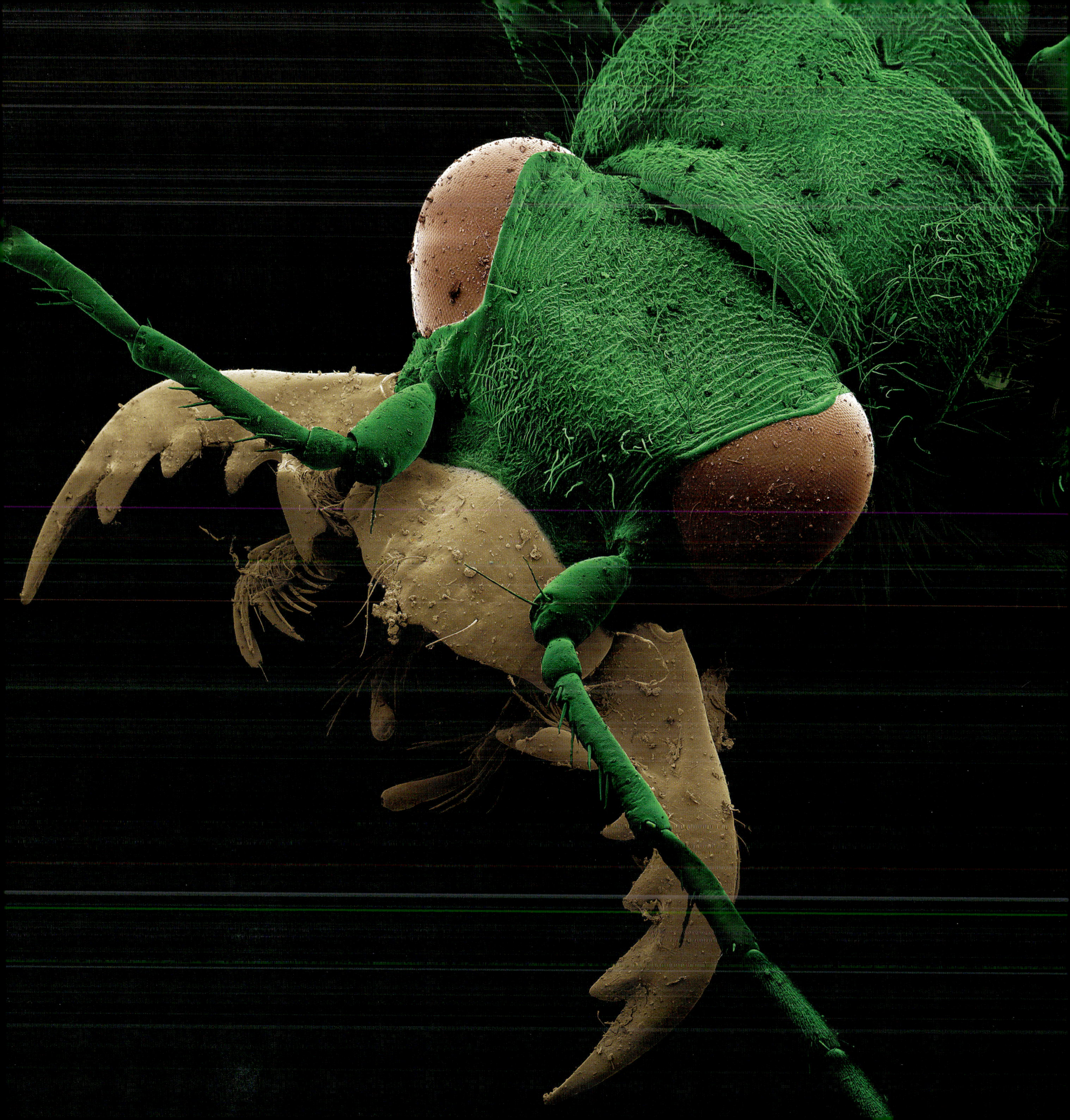

Ground beetles, the family Carabidae, of which there are some 350 species in Britain, are often seen running across the ground or hiding under logs in the garden. Large beetles, up to 3.5 cm long, they are mostly carnivorous, although some feed on vegetable matter, especially seeds. Caterpillars, grubs, and snails are the favorite food for a few species, while others target aphids, springtails and other tiny creatures. Most ground beetles hunt by running or climbing, and while some are capable of flight, others lack wings and their wing cases, or elytra, are fused together, giving them extra protective armour as they scurry around. As a defence mechanism against predators, many species secrete a repellent fluid from a gland at the tip of the abdomen, and some emit an audible squeal. The large species can also bite.

GREEN TIGER BEETLE

Closely related to *Agonum* is the spectacular green tiger beetle, *Cicindela campestris..* Found on heaths and in sandy places, it is fairly common in Britain, favouring open ground where it runs fast and makes short flights. There are five different species, all of them a beautiful iridescent green colour with yellowish spots on the elytra or wing cases. Adults have large eyes and mandibles characteristic of a hunter and both adults and larvae are fierce predators of small invertebrates, but while the adults actively chase their prey, the larvae wait in ambush. In order to create a trap, they dig pits at the base of which they excavate a burrow where they lurk, waiting for a small creature to pass by.

VINE WEEVIL

Weevils belong to the very successful beetle family Curculionidae with more than 50,000 species worldwide. They vary in size from tiny seed weevils, less than 2 mm long, to the large pine weevils, up to 2.5 cm long. Nearly all weevils have a characteristic snout, or rostrum, projecting forward from the head, with mandibles or jaws at the tip, and most species have distinctly elbowed antennae. In some genera, for example *Circulio,* the length of the snout exceeds that of the body, while in others, such as the leaf weevils, genus *Phyllobius*, the snout is very short. The vine weevil (*Otiorhynchus sulcatus*) also has quite a short snout. It attacks a wide range of garden plants and will also attack houseplants given the chance. The adults, up to 1 cm long, are dull black and may have small patches of yellow on the wing cases. Like many weevils, they are slow moving and cannot fly but are persistent crawlers and climbers. Seldom seen, since they are mainly nocturnal and hide during the day in leaf litter or crevices, they crawl up onto plants after dark. Their presence is indicated by irregular notches eaten out of leaf margins or by the death of young shoots on woody plants. The harm done by the adults is not fatal, just unsightly, but the real damage is done by the legless, white larvae, up to 1 cm long, which gnaw away the outer tissues of roots and stem bases of woody plants and bore into tubers. This stunts growth and may even kill plants, especially if potted or young. The first sign is usually yellowing leaves, poor growth or wilting, which no improvement after watering.

The biology of the vine weevil is unusual since adult males are very rare. Nearly all adults are female, which are able to reproduce by parthenogenesis, laying viable eggs without being fertilised by a male. Females can produce several hundred eggs during the summer months in

Iridescent scales of a Diamond Weevil,
Entimus imperialis (right)

The large South American weevil, *Entimus imperialis*, is remarkable for the green, blue and gold iridescent scales (see right) that cover its elytra, or wing cases. The colour comes not from pigments but from the diffraction of light on the microstructure of the scales themselves. The tiny scales, each only about 0.1 mm long, are in rows about 1 mm apart along the 'furrows' of the back of the weevil. In Britain, some weevils, such as *Phyllobius* (overleaf) and *Polydrosus*, also have iridescent scales and are among the most beautiful insects here.

Magnification: x250
Light Micrograph

Vine Weevil,
Otiorhynchus sulcatus (overleaf, left)

Dull black with small patches of yellow, vine weevils are slow moving and cannot fly, but are good climbers, feeding on leaves and young shoots of woody plants. The white larvae of this species feed on the roots and stem bases of the same plants, causing damage that stunts growth.

Magnification: x40
Scanning Electron Micrograph

Leaf Weevil, *Phyllobius sp.*
(overleaf, right)

Most weevils have heads with a long snout, or rostrum, with mandibles at the tip, and the antennae are usually elbowed. *Phyllobius* is a genus of 'short-nosed' weevils, some species of which are characterised by an iridescent scaly coat with short hairs. The adults eat leaves and occasionally the blossoms of trees and woody shrubs, while the larvae are root feeders.

Magnification: x180
Scanning Electron Micrograph

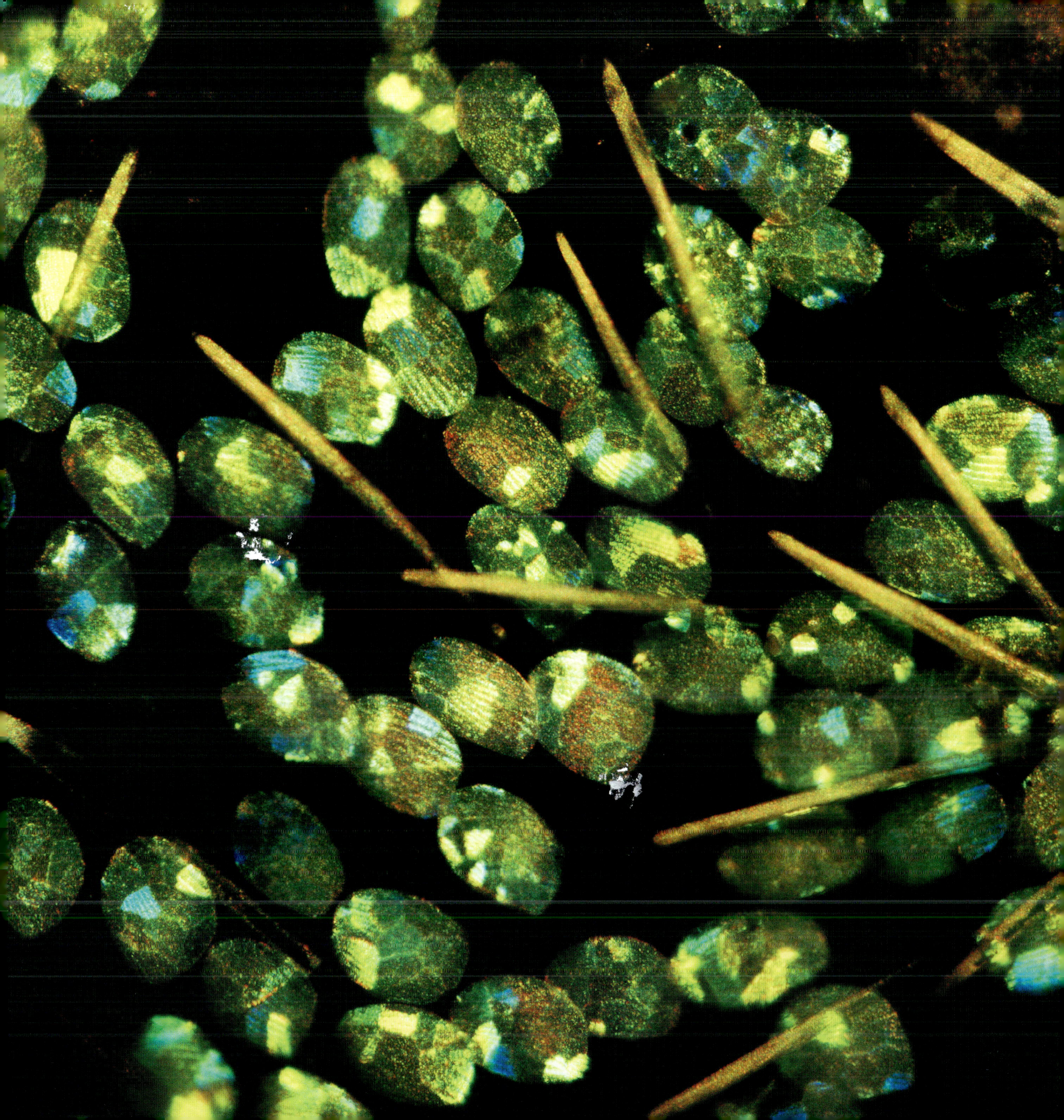

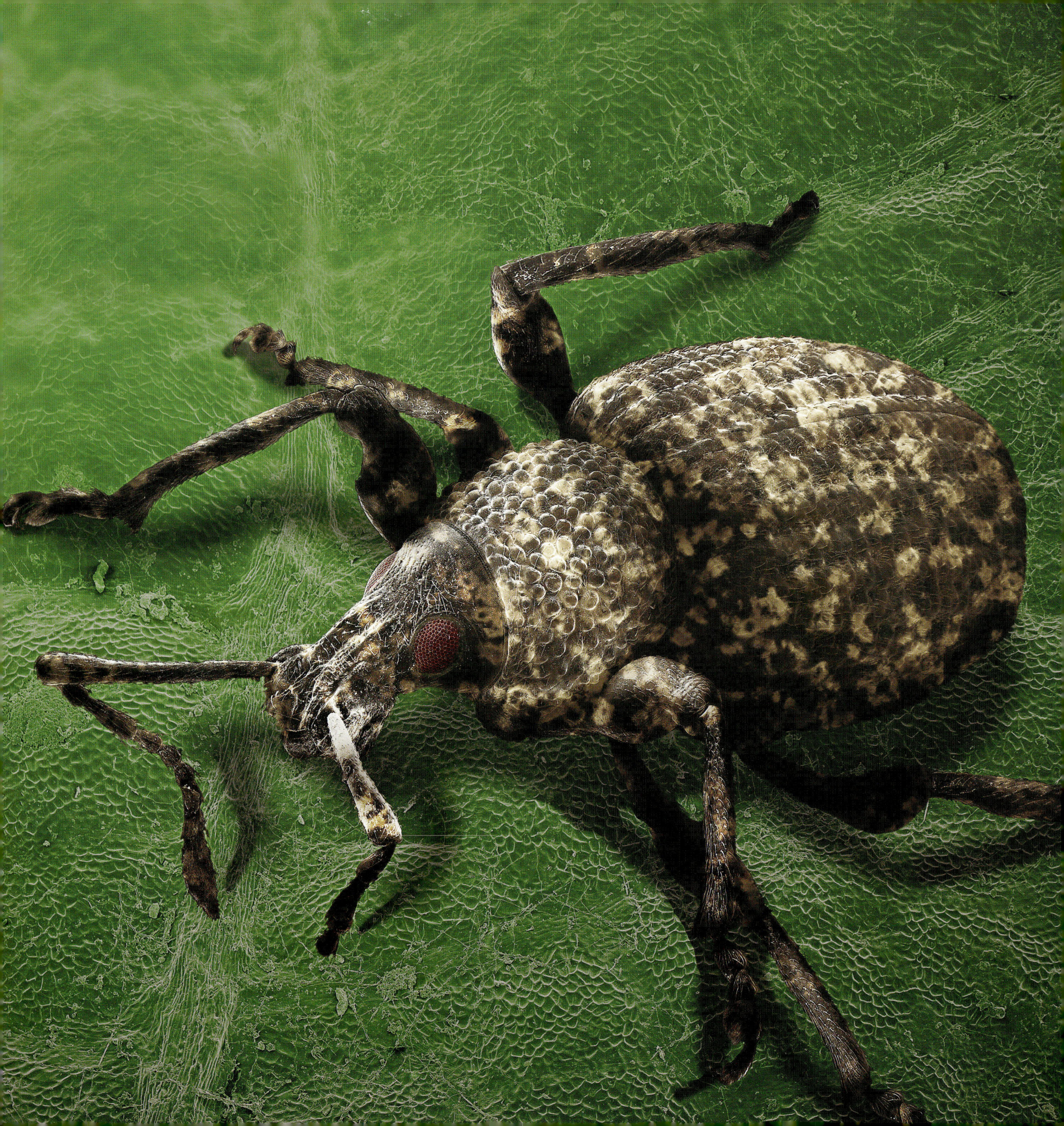

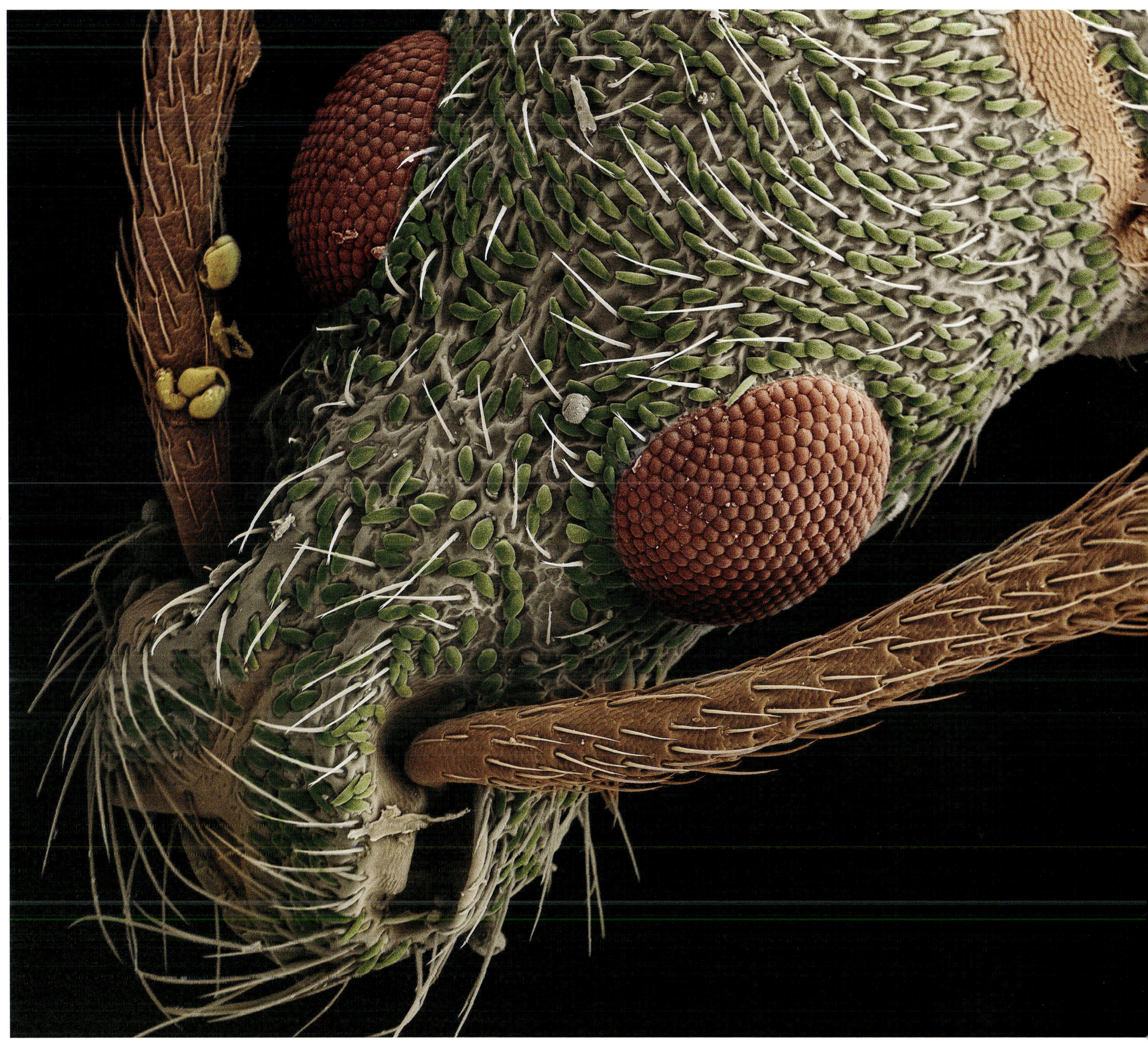

the soil near suitable food plants. The larvae hatch after about two weeks, feeding for several months before pupating in the soil, then the adults may emerge in the autumn, but generally wait until the following spring. A natural predator is the centipede, which eats both eggs and larvae.

BUTTERFLIES AND MOTHS

Among our most conspicuous and showy insects, there are over 70 species of butterflies and 2,000 species of moths in Britain. As a rule, butterflies are active by day and moths by night although this is not always true. Butterflies almost always have knobs on the end of their antennae, whilst moths do not; in fact, some male moths have very intricate, feathery antennae, which are sensitive pheromone detectors, enabling them to locate females by flying along a scent trail. Butterflies and moths have two pairs of wings, which are usually covered with a dense mosaic of tiny chitinous scales forming a myriad of striking colour patterns and designs. Their mouthparts simply consist of a long tube, used to suck up nectar or other liquids, that is normally carried in a coil under the head. Eggs are usually produced in early summer though some species over-winter as eggs. The eggs hatch out into caterpillars, voracious feeders with biting, chewing jaws to ingest the leaves of their food plant. After a few weeks, the caterpillars pupate, and eventually the adults emerge, spreading their wings to dry before flying.

GRASSHOPPERS

The rhythmic chirping of grasshoppers, advertising their presence to each other by rubbing their hind legs against their tough forewings, is one of the familiar sounds of summer. It is called 'stridulation' and is primarily territorial, aimed at rival males. When courting a female, the males make a briefer 'ticking' sound. Their strong hind legs enable them to leap great distances if disturbed. Crickets are similar, though more stout-bodied, tending to run rather than leap. The chirping of crickets, softer than that of grasshoppers, is made by rubbing the forewings together and is a sound characteristic of balmy summer evenings. Both grasshoppers and crickets are well camouflaged, usually only noticed when they move. The common field grasshopper, *Chorthippus brunneus*, is probably the most frequently seen species; it grows up to 2.5 cm long and is very variable in its colour and markings. It is herbivorous, feeding mainly on grasses, and deposits pods of up to 15 eggs in the ground, usually at the base of grass tufts, where they survive the cold winter months. After the nymphs emerge from the ground in the spring, they moult four times before becoming winged adults. *Chorthippus* is a strong flier, active in warm weather when it can frequently be seen sunning itself on walls, or on bare ground in the garden.

SHIELD BUG

Shield bugs are strikingly colourful little creatures with a very distinctive shield-like shape, hence their name. They are also known as stink bugs, since they produce an acrid, smelly secretion from glands in the thorax when disturbed, to deter predators. There are nearly 6,000 species known worldwide, of which 33 occur in Britain. Most shield bugs have long, thin, beak-like mouthparts to stab plant tissue and suck out the sap. A few species attack moth caterpillars and beetle larvae, stabbing them and sucking out their body fluids. Depending on the species, they pass through four or five nymph stages, each increasing in size. The gut of a shield bug is packed with bacteria, and when eggs are laid, the adult female carefully smears them with its own bacterial soup for the newly hatched nymphs to imbibe. It is possible that the bacteria aid digestion or help to metabolise vitamins.

Footman moth, *Family Arctiidae*

The elaborate antennae of a male moth have sensitive chemoreceptors, sense organs that can detect the scent of chemical pheromones from a female moth hundreds of metres away.

Magnification: x30
Scanning Electron Micrograph

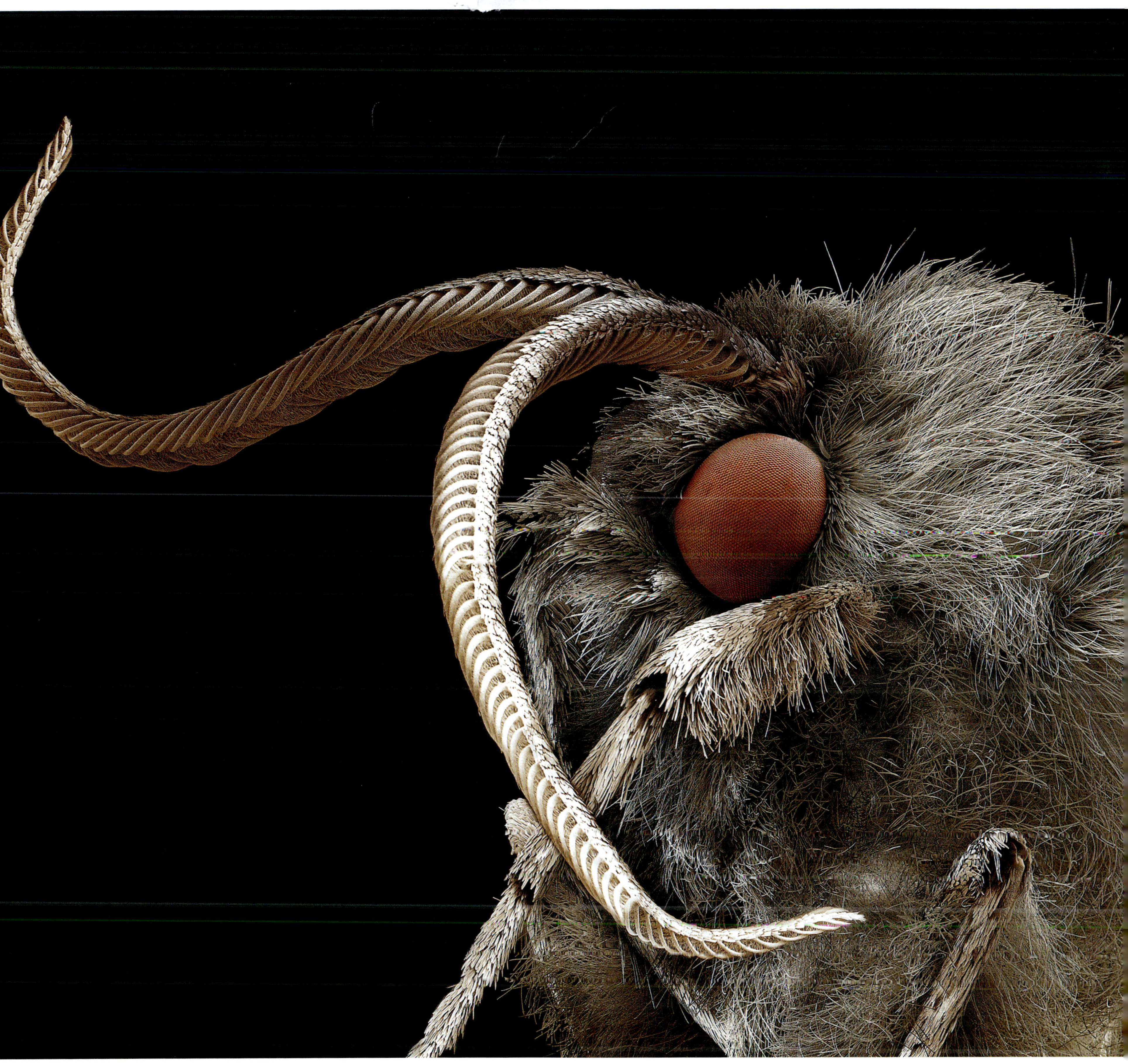

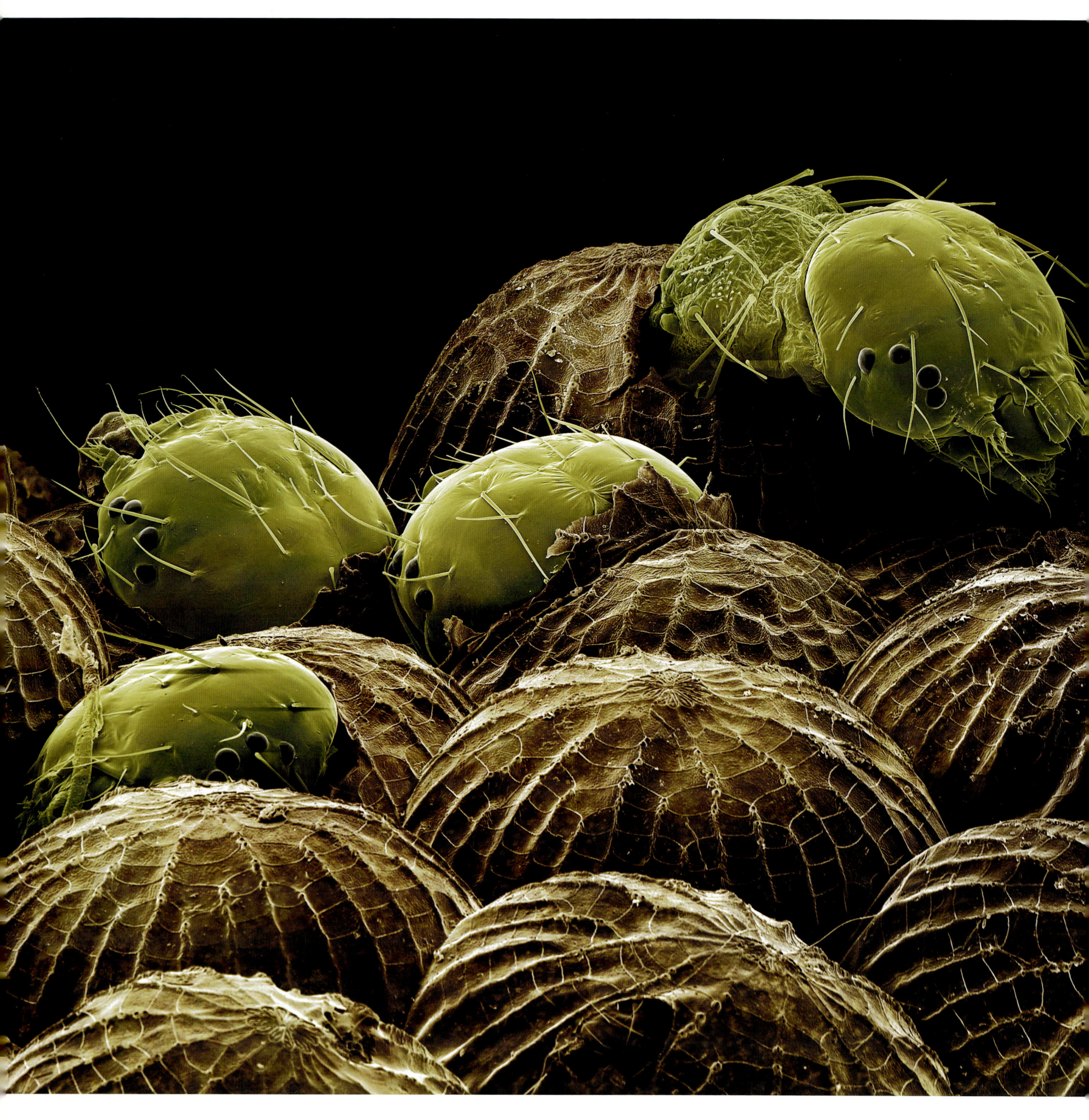

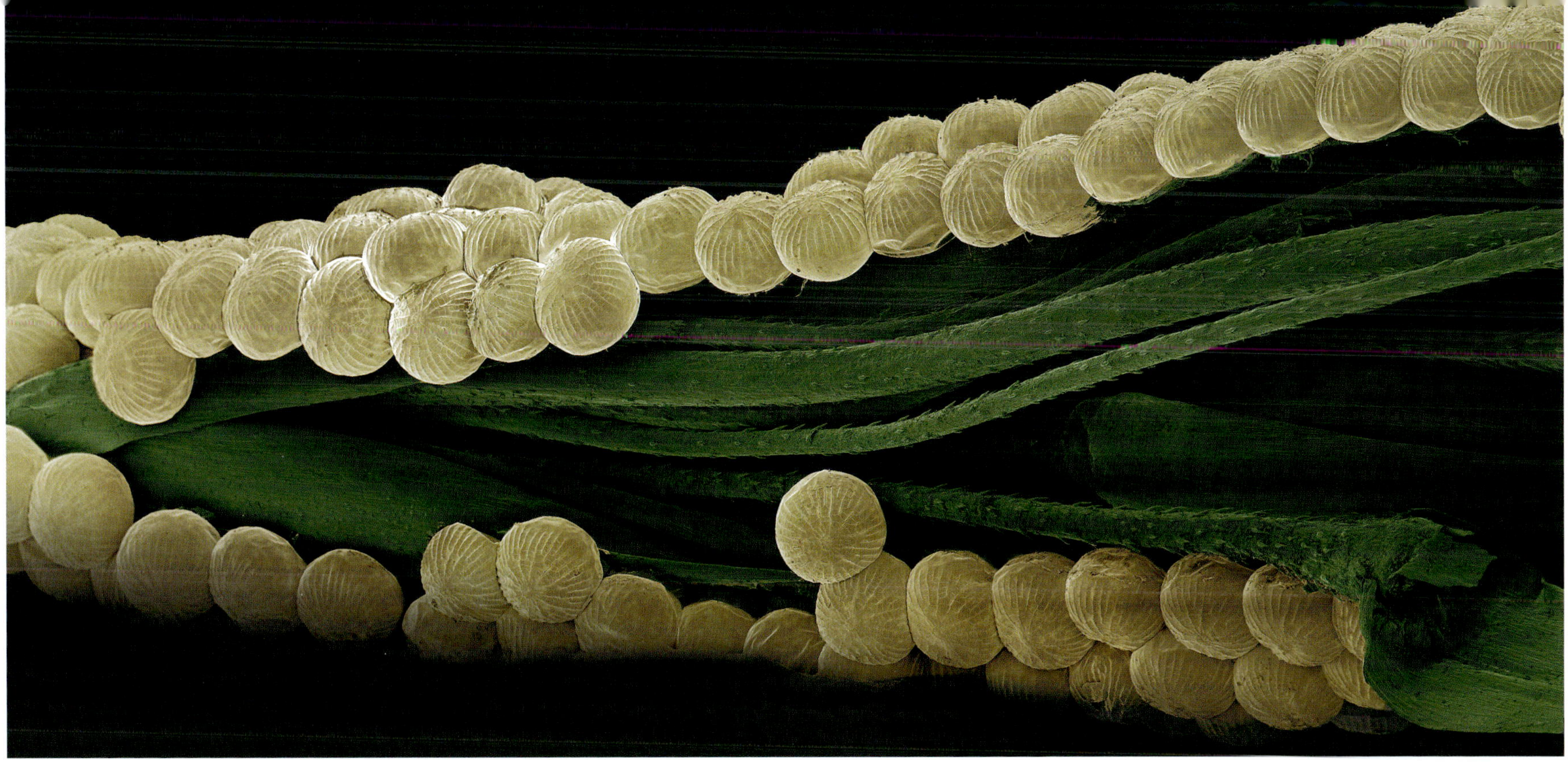

Spectacle Moth, *Abrostola tripartita* (left and above)

The caterpillars of a spectacle moth, family *Noctuidae*, hatch from eggs laid on the surface of a leaf (see above). After chewing their way through the egg case to escape (see left), they will feed on nettles. Spectacle moths favour damp habitats on woodland margins.

Magnification: x300   *Scanning Electron Micrograph* (left)
Magnification: x35   *Scanning Electron Micrograph* (above)

Small Tortoiseshell Butterfly, *Aglais urticae* (below)

The caterpillar of the familiar small tortoiseshell butterfly also feeds on nettles. Many caterpillars, like those of *Aglais*, have protective spines or venomous hairs to discourage predators.

Magnification: x25   *Scanning Electron Micrograph*

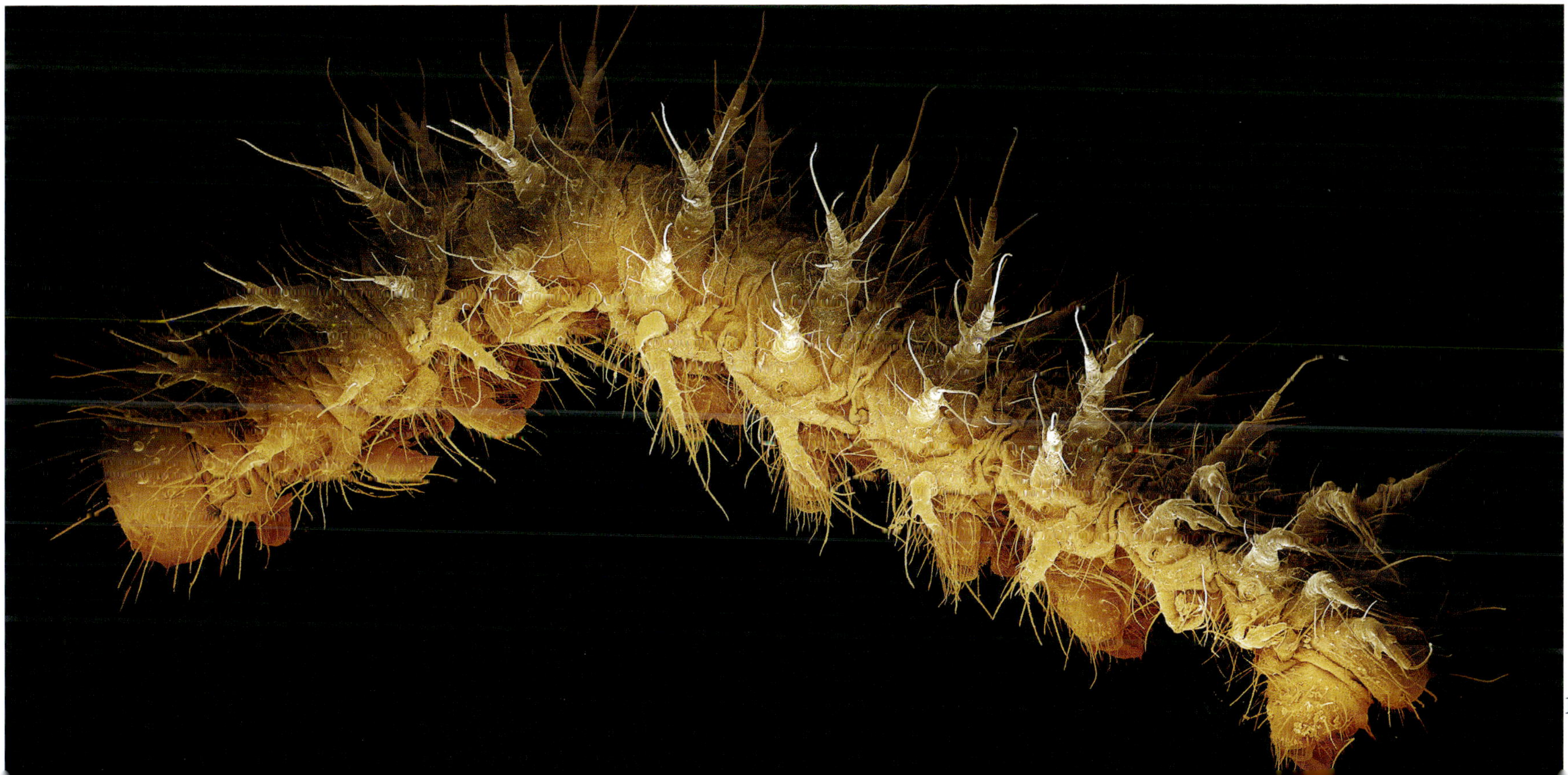

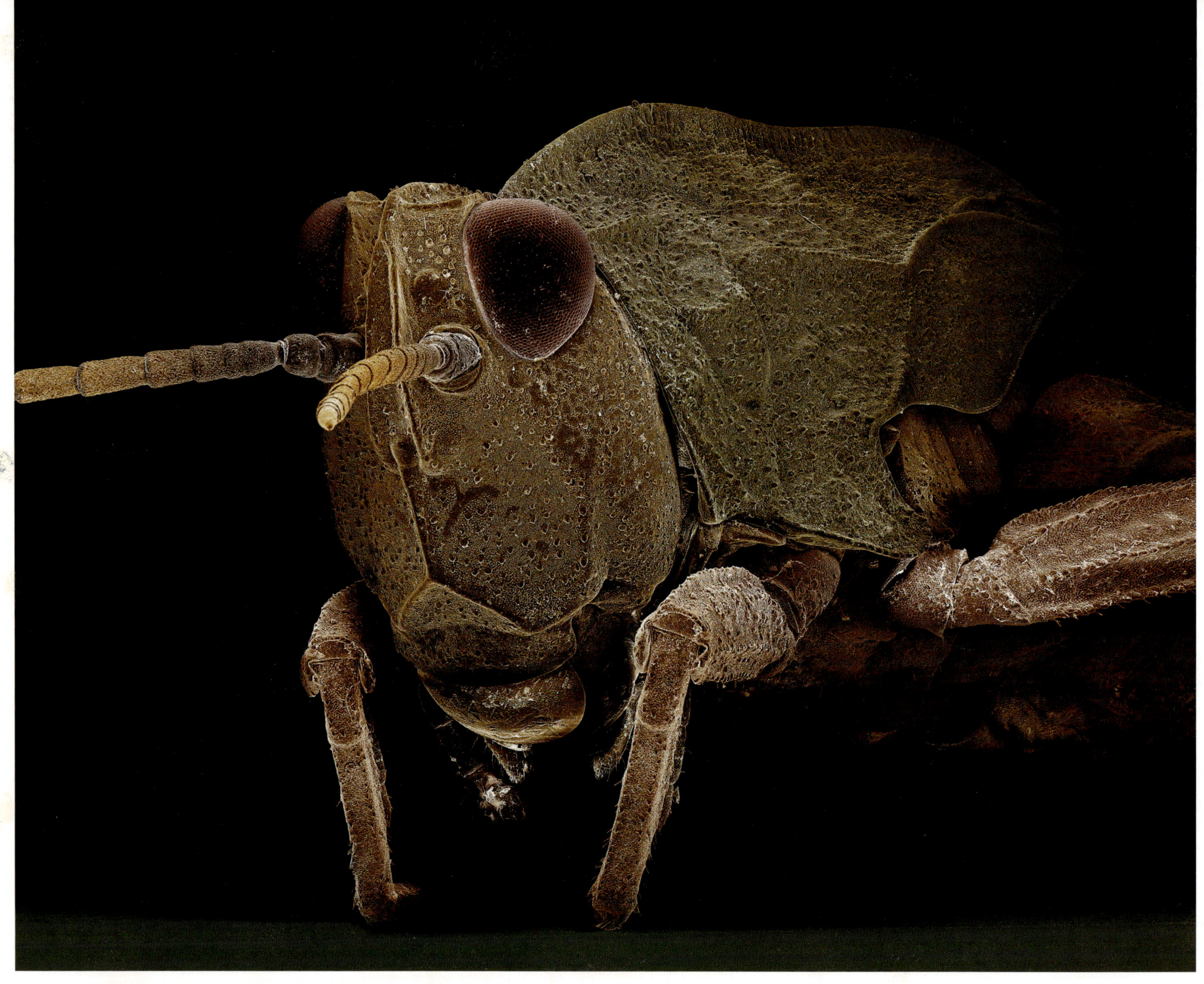

Common Field Grasshopper, *Chorthippus brunneus* (above)

The most common and familiar grasshopper in meadow habitats throughout Britain, the rhythmic chirping of *Chorthippus is* a feature that goes hand in hand with a warm summer's afternoon in the country. The strong hind legs of grasshoppers enable them to leap great distances if disturbed and they can fly quite long distances.

Magnification: x60   *Scanning Electron Micrograph*

Green Shield Bug, *Palomena prasina* (right)

When seen from above, the body of this insect is shield-shaped, hence the name. Bright green during the summer, they turn brown for hibernation over the winter, then lay eggs in the spring. The immature nymphs resemble the adult but are more rounded in shape and lack wings. Green shield bugs feed by sucking sap from a variety of trees and shrubs, especially hazel.

Magnification: x45   *Scanning Electron Micrograph*

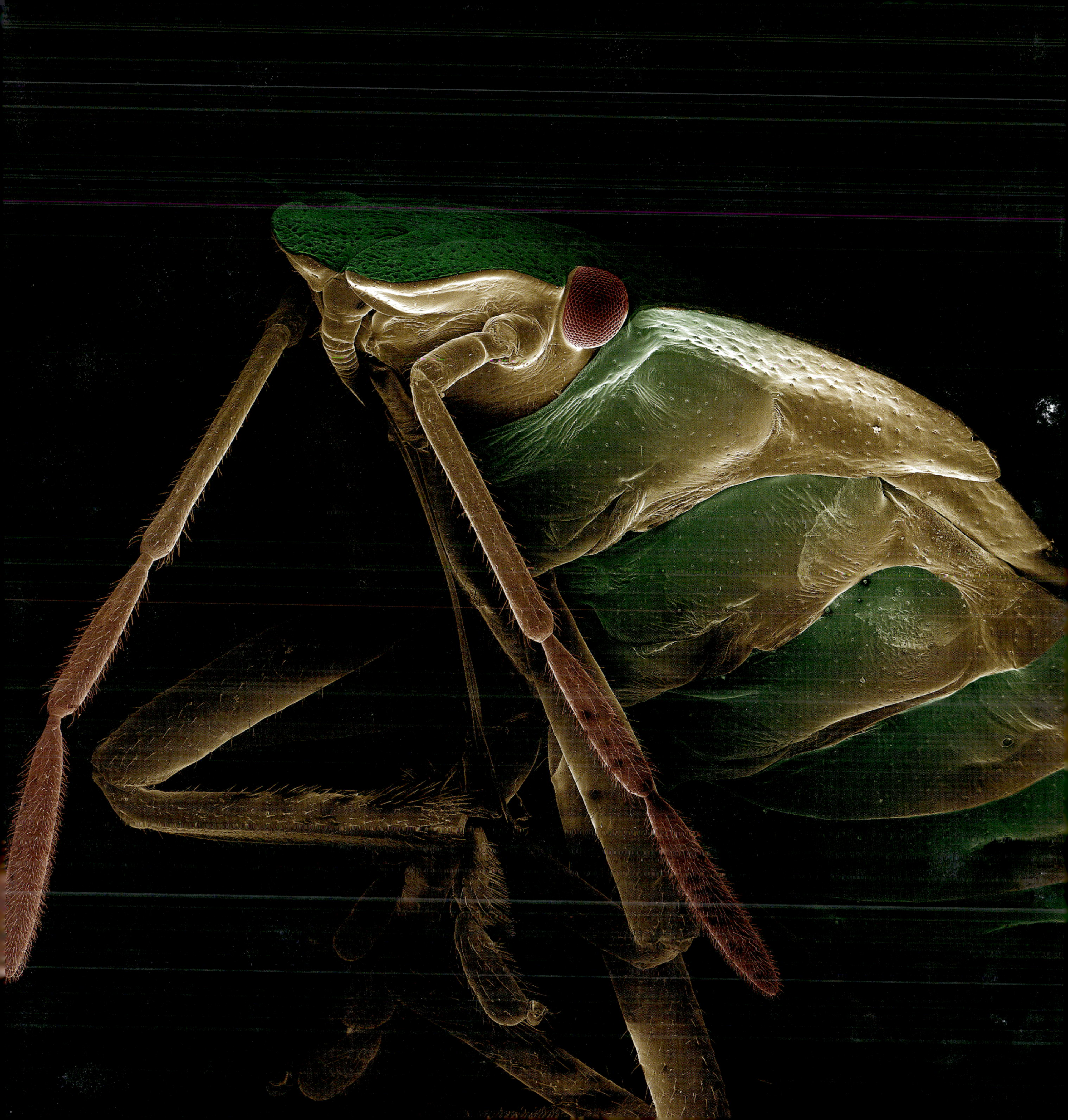

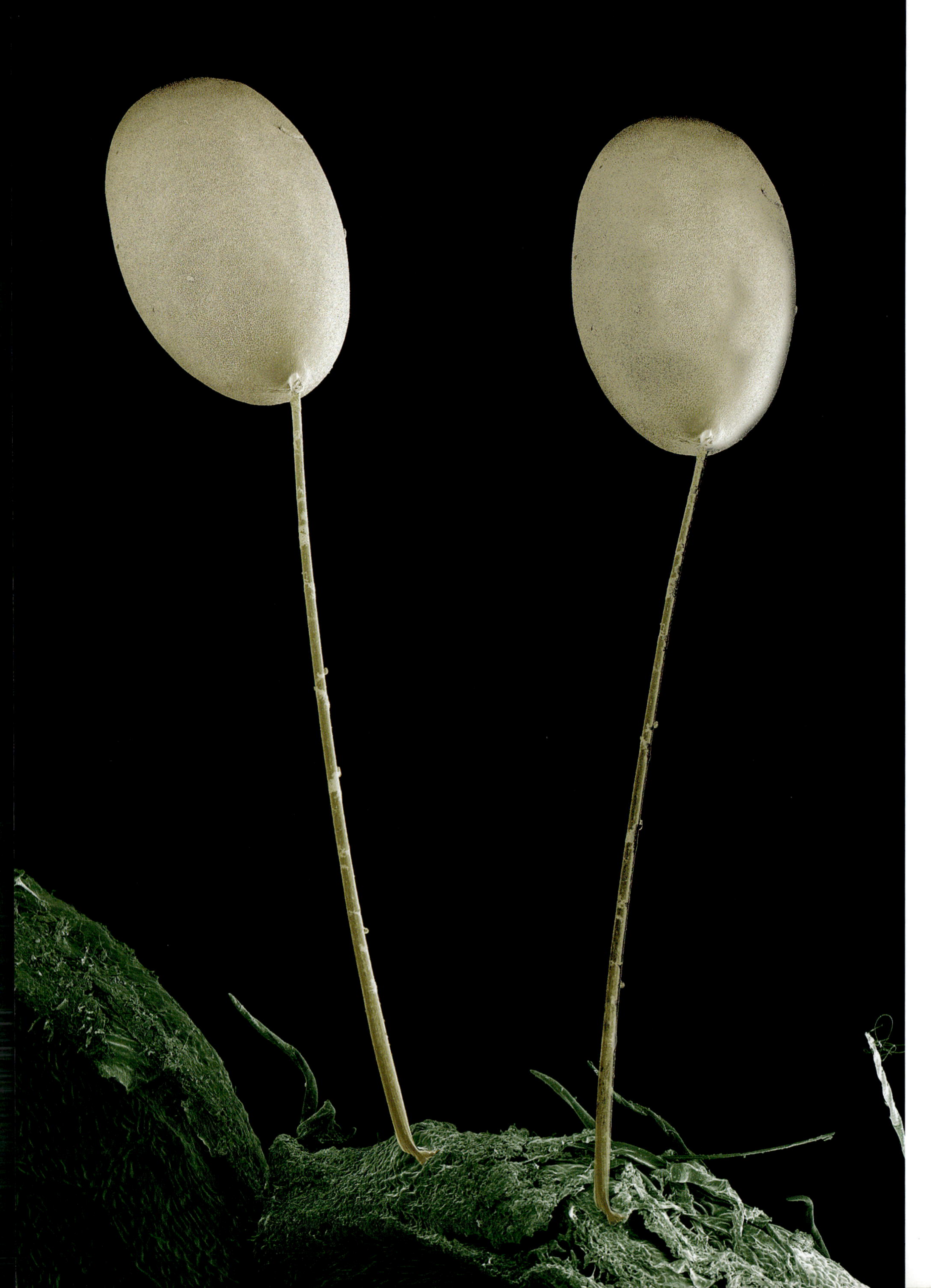

LACEWINGS

Lacewings are insects about 1.5 cm long, with two pairs of filmy wings with a network of fine veins. Their larvae are natural enemies of soft-bodied insects, including aphids, and possess unusual mouthparts, which they use to puncture their prey to suck out body juices. The mandibles are long, slender tubular structures rather like hypodermic needles that curve forward from the front of the head. The adult female lacewing deposits eggs on leaf surfaces, on the tips of hardened mucus strands. In some species, the larvae augment their camouflage by covering themselves with the dried remains of their prey, so they tend to look like a small mass of debris with legs.

Green Lacewing,
Chrysoperla carnea

Beautiful delicate-looking insects with transparent veined wings, lacewings (see right) are friends of the gardener, for their larvae are voracious consumers of aphids. The adult lacewing feeds on nectar and pollen, and is not predatory. The female lays her eggs (see left) on the surfaces of leaves, supported on hardened strands of mucous.

Magnification: x135
Scanning Electron Micrograph (right)
Magnification: x125
Scanning Electron Micrograph (left)

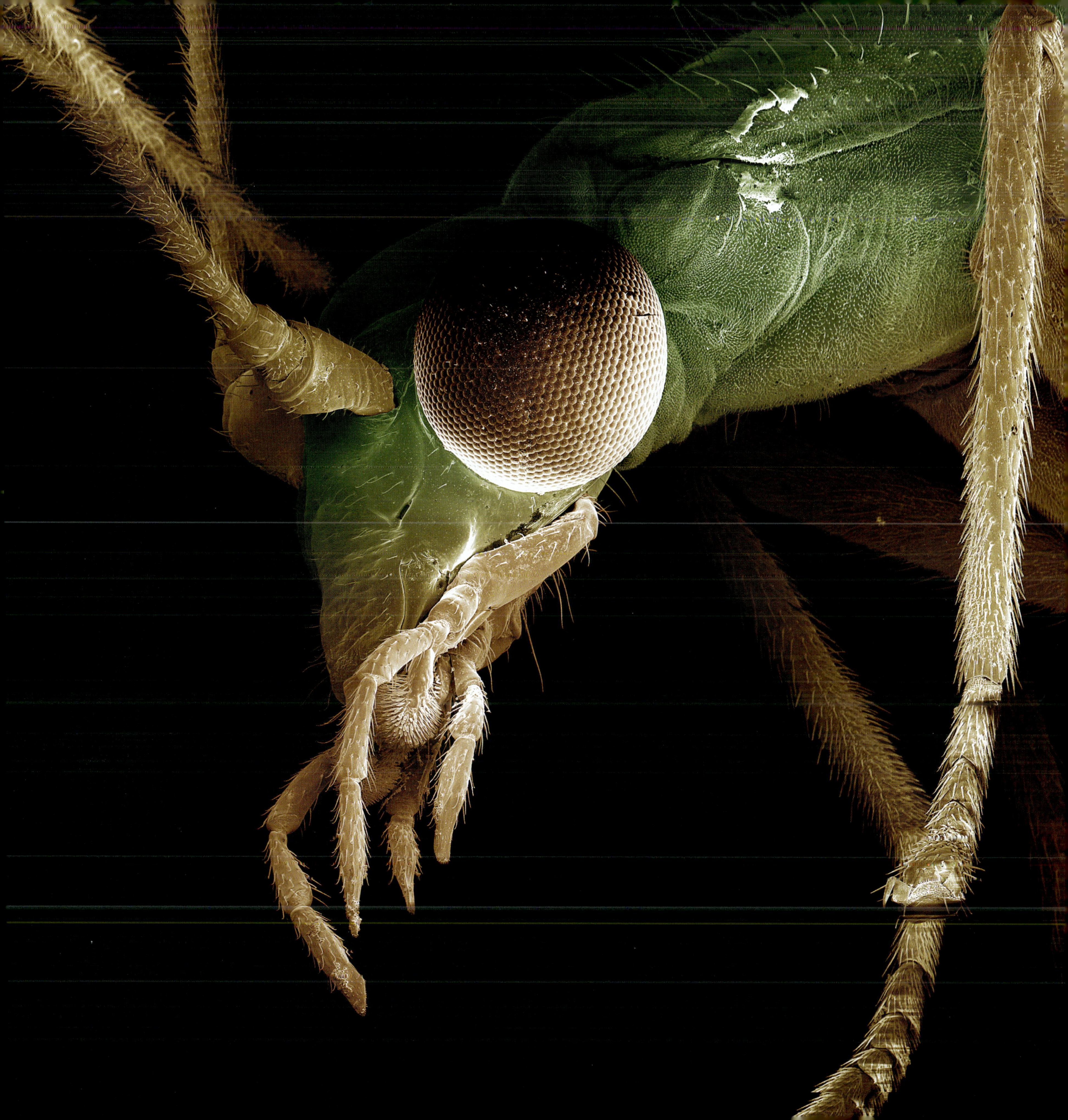

CRANE FLY

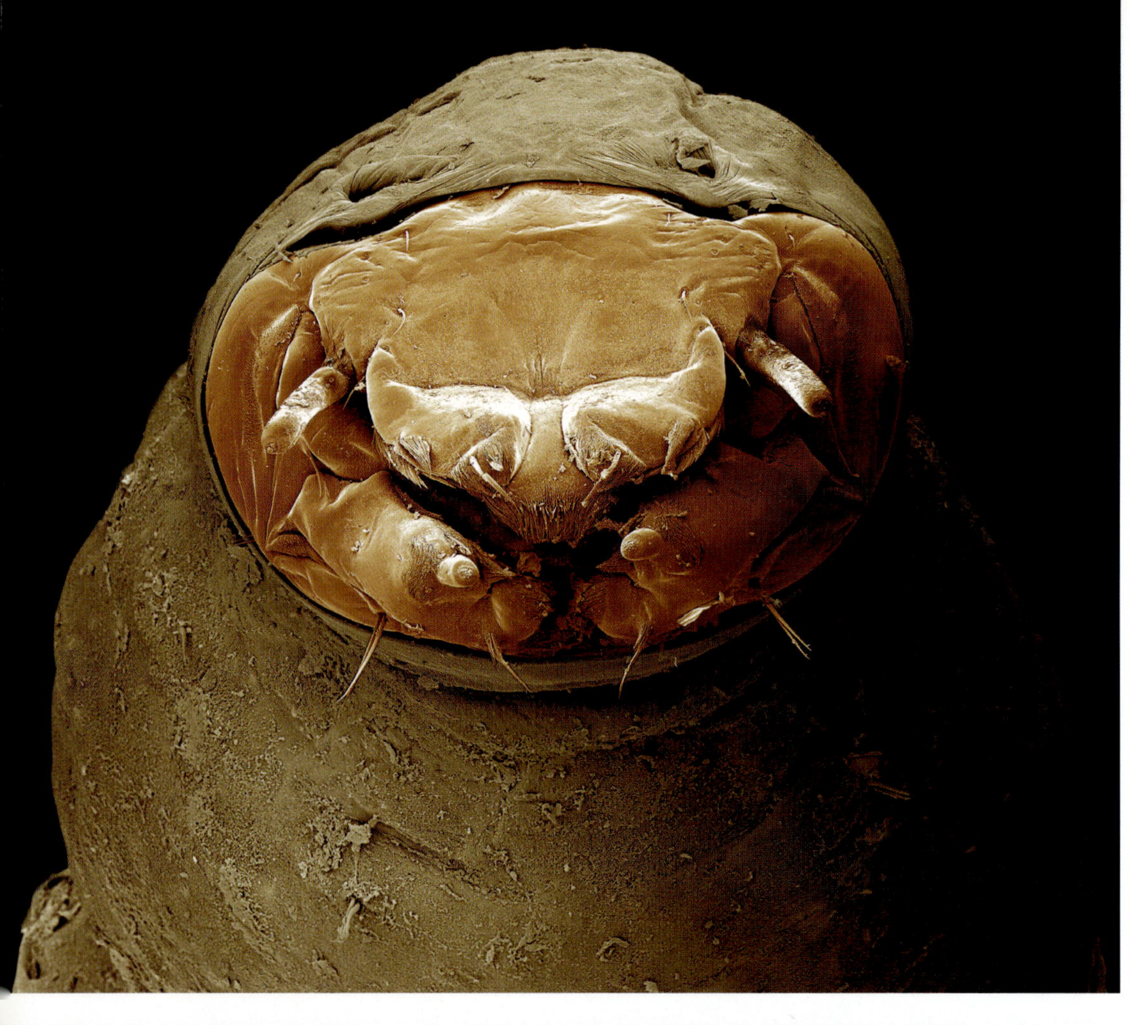

Crane flies or 'daddy-long-legs', such as *Tipula paludosa*, are very common in Britain. Familiar insects, the adults have a slender brown body about 1.5 cm long, thin wings and long dangling legs. Like all flies, crane flies have one pair of membranous wings, with a second pair modified to drumstick-like appendages known as 'halteres', which aid in balancing. The larvae, known as 'leatherjackets', grow up to 4 cm in length, feeding on the roots and the bases of plant stems, especially grass, and can cause damage to lawns. They spend the winter in the soil and can continue to be active in temperatures as low as 5°C, but as spring approaches they become increasingly active, eventually heading towards the surface to pupate. The adults emerge between June and September, although emergence reaches a peak from mid-August to mid-September when they can be seen frequenting the surface of lawns. Shortly after emergence, the adults mate and females lay one batch of eggs amongst grass or other vegetation. Another crane fly, *Lipsothrix nigristigma*, which is widely distributed in continental Europe, is rare in Britain having been found only at a few sites in Shropshire. It has a yellowish body and is much smaller than its more familiar cousin.

WEEDS

Weeds are simply plants that interfere with the planned aesthetic order of a garden by growing in the 'wrong' place. They are usually native plants, which are best adapted to the environment where they grow, and the seeds they produce can lie dormant in the soil for many years. Since they compete very successfully with garden plants for space, water, light and nutrients it is necessary to keep them under control.

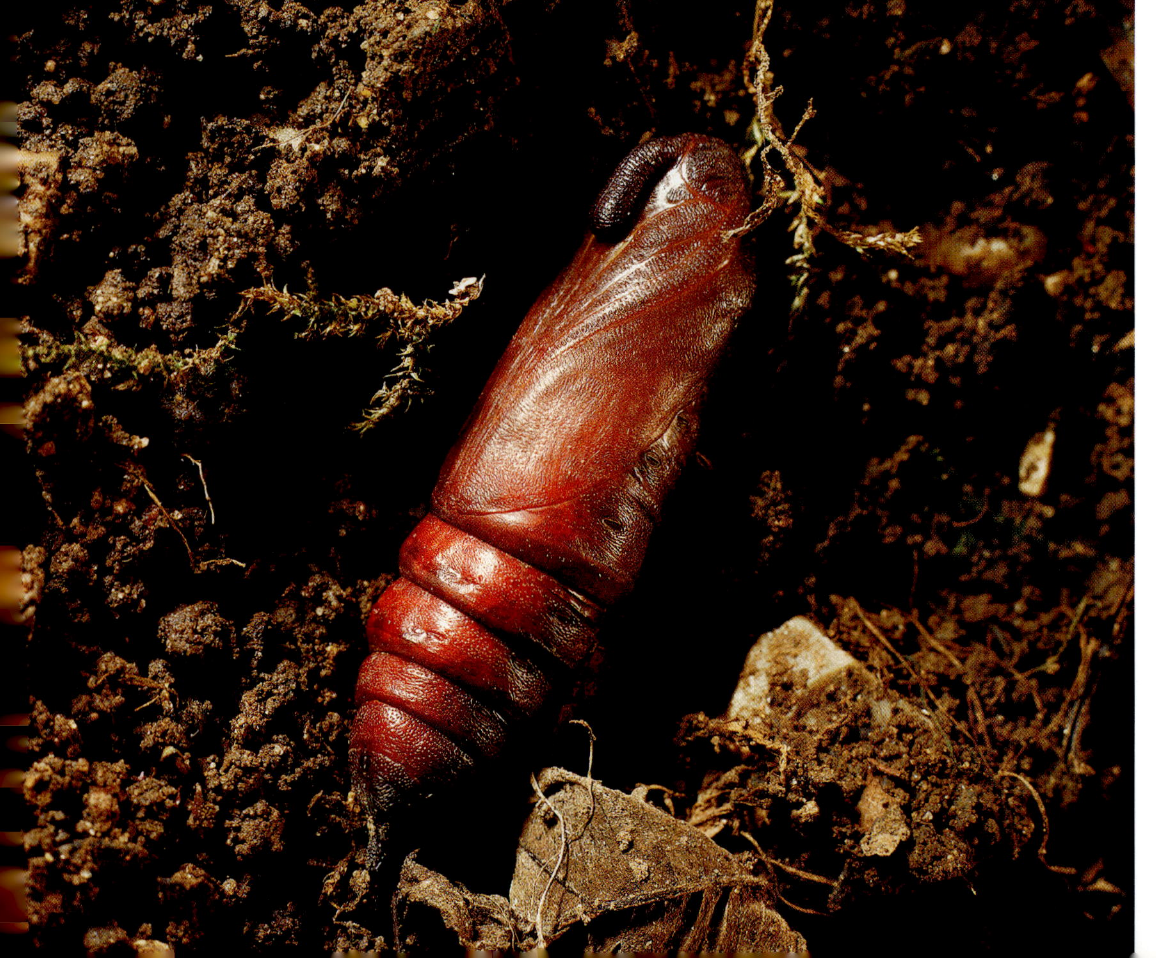

Crane Fly or 'Daddy-Long-Legs', *Tipula paludosa*

The larva, or 'leatherjacket', of the crane fly (above left) chews the roots and stem bases of plants, especially grasses, and can cause damage to lawns. The larva pupates (below left) just under the surface of the soil. The adult crane fly (right) emerges in late summer.

Magnification: x40
Scanning Electron Micrograph (above left)
Magnification: x3
Macro-photograph (below left)
Magnification:x100
Scanning Electron Micrograph (right)

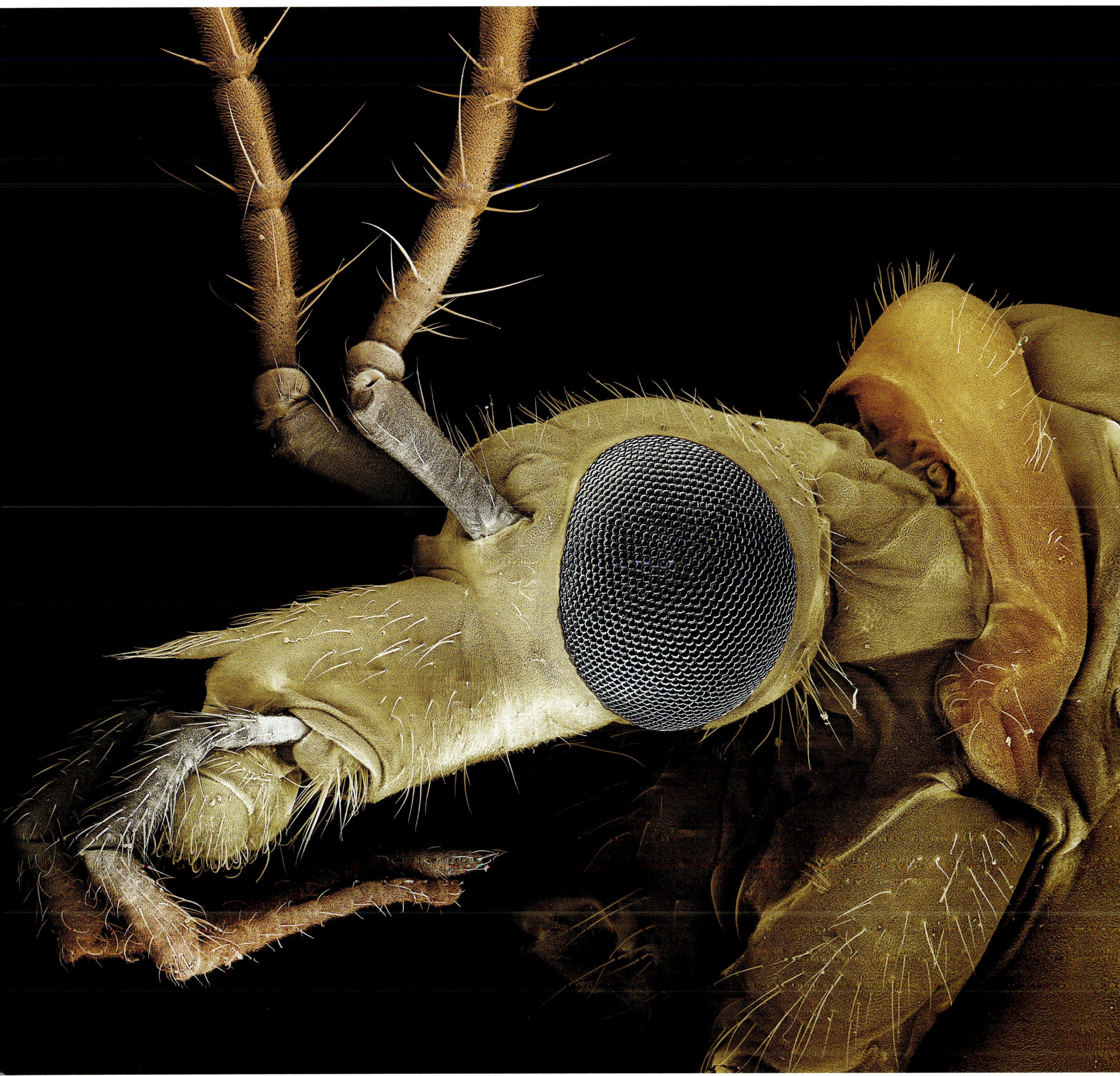

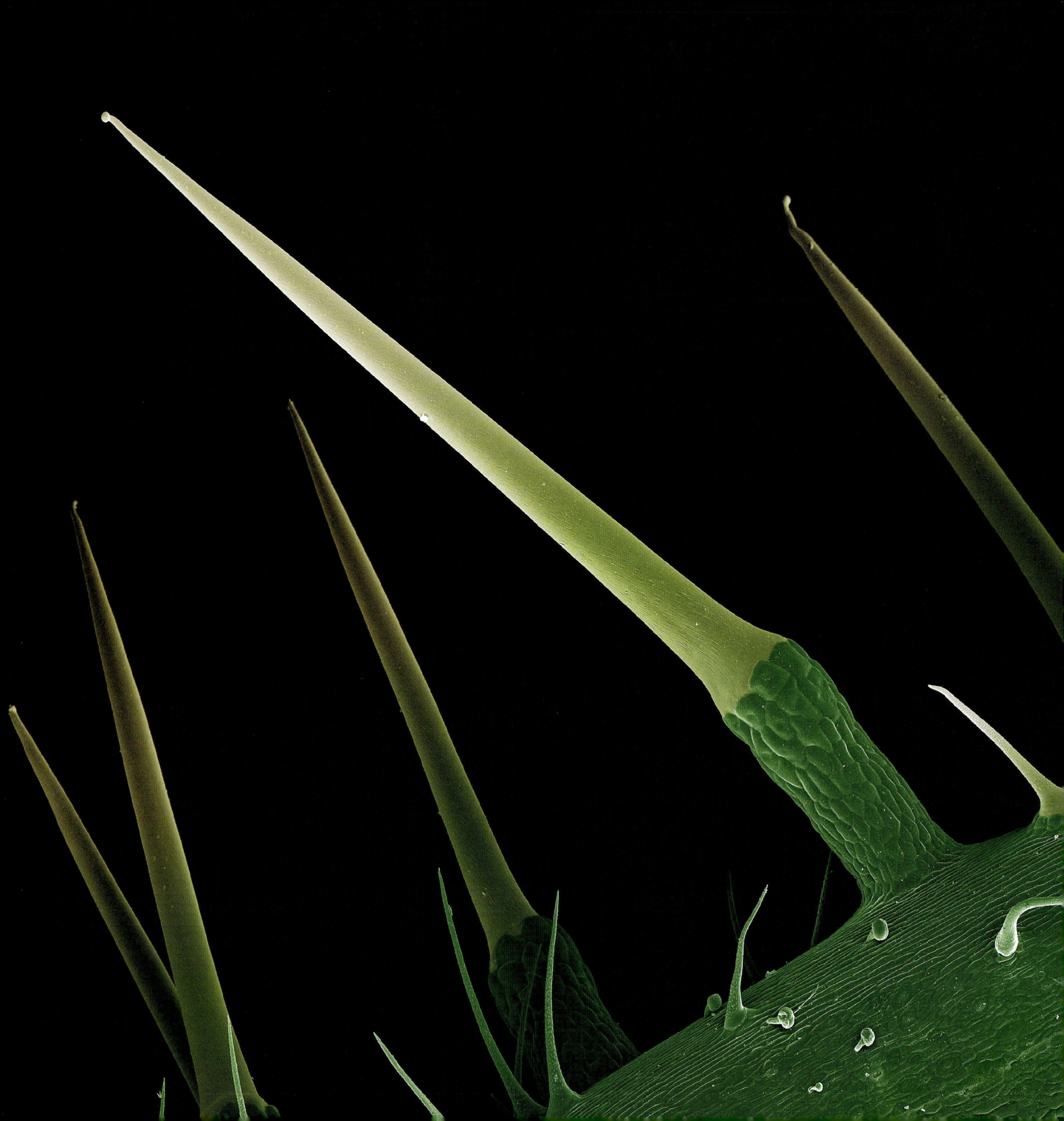

STINGING NETTLE

Stinging nettles, *Urtica dioica*, are familiar weeds, which provide an important source of food for the caterpillars of many species of butterflies, some of which feed exclusively on the leaves. A natural feature of the British countryside, nettles are found in wooded areas and on disturbed ground. They are hardy perennials, growing up to about a metre in height, with male and female flowers arranged in long catkins, coloured green, red or white. For a butterfly rich garden, it is a good idea to allot an area to nettles in order to help them breed. Nettles are also good for the garden as they can act as an accelerator for compost, and are a rich source of nitrogen and minerals. When cooked, the leaves can be eaten, and various potions can be prepared from nettles, which are claimed to be anti-inflammatory, anti-parasitic, antiseptic, a digestive stimulant, and a menstrual promoter. In Roman times, nettle stings were used as a cure for chronic rheumatism.

Unfortunately, nettle stings can be quite unpleasant. The toothed leaves and stem are covered in tiny sharp-pointed hairs, which are hollow tubes with walls of silica, like glass needles. The tips are very easily broken when touched, leaving a sharp point to pierce the skin. The mechanical force on contact causes the hair to bend, squeezing the stinging liquid from a sac at the base, up the tube and into the skin. The liquid is a cocktail of histamine, acetylcholine and 5-hydroxytryptamine (serotonin). The mechanism provides effective protection for the plant from large herbivores.

Stinging Nettle, *Urtica dioica*

A deterrent to large herbivores such as cattle, the leaves and stem of the stinging nettle are covered in venomous hairs. At the slightest physical contact, the tips of the hairs break, leaving a sharp point that penetrates the skin releasing a potent stinging cocktail of histamine, acetylcholine and 5-hydroxytryptamine (serotonin).

Magnification: x125
Scanning Electron Micrograph (left)
Magnification: x25
Scanning Electron Micrograph (right)

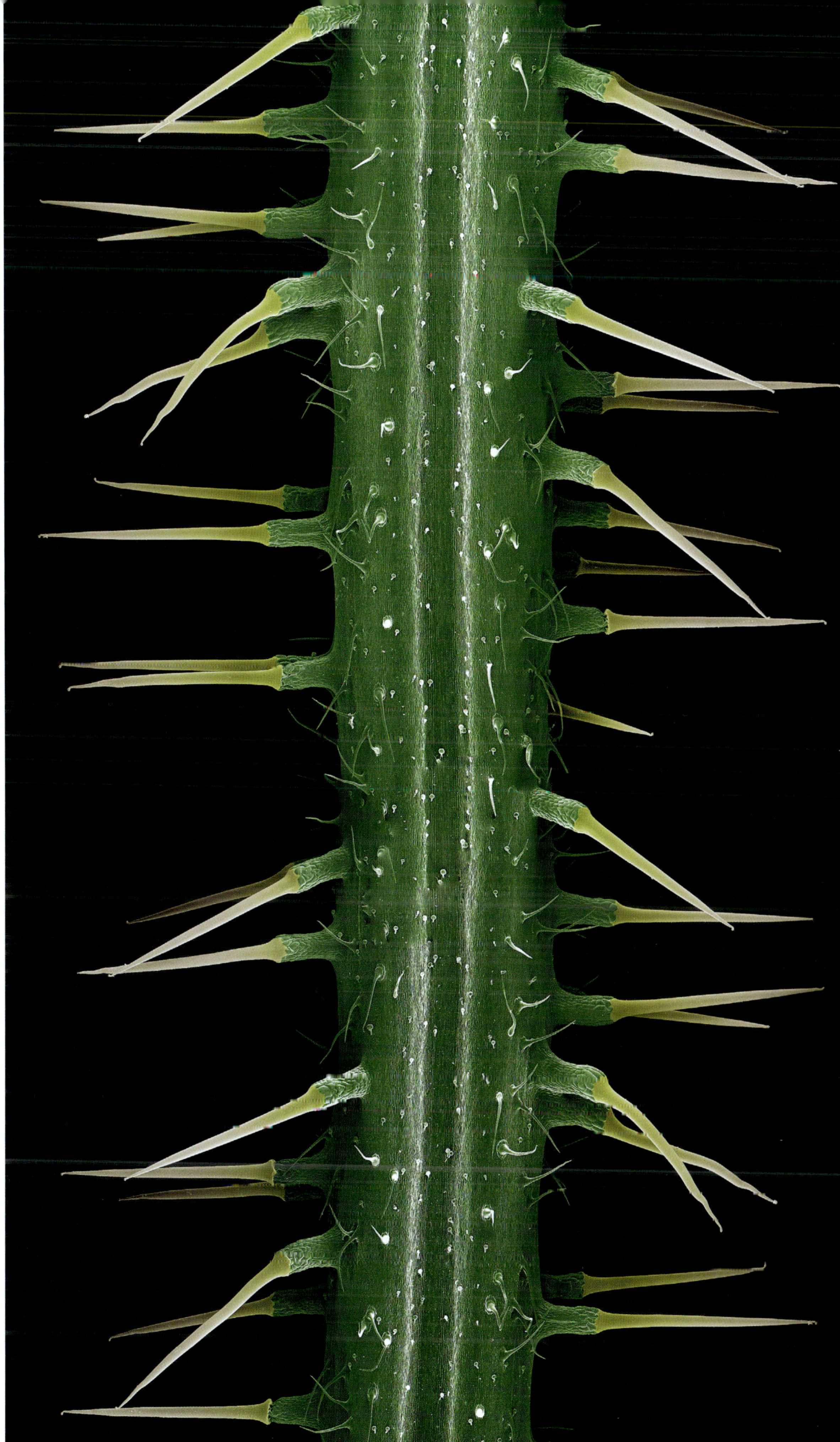

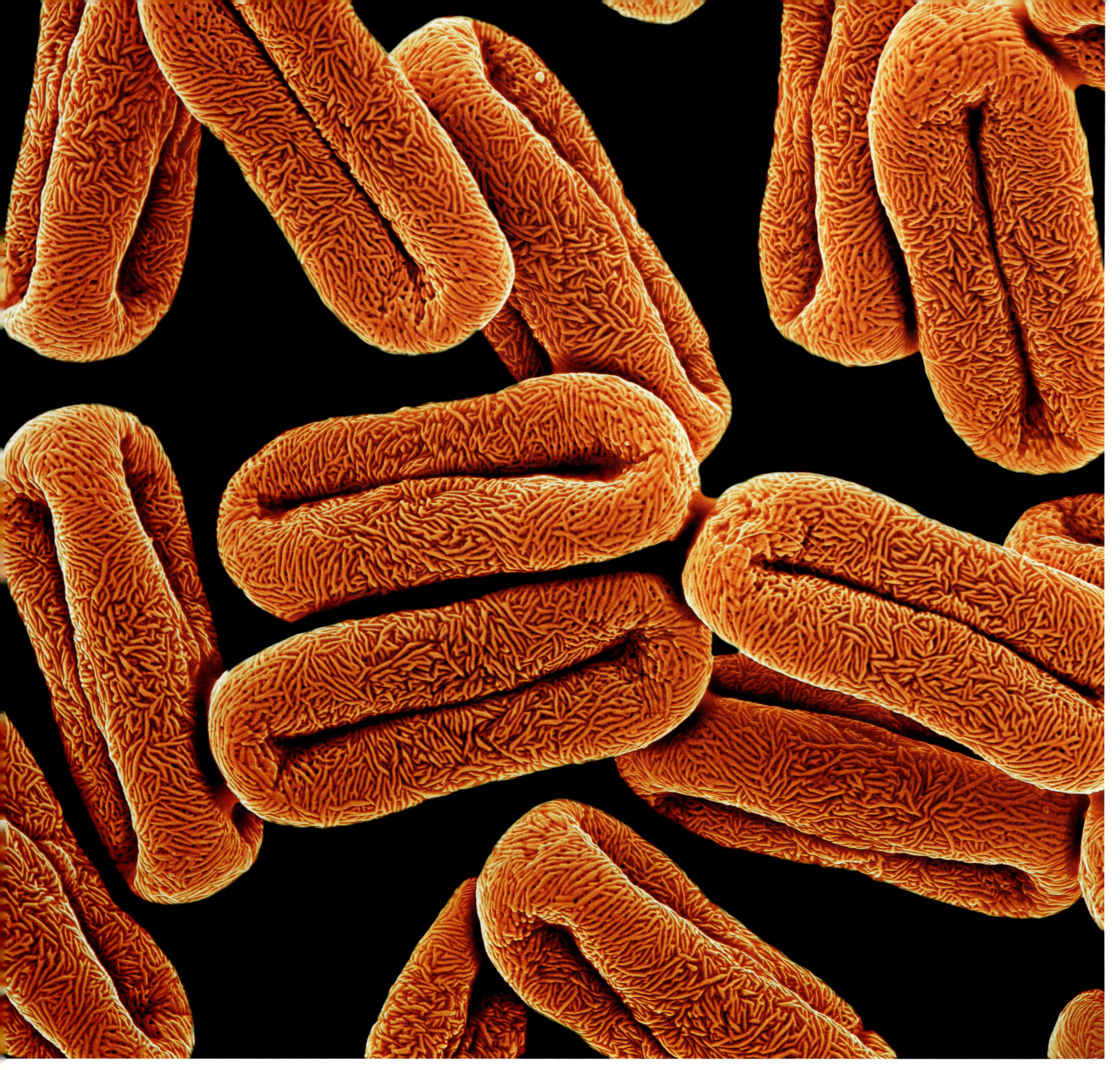

Pollen from an umbelliferous flowering plant (insect-pollinated)
*Magnification:*x5000
Scanning Electron Micrograph

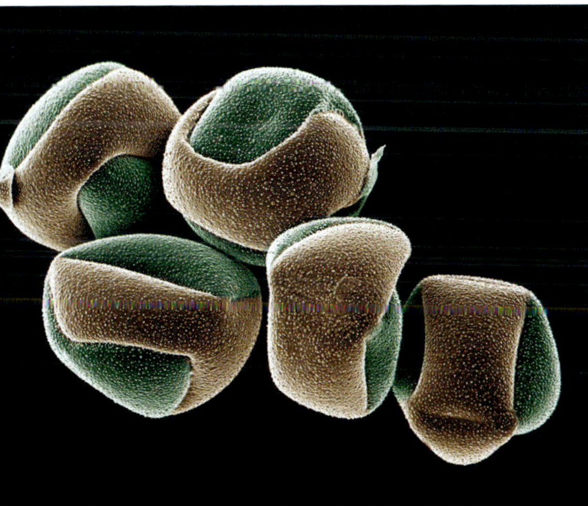

Cactus pollen (insect, or other animal-pollinated)
Magnification: x450
Scanning Electron Micrograph

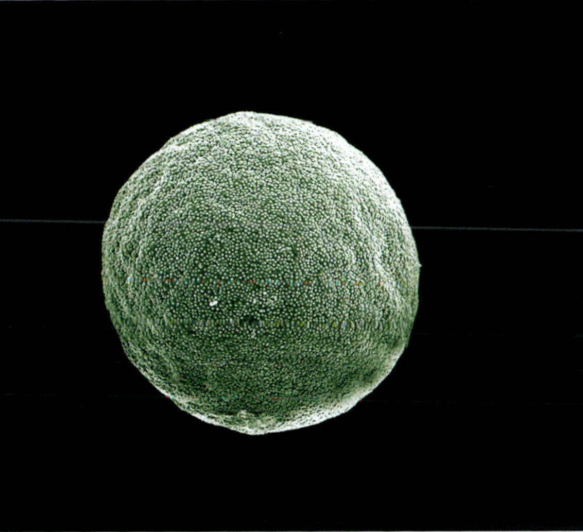

Grass pollen (wind-pollinated)
Magnification: x1500
Scanning Electron Micrograph

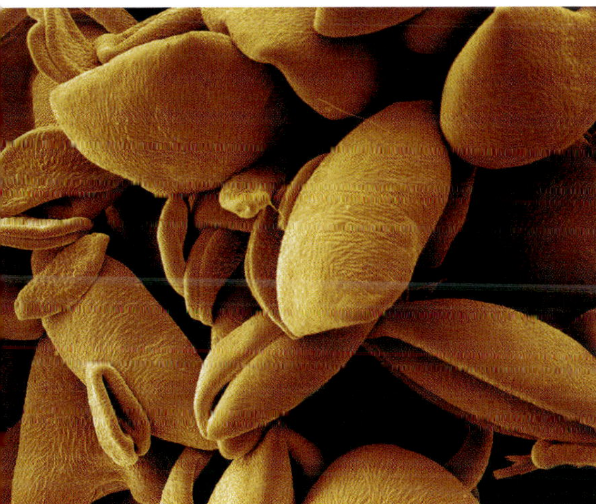

Lily pollen, (insect-pollinated)
Magnification: x340
Scanning Electron Micrograph

Pollen and Pollination

The problem for plants to overcome is to ensure that pollen, containing a small packet of genetic information from a male plant, reaches the female receptive organ, or stigma, to achieve fertilisation. Many flowers are hermaphrodite so self-fertilisation is always a safe option, but for genetic mixing it is necessary to transfer pollen to a flower on another plant which might be far away. Pollen is therefore dispersed, either by wind, by animals, or in the case of aquatic plants by water.

Pollen grains are produced in many different sizes and shapes, often with intricate patterns and textures on the outer coat, or exine, which is made of one of the toughest materials in nature. The shape and form of a pollen grain is unique to each plant species, rather like fingerprints are unique to each of us. They are extremely durable, able to resist extreme environmental conditions, remain viable for a long time, and can still be identified after thousands of years. Because of this, the study of pollen has proved an invaluable tool in forensic science and in understanding evolutionary history. The structure of a pollen grain is adapted to its mode of dispersal. The pollen of wind pollinated plants such as trees or grasses is either extremely small, or has structural extensions like tiny wings or spines, to increase the surface area so it can stay airborne. Dispersal by the wind has such a small chance of success that the dust-like pollen is produced in vast quantities during the heat of summer. Airborne pollen, if inhaled or comes into contact with the eyes, can cause hay fever, an allergic response affecting up to 20% of the population in Britain. For the sufferer, it tends to be a regular event, varying from a mild attack to an acute allergic reaction that can make life very unpleasant.

Plants that employ animal vectors, such as insects, birds or bats, produce pollen grains that are sticky or with sculpted shapes that will adhere to the vector's body. Animal-aided dispersal is quite efficient, the flowers luring their potential agents with conspicuous colours and perfume, and with nectar as a reward. The pollen grain itself can also afford to be larger and packed with nutrients – an extra reward for insects, which might eat a few in return for transferring others to another flower. Bees transport pollen on their hairy bodies as well as gathering it to feed to their larvae. Many other insects, including butterflies and moths, pollinate our garden plants, an activity essential to ensure the growth of fruit and vegetables, and the production of seeds for future generations of decorative plants.

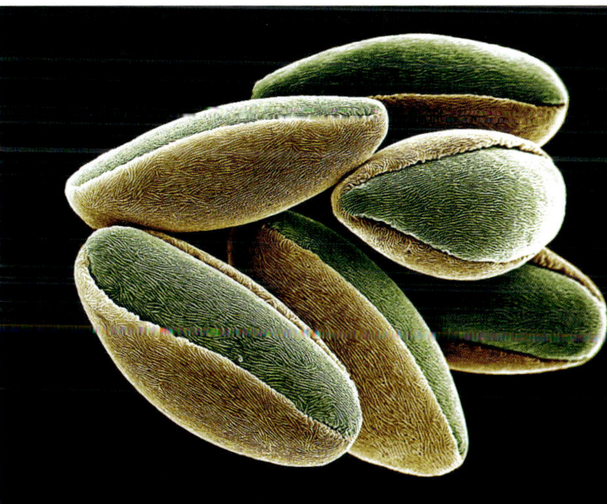

Speedwell pollen, *Veronica sp.* (insect-pollinated)
Magnification: x1100
Scanning Electron Micrograph

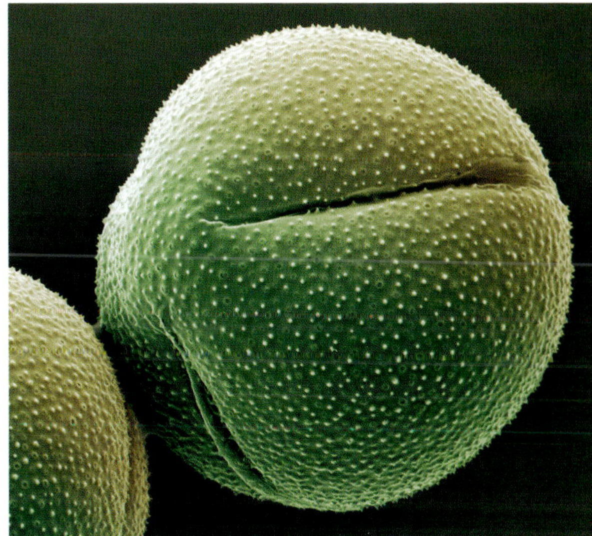

Cactus pollen (insect, or other animal-pollinated)
Magnification:x1500
Scanning Electron Micrograph

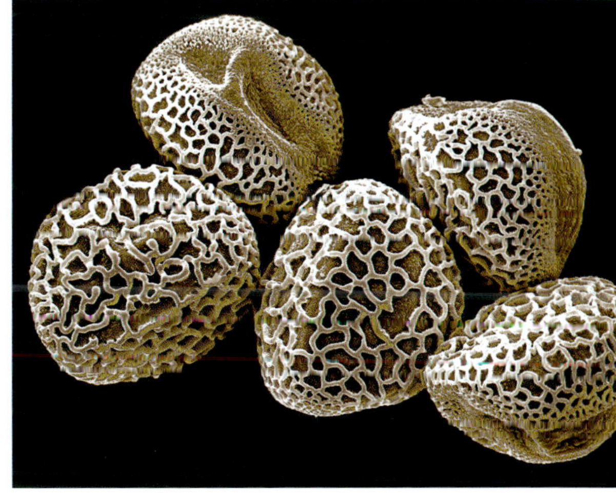

Lily pollen (insect-pollinated)
Magnification: x550
Scanning Electron Micrograph

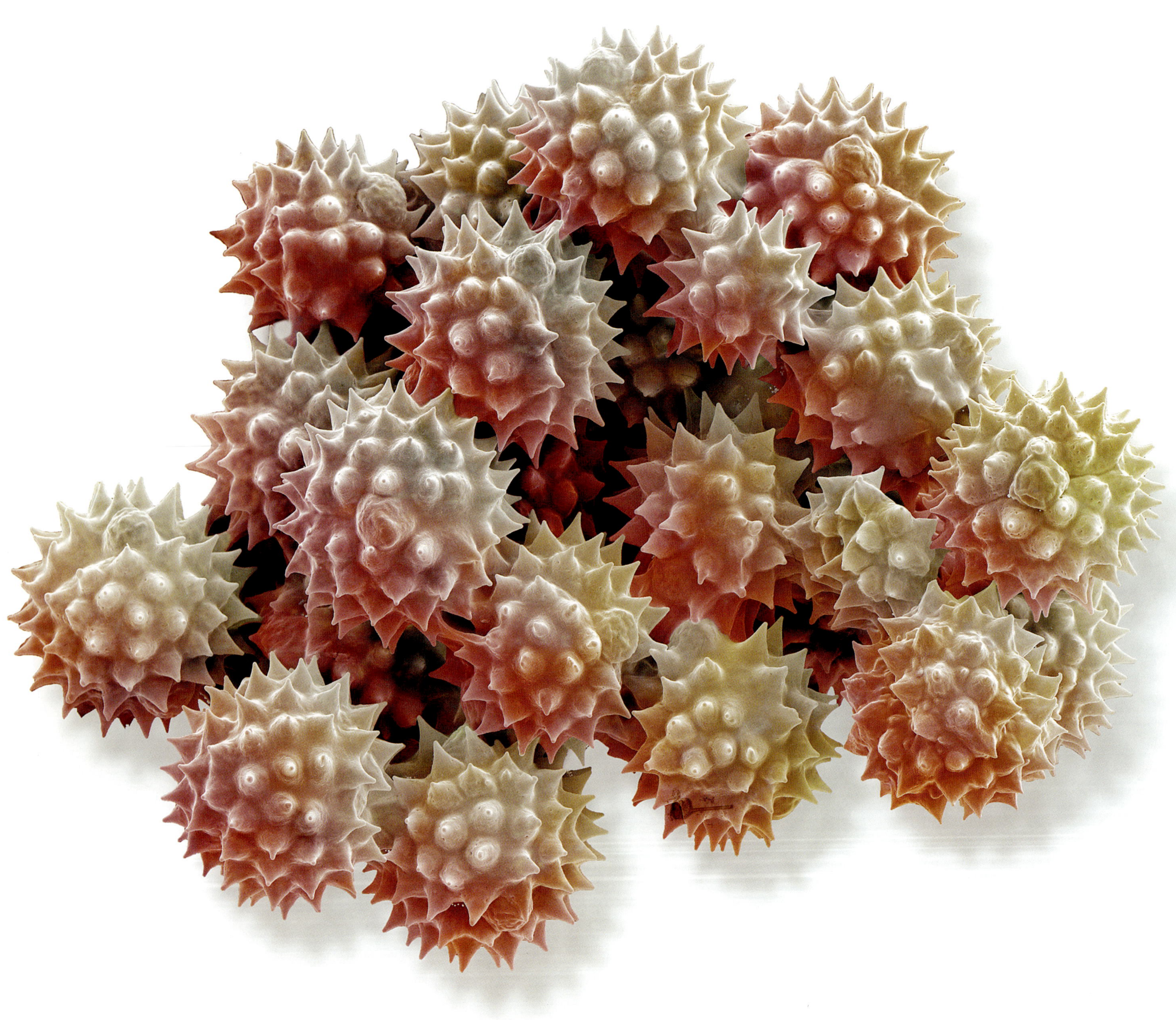

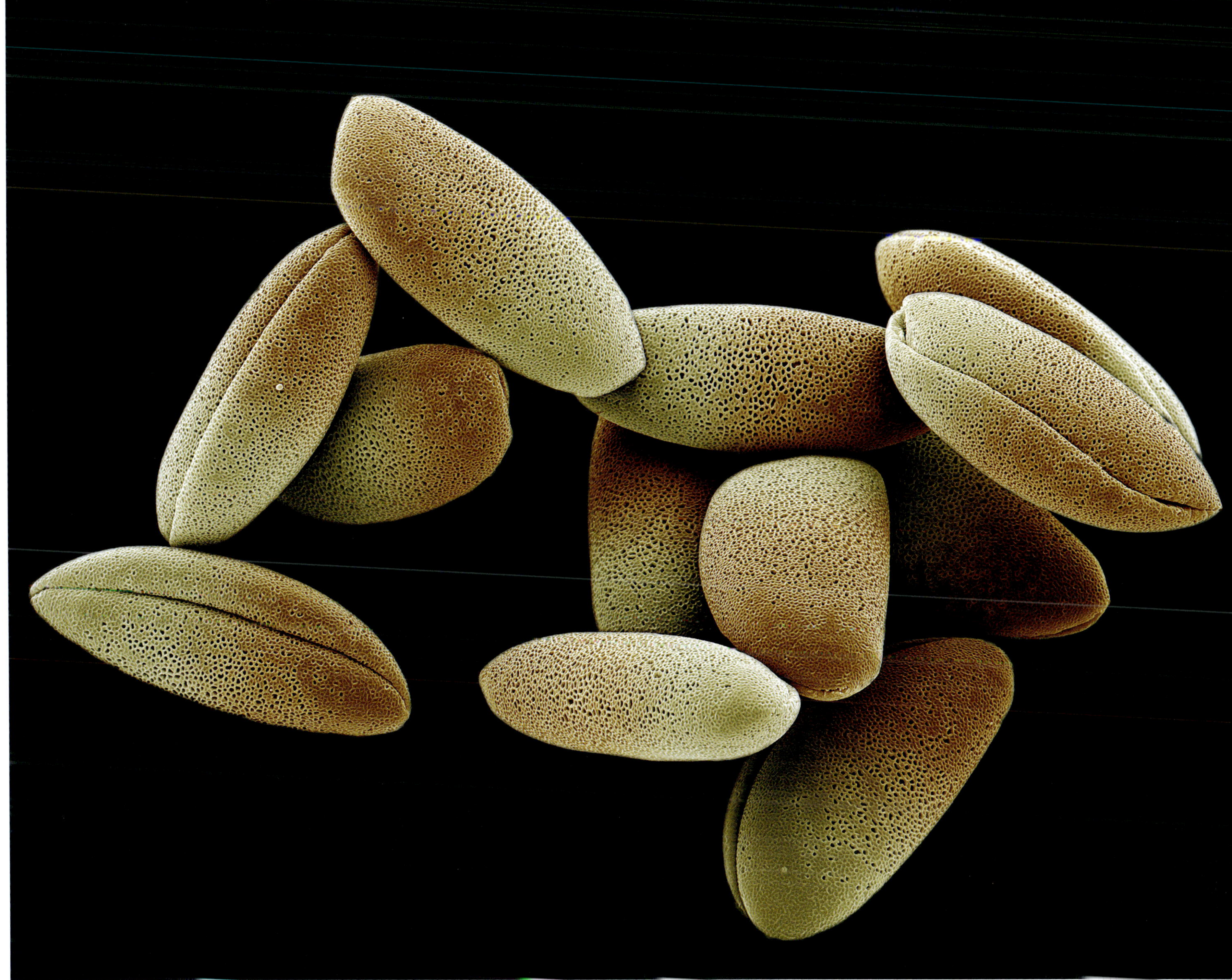

Pollen

The pollen of two insect pollinated plants, a lily, *Lilium sp.* and a groundsel, *Senecio vulgaris*, show different strategies. The pollen of the lily (above) is sticky, so it adheres easily to an insect's body. The pollen of the groundsel (left) is spiky, to adhere to hairy insects.

Magnification: x2800	*Scanning Electron Micrograph* (left)
Magnification: x1100	*Scanning Electron Micrograph* (above)

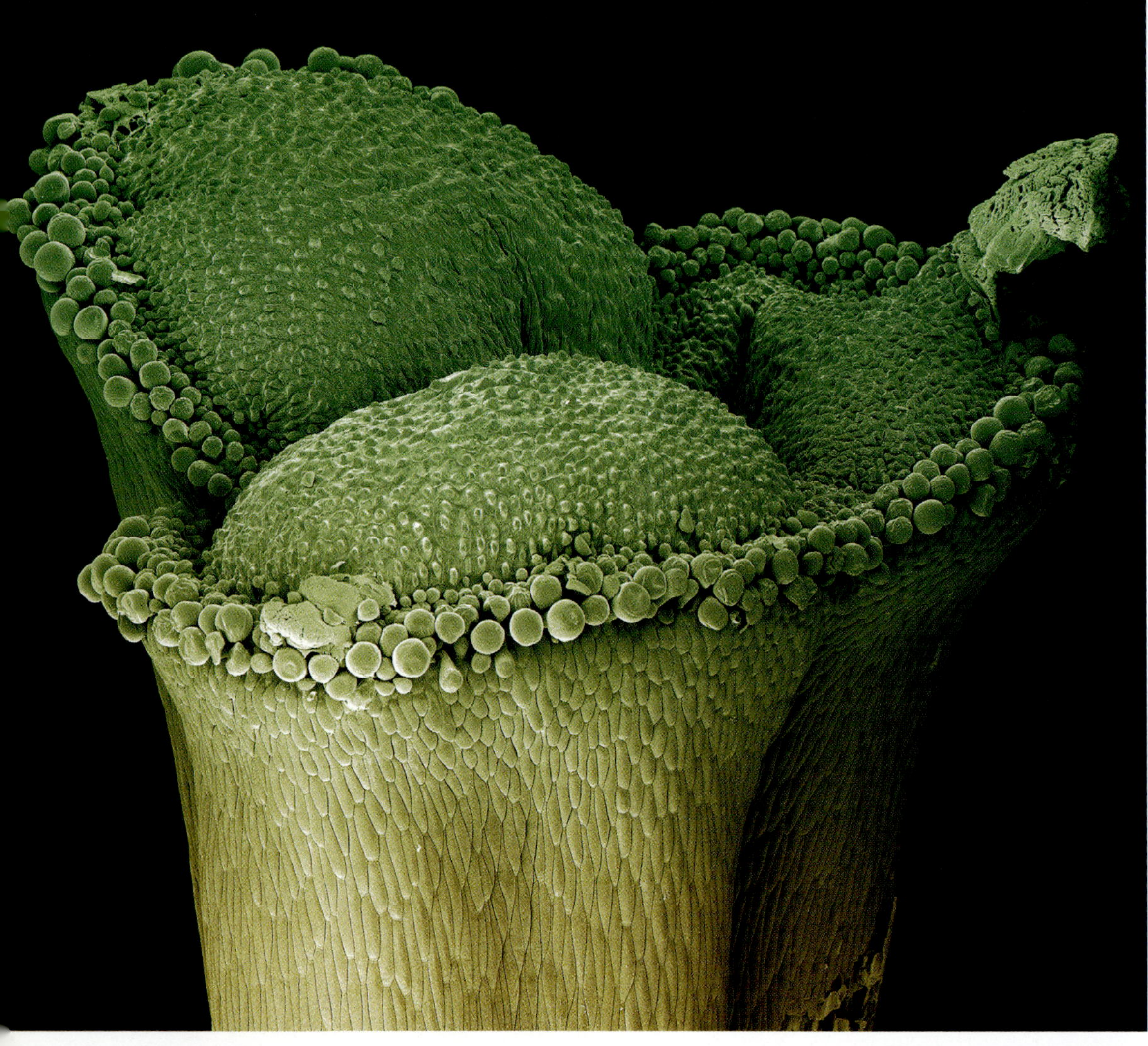

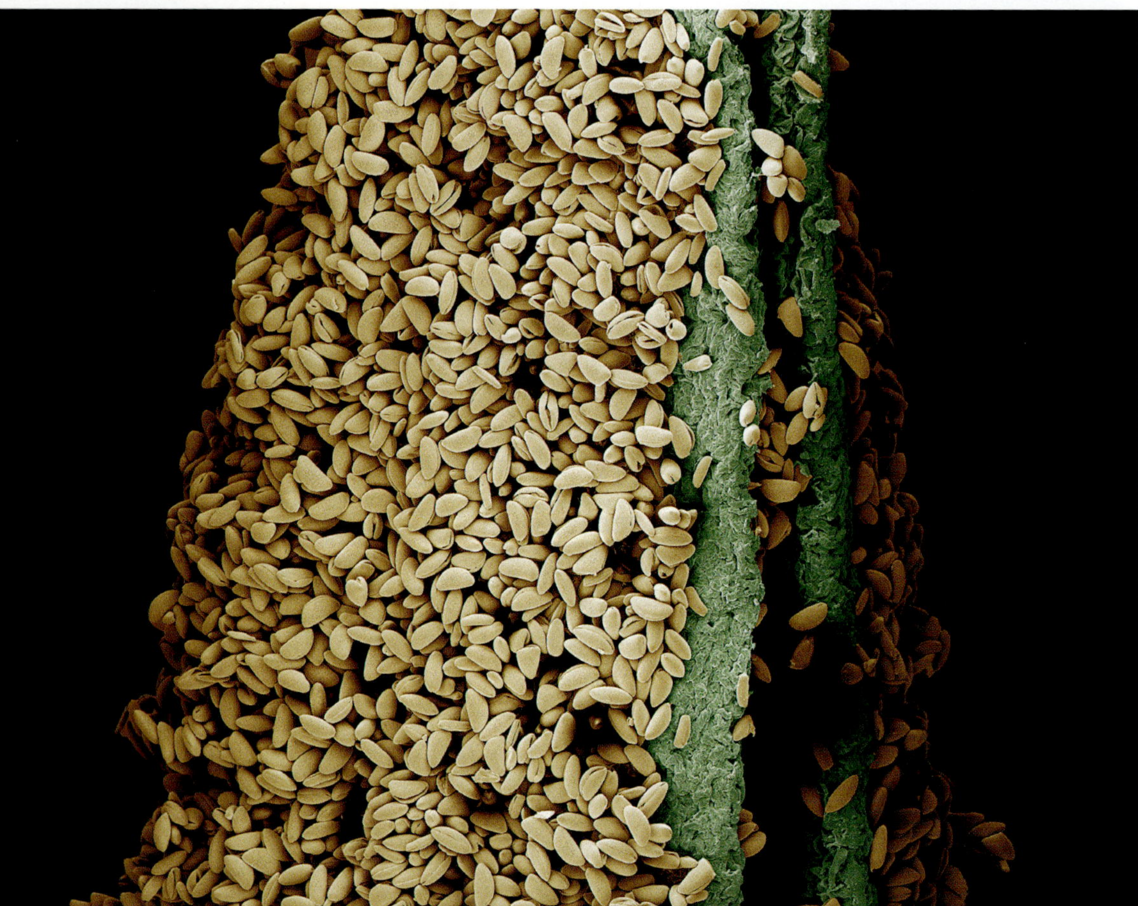

Lily flowers possess both male and female sex organs. The male stamens rise from the centre, each bearing a yellow, pollen-bearing anther at its tip. The pistil also rises from the centre, extending beyond the stamens, supporting the female receptive organ, the stigma, which is divided at the tip into two lobes that are sticky at the surface. The pollen is also sticky, and heavy, so it cannot be blown by the wind, but must be gathered and distributed by insects or other active pollinators. Lily flowers are often fragrant, a device to attract pollinating agents such as bees, and produce copious amounts of nectar as a food reward in exchange for the work of transferring pollen to a lily stigma. The pollen grains naturally dry to about 20% moisture before being released, making them lighter. Once on a stigma, lipids on the surface stimulate the pollen grain to hydrate and germinate, as a pollen tube that grows down the style, through the wall of the ovary and into an ovule. As the pollen tube develops, two sperm are produced, one of which eventually fuses with the ovum, a process called syngamy. The second sperm fuses with the central cell of the ovule, and this union produces the endosperm of the seed, which is composed of starch, proteins and oils that provide nourishment for the developing embryo.

The stigma of a lily (above left) is the female receptive organ; the male anther (below left) is laden with pollen. When a pollen grain comes to rest on a stigma, it grows a pollen tube (see right) towards the ovule, carrying the male gamete to fertilise the egg cell.

Magnification: x50
Scanning Electron Micrograph (above left)
Magnification: x60
Scanning Electron Micrograph (below left)
Magnification: x850
Scanning Electron Micrograph (ight)

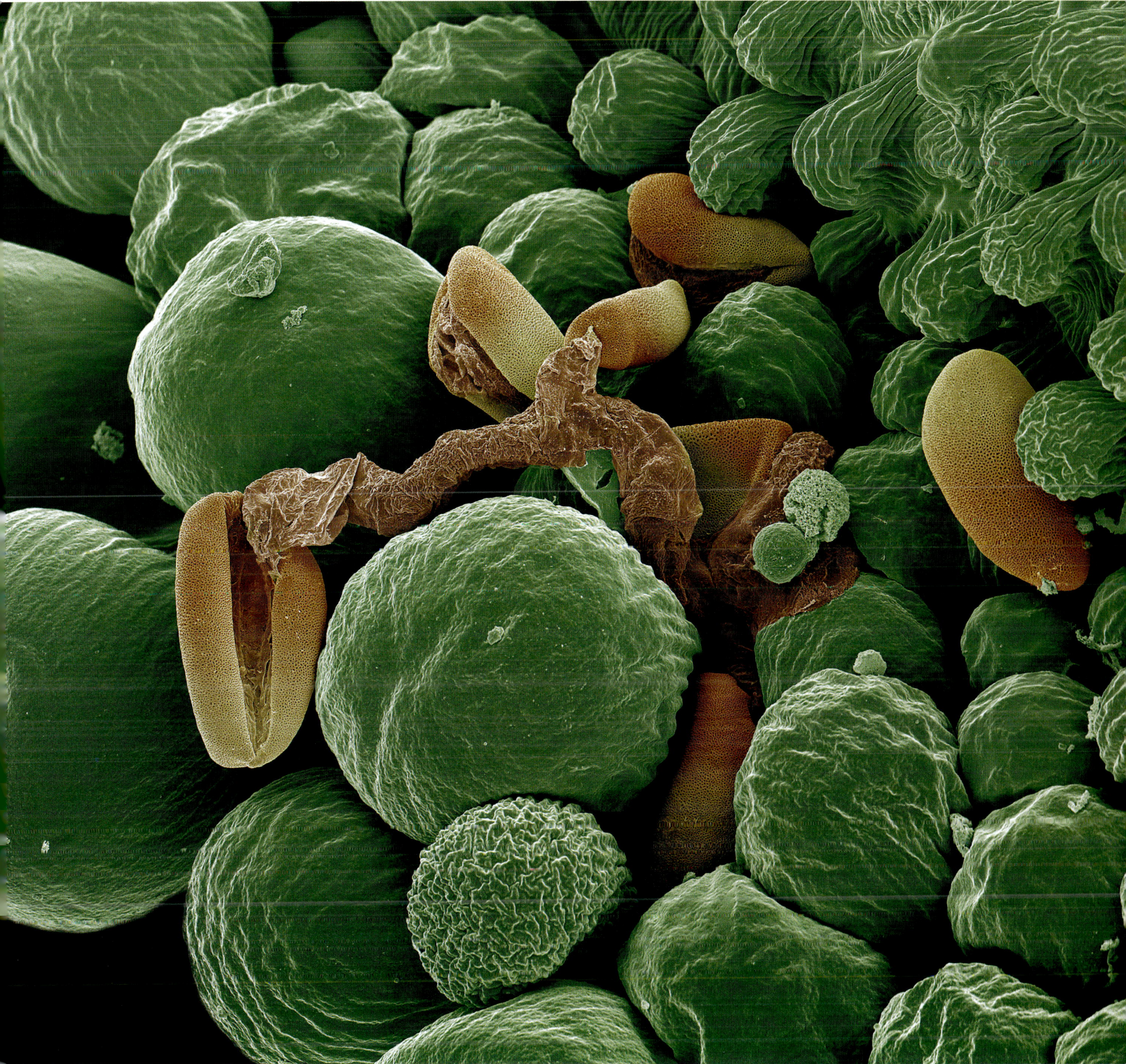

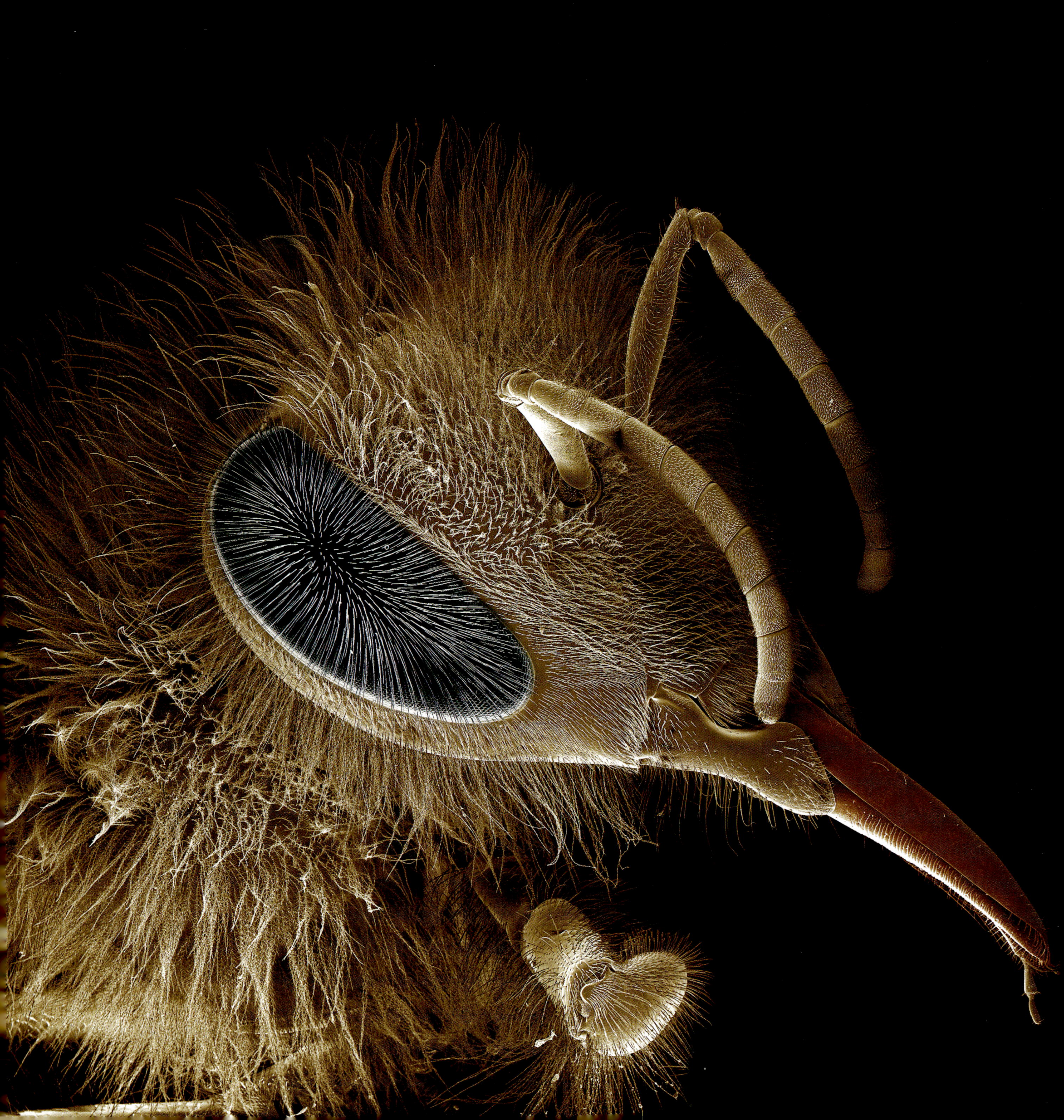

BUMBLEBEES

Bumblebees have immense value as pollinators of many wild and cultivated plants. Pollination of some crops, like runner beans, depends almost entirely on the foraging activity of bees. Likewise, many varieties of apples, pears and plums produce more fruit when bees are plentiful on the blossom. Like ants and their smaller cousins honeybees, bumblebees live in colonies with a complex and cooperative social organisation, consisting of different types of individual; queens, female workers and male drones, each with their own task to accomplish for the benefit of the colony as a whole. While honeybee colonies are large and perennial, bumblebee colonies are small and renewed each year, dying out in the autumn, leaving only young mated queens to survive the winter and start new colonies in the spring. Queen bumblebees emerge from hibernation in early spring, busily feeding and searching for possible nest sites in grass tussocks, moss clumps or cool, dark places such as old mouse burrows, or underneath garden sheds. Once in her chosen location, the queen makes a small cake of nectar and pollen and lays up to twelve eggs on it. Around this, she creates a wax cell, and, after about two weeks, the first of the new season workers hatch. She rears this first brood by herself, but the worker bees will eventually take over the duties of collecting food, rearing the young, and building and maintaining the nest, leaving the queen to devote herself entirely to egg-laying. The waxy material used to produce the nest comb and brood-cells is made from special glands, and the whole nest is usually covered and protected inside a ball of dead, finely shredded grass, moss, animal fur or similar material. Bumblebees feed on pollen and nectar, and rear their grubs on the same diet. Towards the end of summer, bumblebee colonies produce males (drones) and new queens. The males, which are stingless, do no work in the natal colony and quickly leave the nest to search for and mate with new young queens from other colonies. After mating, the males will die, and the fertilised young queens abandon their nests to find suitable places to hibernate over the winter in small underground chambers, under stones, in logs or grassy banks. Come the first sharp drop in temperature and frost the old queen, her workers and any surviving males also die.

Honeybee, *Apis mellifera*

Their origins are in South-East Asia but honeybees have been introduced worldwide by humans, and nurtured everywhere for the honey they manufacture from pollen and nectar. They are also important pollinating agents. Worker bees guard the entrance to their nest; if threatened, they will sting, and the venom contains an alarm pheromone that signals other workers to attack. Unlike wasps, bees can only sting once, because the stinging apparatus is attached to internal organs. As the bee pulls away from an attack, the barbed sting (right) remains in the victim but is wrenched, with fatal consequences, from the body of the bee.

Magnification: x40   *Scanning Electron Micrograph* (left)
Magnification: x75   *Scanning Electron Micrograph* (right)

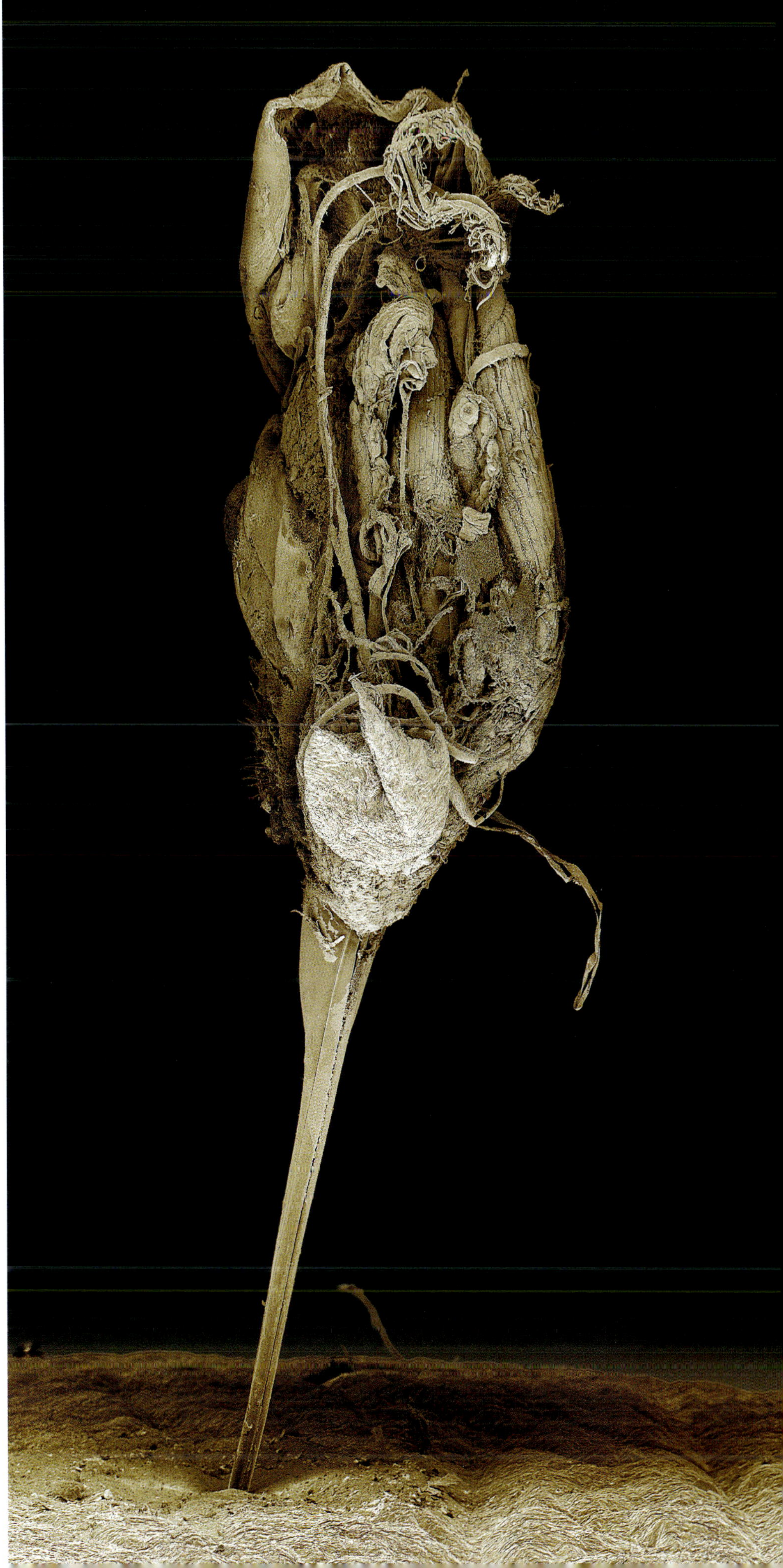

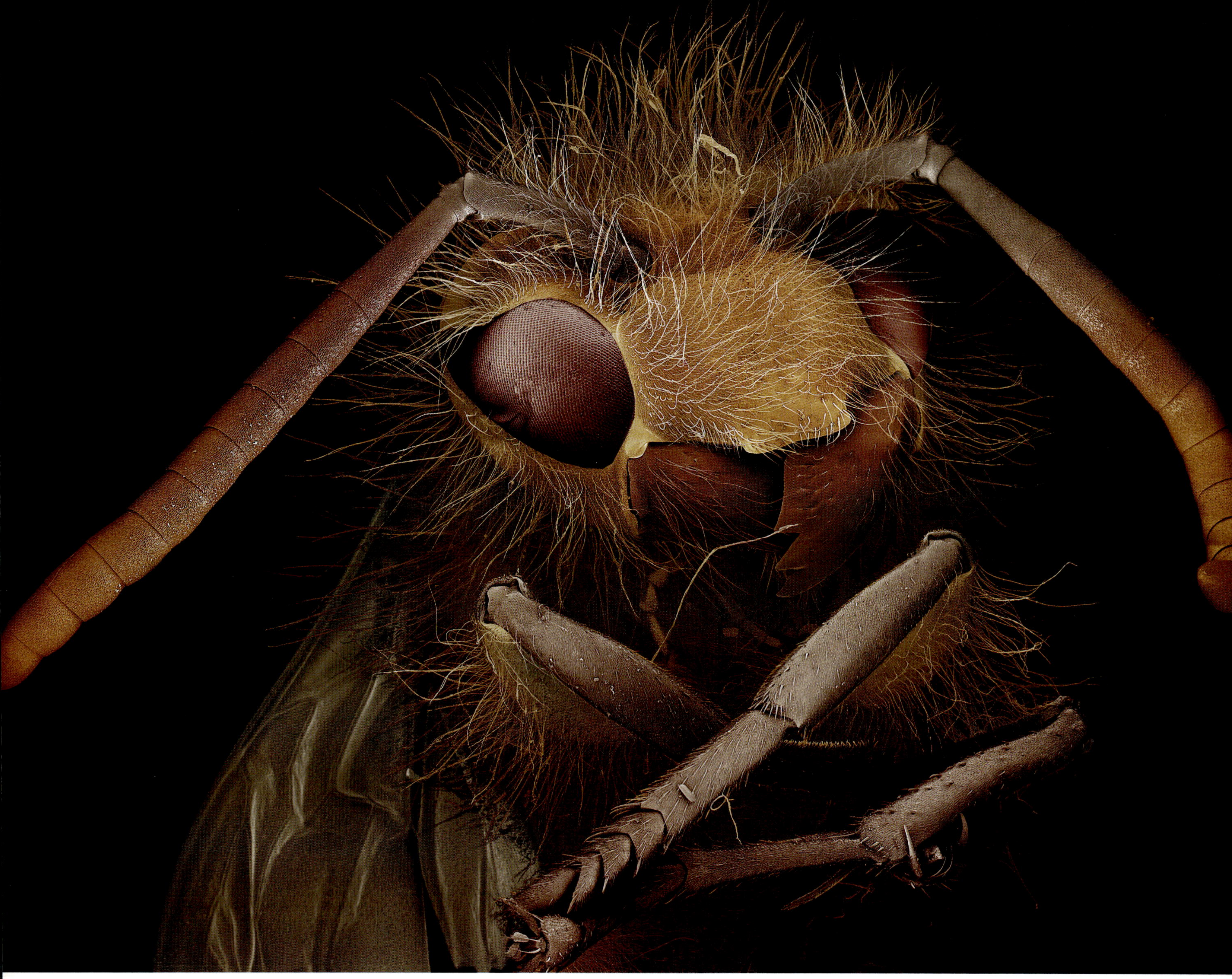

Common Wasp, *Vespa vulgaris* (above)

Related to bees, wasps are social insects feeding on fruit and nectar, but they also hunt other insects to feed to their larvae. They use their powerful jaws to catch insects sometimes much larger than themselves, and can strip wood fibres from trees to be chewed up as a nest building material.

Magnification: x45   *Scanning Electron Micrograph*

Wasp Sting, *Vespa vulgaris* (right)

The wasp sting (right) is finely barbed but can be extracted easily from the victim, so it can be used repeatedly, not only for protection, but for immobilising prey; wasps are hunters.

Magnification: x400   *Scanning Electron Micrograph*

WASPS

The bright yellow and black striping of wasps is a warning pattern to prospective predators, such as birds, that this insect is venomous and best left alone. Wasps are social insects related to bees, feeding on plant material especially fruit and nectar. However, they are also hunters, killing other insects to feed to their larvae, using a retractable sting at the tip of the abdomen to immobilise their victim. Unfortunately, wasps can be quite aggressive to humans and their sting is painful; they will certainly attack if threatened or if their nest is disturbed. However, they are useful pest control agents in the garden and only become a nuisance in the late summer when they feed on soft fruits and the cooler weather slows them down. The wasp most often encountered in Britain is the 2 cm long common wasp (*Vespula vulgaris*), but the large and impressive hornet (*Vespa crabro*), which can be up to 4 cm long, also occurs here though it is relatively uncommon and largely confined to the south of England.

In the spring, after emerging from hibernation, a young queen, who is already carrying fertilised eggs, starts to build a nest in a hole underground, inside a hollow tree or in the roof cavity of a building. When finished, the nest is round in shape, created from a paper-like material, made from wood chewed and mixed with saliva. Inside, the queen builds a few cells, or small compartments, where she deposits eggs. After hatching, these first larvae are fed on a diet of chewed-up insects by the queen and will become adult workers, which take over the tasks of nest building, maintenance and food collection, leaving the queen free to devote herself to producing more eggs. By late summer, as the colony grows, a nest may reach the size of a football or even larger. Inside, it consists of several horizontal layers of combs, in which the larvae are reared. When autumn comes, new queens mate and go into hibernation as winter approaches.

SPIDERS

Britain has more than 600 species of spiders, all of them harmless to humans although a few can deliver a nip if threatened. All spiders are predators; they possess powerful fangs on their chelicerae, which inject venom and enzymes then suck out the juices of their prey. The spinning of silk, a protein material produced by spinneret glands near the tip of the abdomen, is a characteristic of spiders. Diversity in construction of webs is fascinating, from the familiar spiral orb web to irregular sheets such as cob-webs, funnel-shaped webs, and three-dimensional webs, all of them ingenious snares for trapping prey. Spiders use silk for other purposes too, such as the protection of egg sacs, for wrapping and immobilizing prey, the construction of retreats in which to shelter, or as safety lines. Large spiders move from place to place by walking, but newly hatched spiderlings, and even the adults of small species, can disperse aerially, using a fine thread of silk as a sail. The aeronaut first crawls to the tip of a leaf or grass blade, stands on 'tiptoe' with the abdomen pointing skyward, and then releases a strand of silk from its spinning organs. The fine thread is caught in the breeze and lifts the spider upwards on rising air currents. Called 'ballooning', this aerial form of dispersal has disadvantages however; the dangling spiders are helpless against attack by insectivorous birds, and the wind may transport them to unsuitable environments.

Jumping spiders and the zebra spider, common in British gardens, are active hunters with huge eyes and excellent stereoscopic vision, able to leap onto their prey from a distance of up to 15 cm with great precision, anchored by a safety line in case they miss their target. Quite small, they are very active in warm, sunny weather and can often be seen on stone walls or paths, where they stalk their prey, usually small flies. They belong to the biggest family of spiders, the Salticidae, with over 5,000 species worldwide.

There are many other species of spiders found in the garden; some are orb spinners whose immaculate web traps become visible when covered in dewdrops in the autumn. Crab spiders are often cleverly camouflaged to match the flowers in which they hide with their jaws ready to strike at any unwary landing insect. The woodlouse spider has jaws powerful enough to pierce the tough armour of woodlice, and strong enough to pierce human skin. All garden species, particularly money spiders and wolf spiders, are beneficial since they prey on undesirable pests of garden plants such as aphids, leafhoppers, gall midges, and flies.

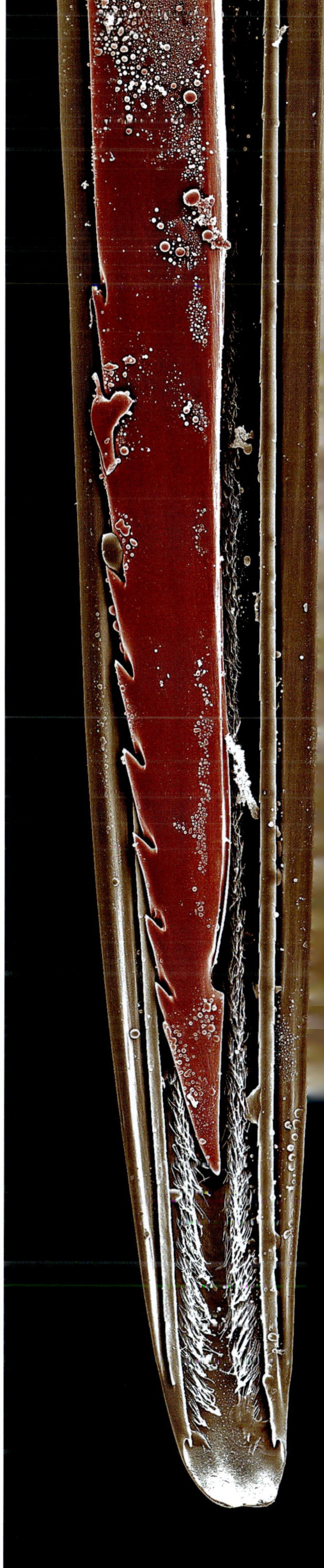

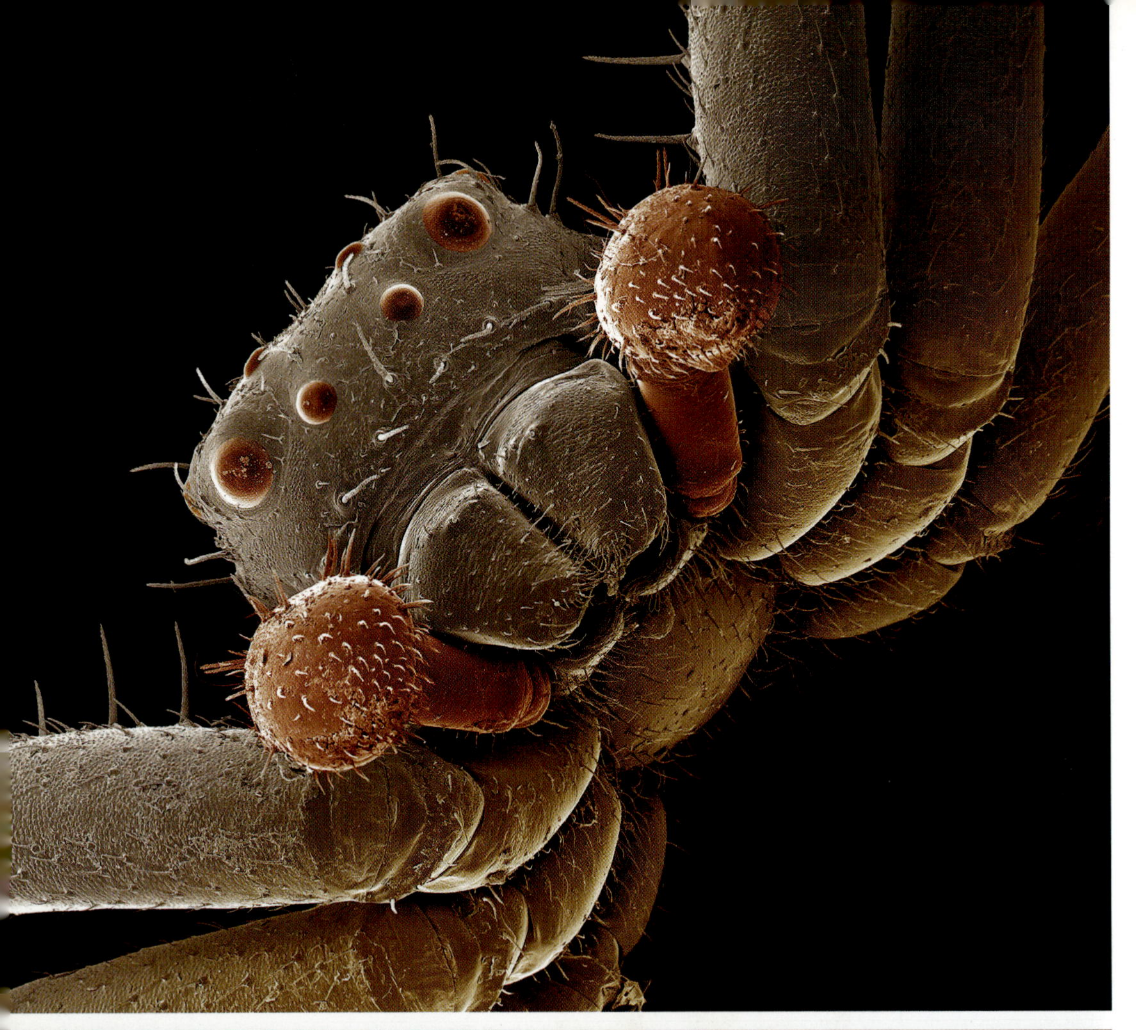

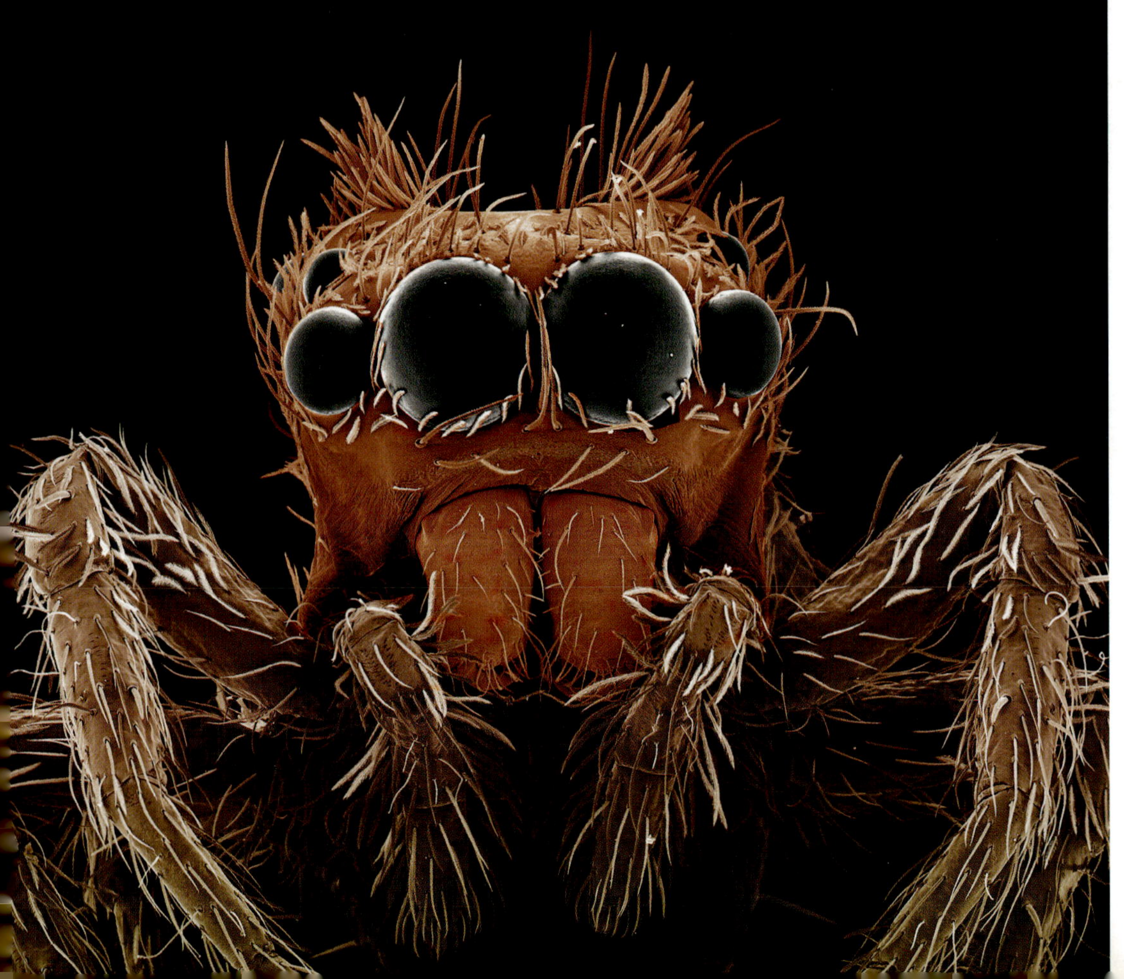

HARVESTMEN

Harvestmen are often confused with spiders since they possess eight legs, but they are distinguished by a globular body that has no division between the head and abdomen. On the top of the body, there is a small raised turret on which there are two simple eyes. Most species have very long legs, with the second pair always the longest, and the jaws, or chelicerae, are quite slender, with delicate pincers. Harvestmen produce no venom, and are therefore harmless. They also lack silk-producing organs. Most have an annual life cycle, usually passing the winter as eggs and maturing in late summer - hence their name. The harvestmen most frequently seen in the garden, for example *Phalangium*, *Odiellus* and *Leiobunum*, are quite large, with a brownish body up to several millimetres long. They are omnivorous, feeding on a wide range of living and dead animal matter, such as insect larvae, mites and young spiders, or juicy plant matter, primarily for its water content although they also obtain water from dewdrops. Most species are nocturnal, but can be seen resting on walls or fences during the day, with their legs spread widely. Their colouring often matches

Crab Spider, *Xysticus sp.* (above left)

Crab-like in appearance, these spiders have the first two pairs of legs longer than the rest. They are active hunters, waiting in ambush for prey and are often well camouflaged, their colour matching their surroundings. Some species wait inside flowers to grab unsuspecting nectar-seeking insects. The male *Xysticus* (see above left), recognized by his boxing-glove style palps, does not risk life and limb as many spiders do when courting a female. He strokes her until she falls into a torpid state, after which he ties up her front end with silk before using his palps to transfer his sperm to her epigyne, or genital opening.

Magnification: x65   *Scanning Electron Micrograph*

Jumping Spider, *Portia sp.* (right, below left and page 216)

Like all jumping spiders, *Portia* (see below left, and right) possesses huge eyes and excellent stereoscopic vision. It stealthily stalks its prey before leaping with great precision to grab it with its fangs. This genus, which occurs in Africa, South East Asia and Australasia, is unusual among jumping spiders in that its favourite prey animals are other spiders. It risks life and limb with dangerous forays on to their webs, but it manages this with considerable cunning, only rarely missing a kill. Most jumping spiders do not build webs, but *Portia* does so, not only to catch insects to supplement its spider diet, but also as a place for courtship rituals.

Magnification: x70   *Scanning Electron Micrograph* (below left)
Magnification: x90   *Scanning Electron Micrograph* (right)
Magnification: x110   *Scanning Electron Micrograph* (page 216)

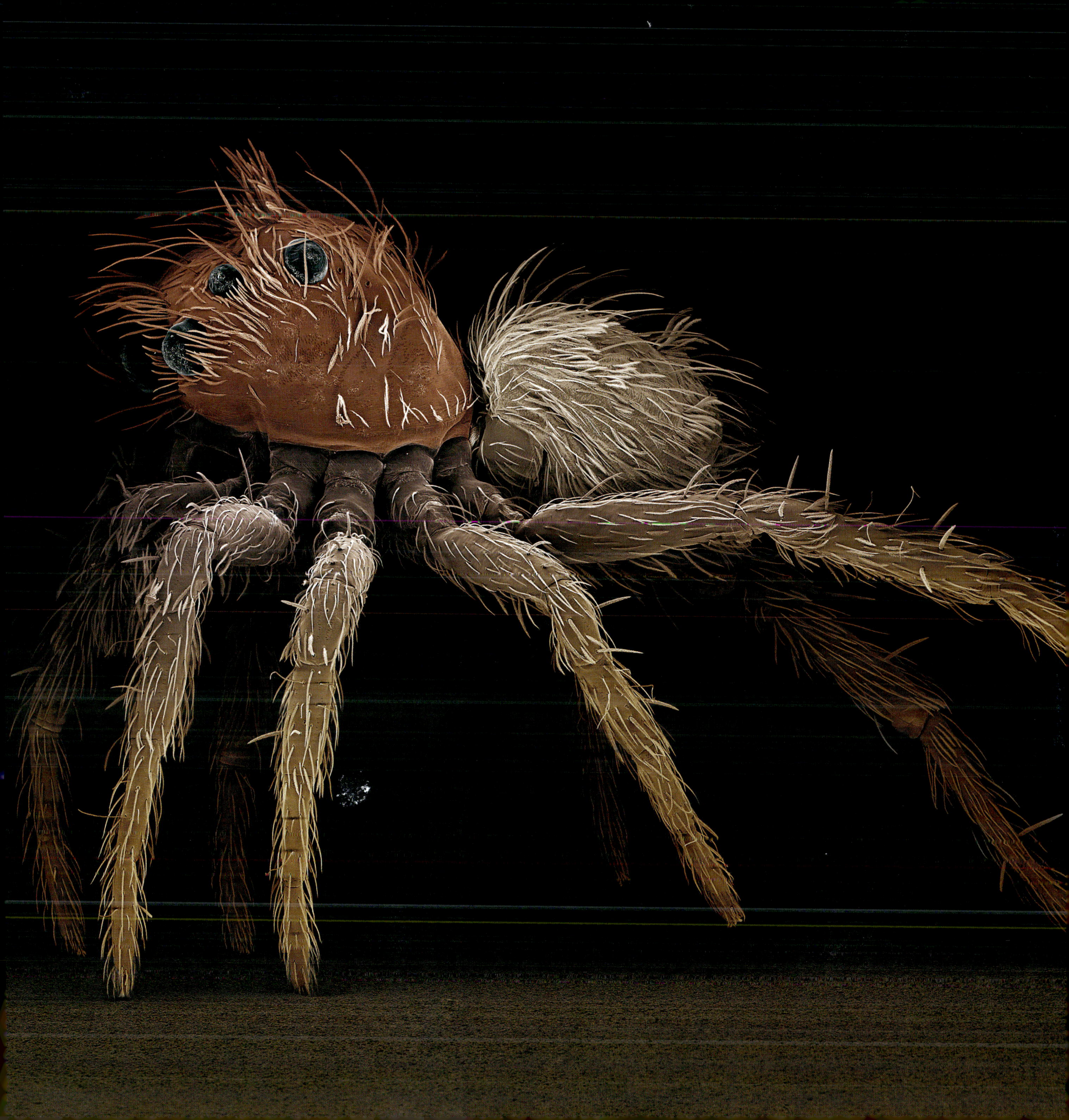

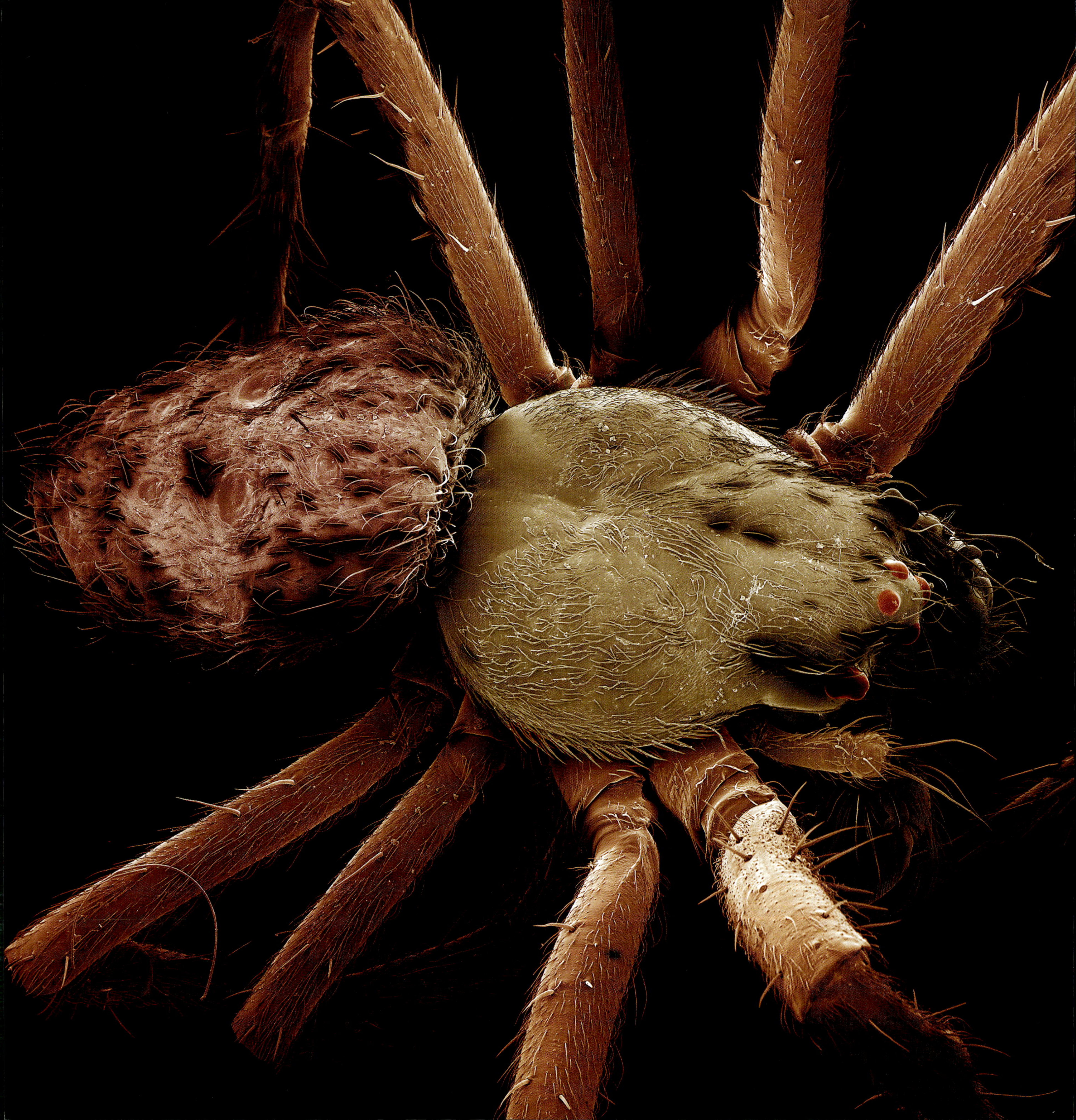

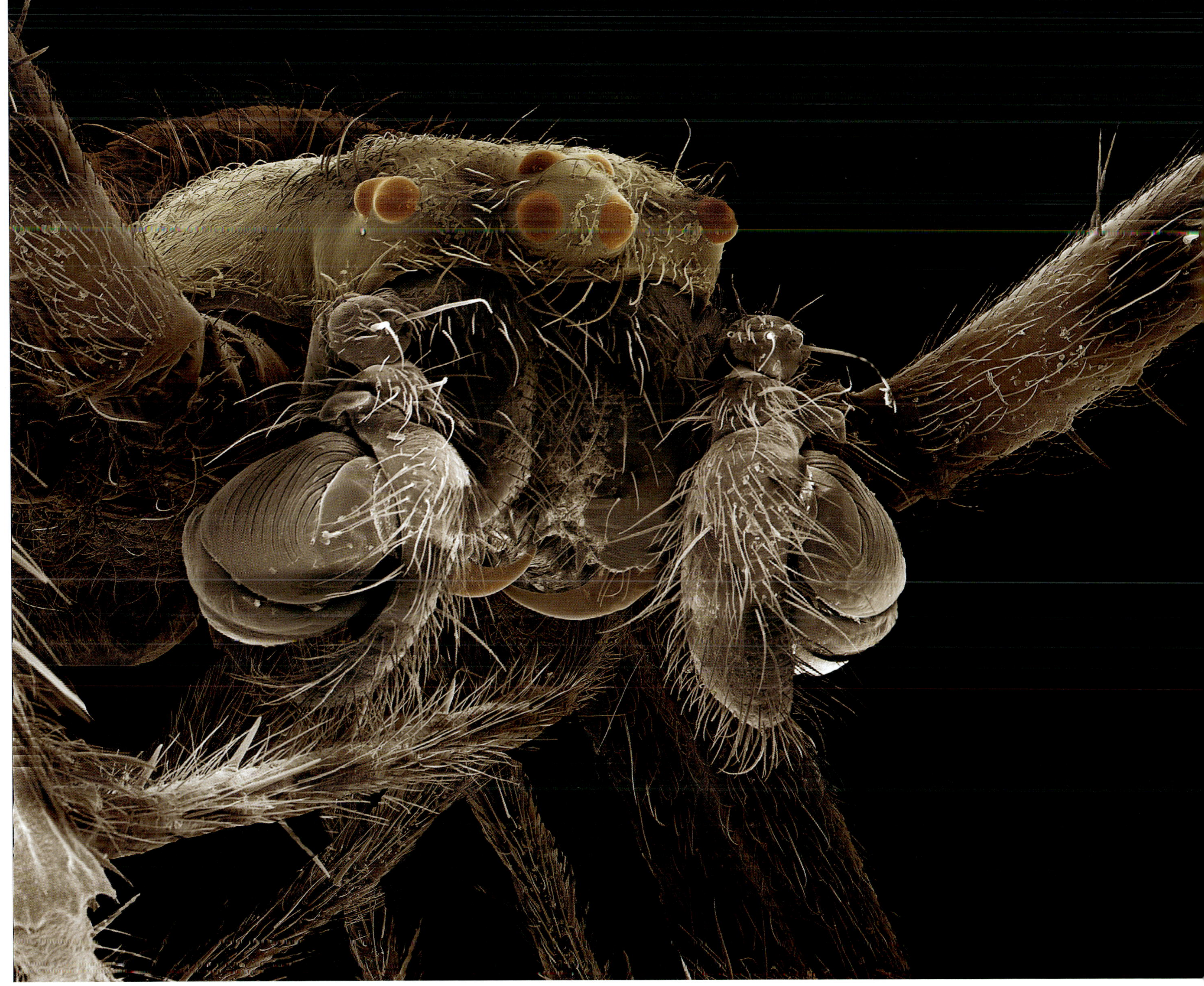

Garden Spider, *Araneus diadematus* (left, above and overleaf)

A spider's body is divided into two sections, the prosoma from which the legs and mouthparts originate, and the abdomen, or opisthosoma, containing the lungs, reproductive organs and silk glands. Garden spiders spin orb webs with a signal line leading to the spider's retreat. Vibrations from a trapped victim will bring the spider in a rush to investigate. The chelicerae are tipped with sharp jaws, used to chew prey to a pulp mixed with digestive juices. Either side of the mouthparts are the leg-like palps, which in the male (see aabove) are large and bulbous, used for sperm transfer to the female (overleaf, left).

Magnification: x30   *Scanning Electron Micrograph* (left)
Magnification: x50   *Scanning Electron Micrograph* (above)
Magnification: x30   *Scanning Electron Micrograph* (overleaf, left)

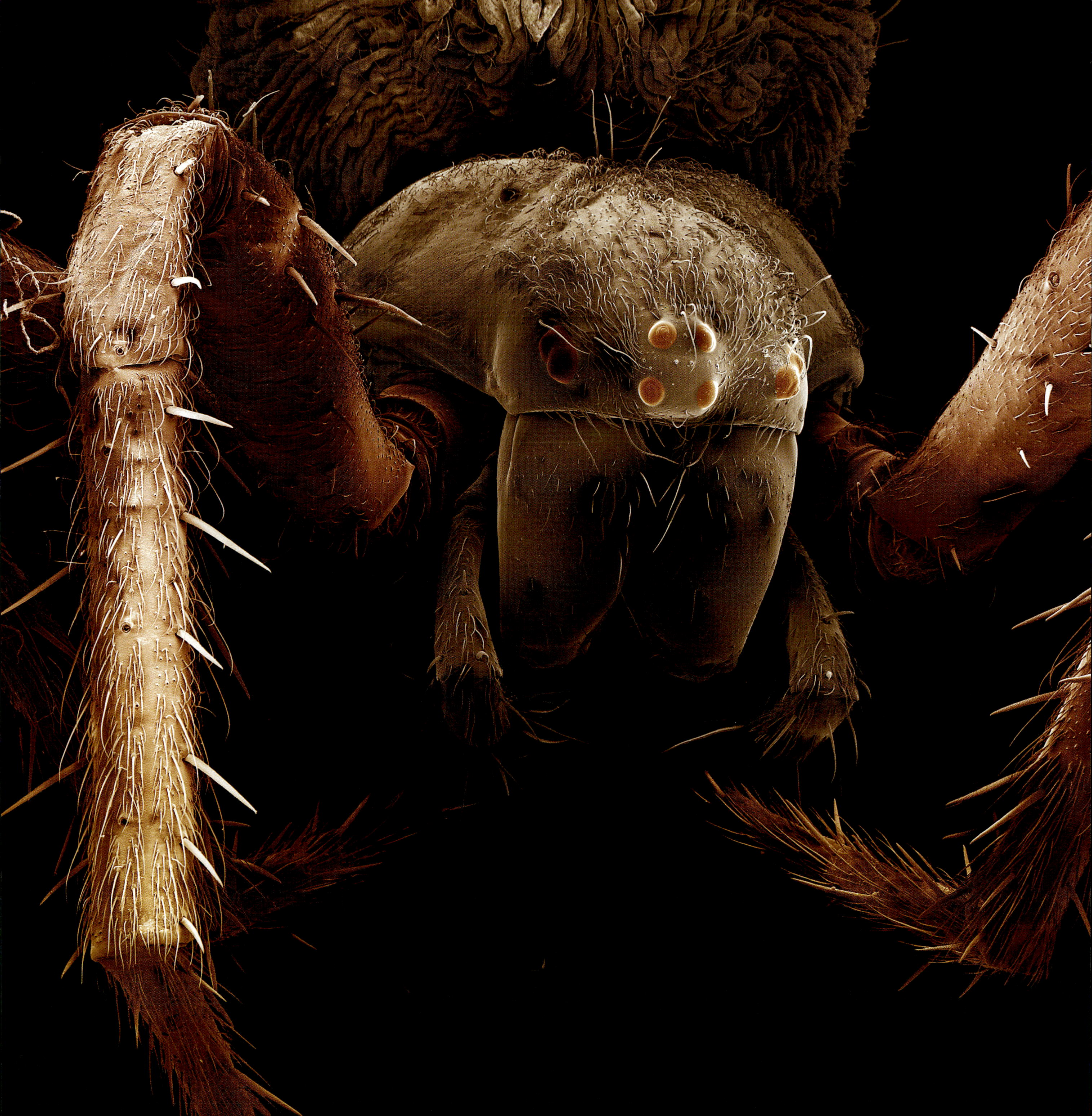

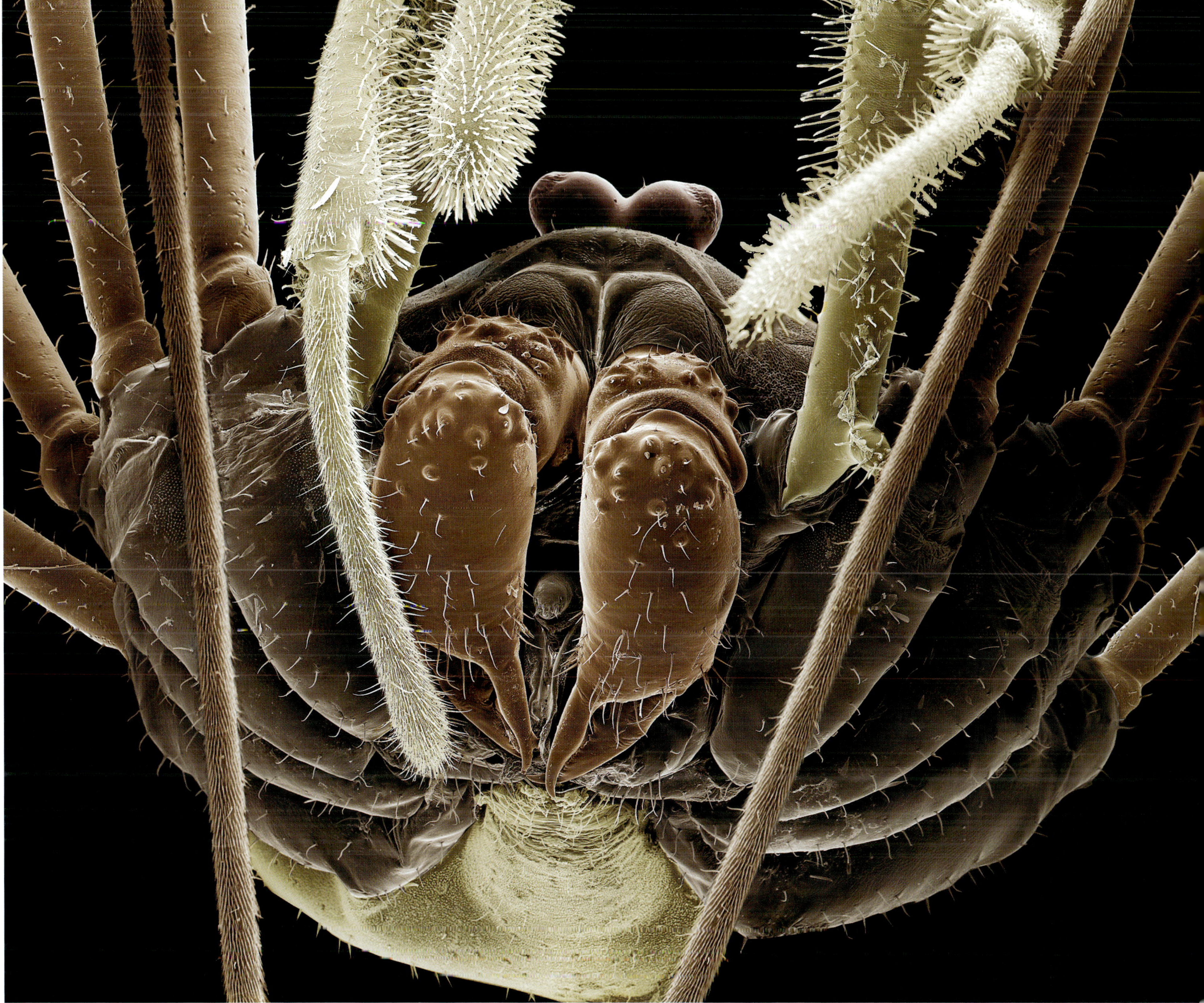

Harvestman, *Phalangium opilio* (above)

Harvestmen are related to spiders having eight legs, but the body is not divided into two parts. They have no silk glands, no fangs, no venom and only two eyes. They have very long legs and can run quickly, feeding on small creatures, as well as rotting organic matter, using tiny claws on the tips of their pedipalps to transfer food to the mouth.

Magnification: x75   *Scanning Electron Micrograph*

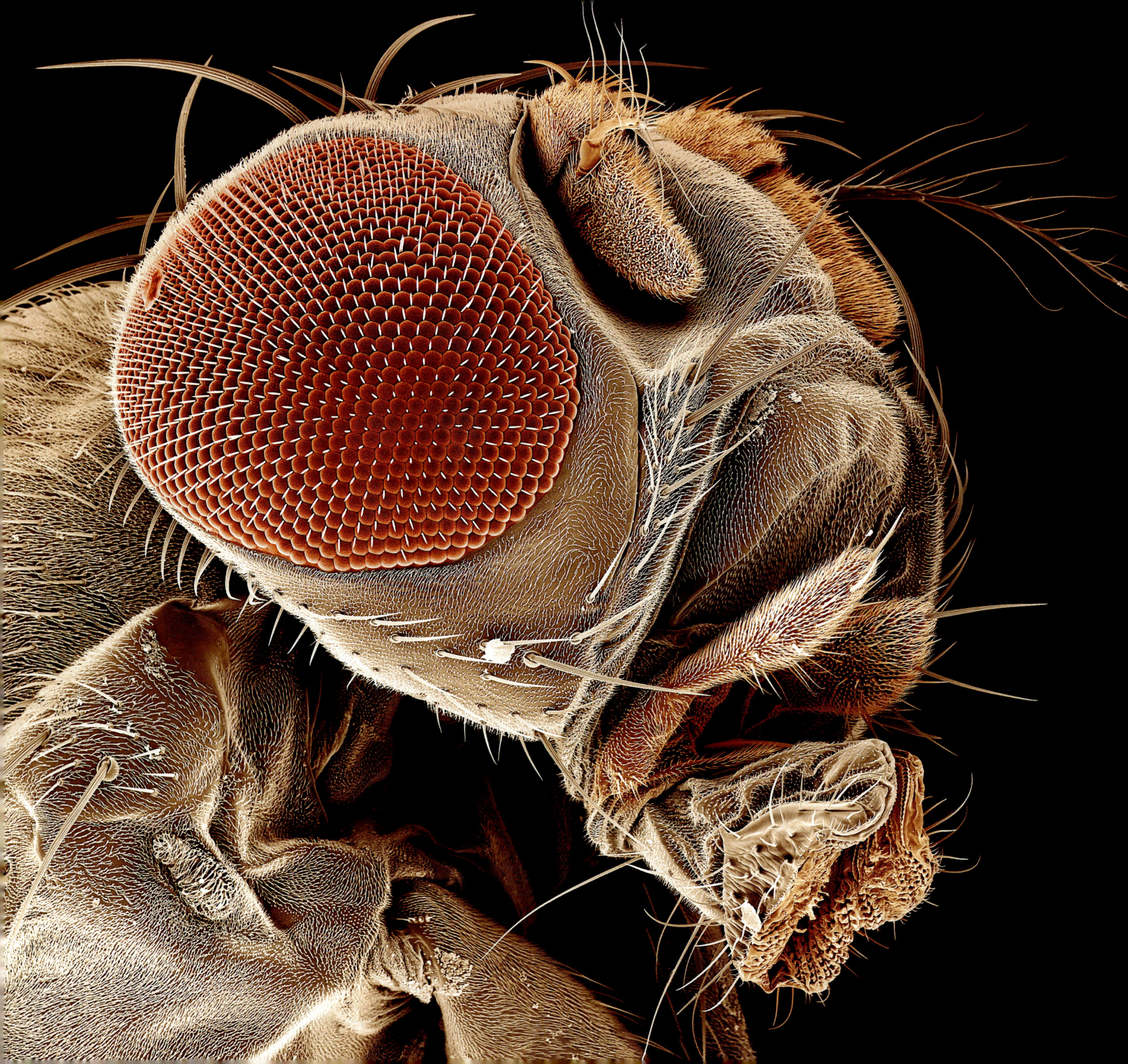

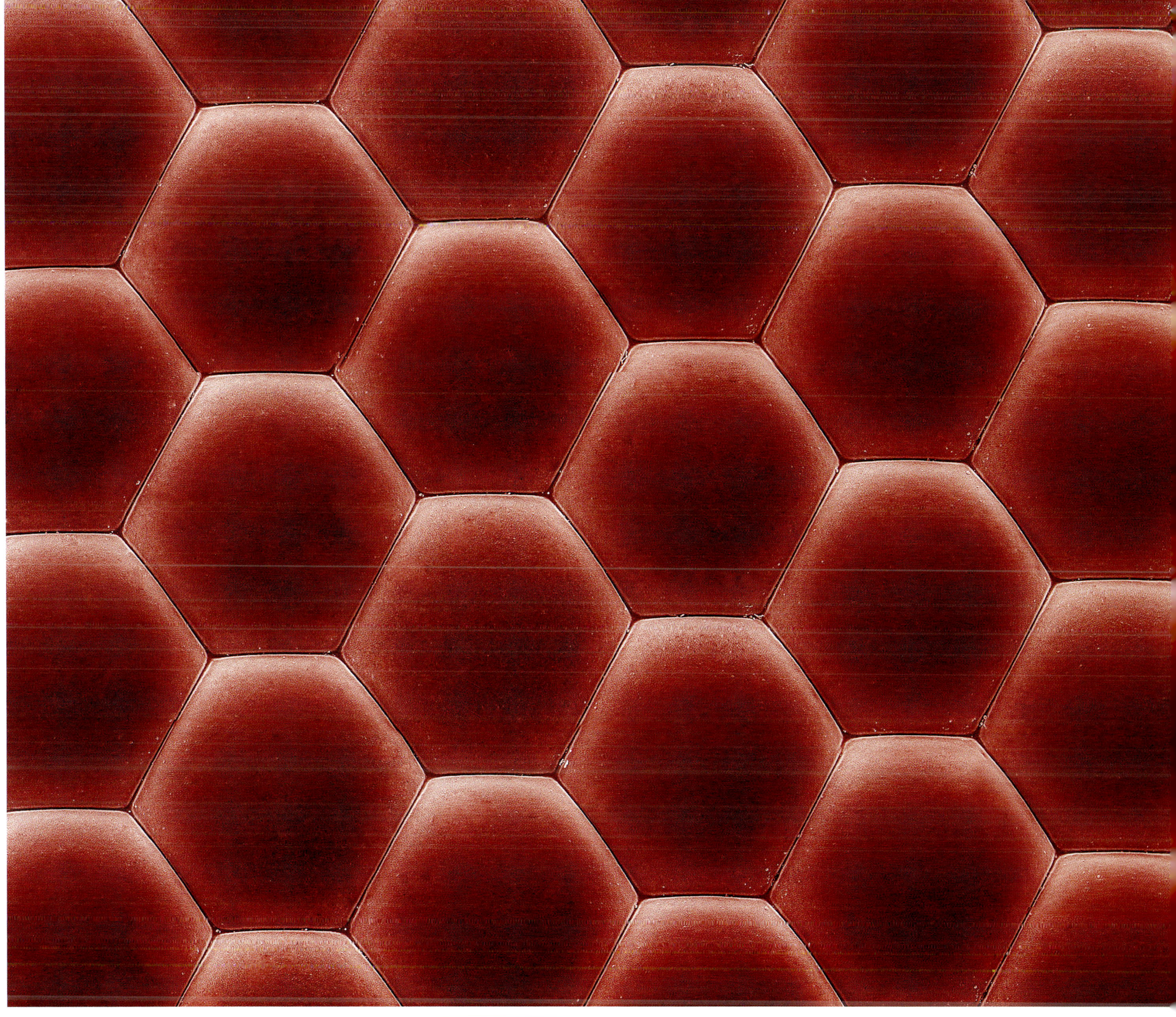

Fruit Fly, *Drosophila sp.* (left)

Attracted to ripe or fermenting fruit and vegetables, fruit flies are primarily nuisance pests. They can, however, contaminate food with bacteria and reproduce rapidly, the female producing up to 500 eggs, with a life cycle that can complete in little more than a week.

Magnification: x300   *Scanning Electron Micrograph*

Eye of a House Fly, *Musca domestica* (above)

A close up view of the mosaic of a house fly's compound eye shows the facets, or external lenses, to the light detecting organs, which are called ommatidia.

Magnification: x2600   *Scanning Electron Micrograph*

their background, so they can be almost invisible, but when camouflage fails, they rely on speed to escape danger, at the same time releasing an odorous and distasteful fluid from glands situated near the base of each front leg.

FRUIT FLY

Fruit flies of the genus *Drosophila* are especially common during late summer when they are attracted to ripe or fermenting fruits and vegetables. They are primarily nuisance pests but they have the potential to contaminate food with bacteria. About 3 mm long, with red eyes, they congregate in large numbers and reproduce rapidly, an adult female capable of producing up to 500 eggs with a life cycle that under ideal conditions completes in little more than a week. The courtship behaviour of *Drosophila* is a complex and fascinating ritual. After approaching a female and drumming on her head with his front legs, the male faces her and they 'dance', shuffling from side to side. The dance reaches a climax when the male spreads and twists his wings in a final display before mounting her. After mating, the female deposits her eggs near the surface of fermenting foods, garbage containers, mops, cleaning rags, or wherever there is a moist film of fermenting organic material. The eggs are about 0.5 mm long and hatch within a day. The tiny worm-like larvae live on the surface of the fermenting mass, growing continuously, and moulting four times before forming an immobile pupa. The adult fruit fly emerges two to four days later.

SCALE INSECTS

Closely related to whitefly, scale insects, for example *Parthenolecanium corni*, have a very strange lifestyle. The males are winged insects about 1 mm long, and look like very tiny midges, but the adult females are wingless, legless, and parasitise trees and shrubs, especially fruit and ornamental trees, feeding on sap. Covered by a protective brown scale about 3 mm long, they remain anchored to their host plant by their proboscis. The females can produce a thousand eggs at a time, producing five or six generations in a year, so infestations can be severe and can seriously impair plant growth. Eggs that are produced in late summer overwinter and hatch in the following spring. Young female nymphs are mobile, seeking a new site to feed, but after the first moult, their legs disappear and they spend the rest of their lives rooted to one spot.

WHITEFLIES

Whiteflies, *Trialeurodes vaporariorum*, are very small winged insects, just over 1 mm long, recognised because they appear to be dusted with a white powdery wax. They are natives of Central America, but have arrived in Britain and other temperate regions with trade goods, and are now established since they can survive harsh winters in heated greenhouses or indoors. Whitefly can infest many greenhouse and indoor plants, for example tomatoes and cucumbers, and they sometimes occur on outdoor plants and weeds during the summer. Breeding, which is mainly asexual, can continue throughout the year in warm conditions. The adult flies may live for a month or more and each female can produce up to 200 eggs, which she places on the underside of leaves. The young nymphs crawl about on plants for a few hours after hatching, before settling in one place. Curiously, their legs and antennae degenerate, and they become immobile scales, which feed on plant sap for about 2 weeks before becoming a non-feeding phase from which the adults emerge. Development from the egg to adult takes about 3-4 weeks.

Whiteflies occur in large numbers, with usually hundreds of them together, and can be seen as clouds of pale flying dust around the host plant if disturbed. A severe attack can severely weaken a plant. They excrete large amounts of sticky sugar-rich exudates, which lands on leaves below the whitefly, and then becomes covered in a sooty mould. Although this does no direct damage to the host plant, it may stop the leaves from photosynthesizing efficiently. Whitefly can be difficult to control because they have a very fast lifecycle, but in the greenhouse biological methods can be effective, using other insects that eat or parasitise them, such as a small wasp, *Encarsia formosa*, which attacks and kills the whitefly scales, a small black ladybird, *Delphastus sp.,* or the predatory bug, *Macrolophus* sp.

FUNGUS GNAT

Sometimes mistaken for midges, fungus gnats such as *Bradysia paupera or Sciara thomae* are small flies around 2 mm long that run over the soil surface or fly slowly around potted or border plants. They are responsible for the spread of some fungal diseases of plants. The adult females deposit eggs in moist soil where they hatch after five to seven days. When they first emerge, the translucent white larvae are no more than 1 mm long, but can grow to ten times

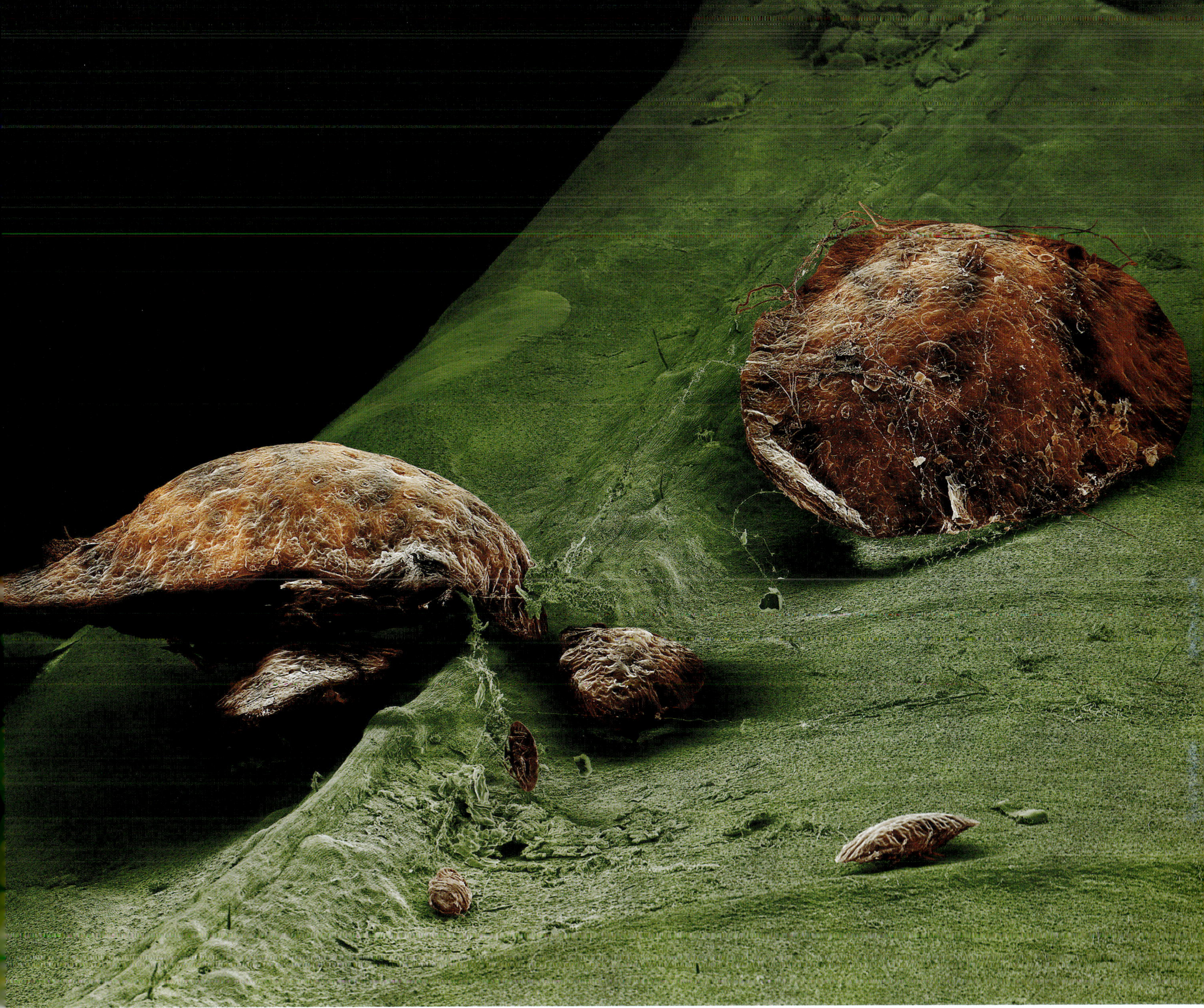

Brown Scale Insect, *Parthenolecanium corni*

Scale insects have a very strange lifestyle, the adult females are wingless and legless, parasitising trees and shrubs, to feed on their sap. Covered by a protective brown scale, they remain anchored to the host plant by the proboscis.

Magnification: x60   *Scanning Electron Micrograph*

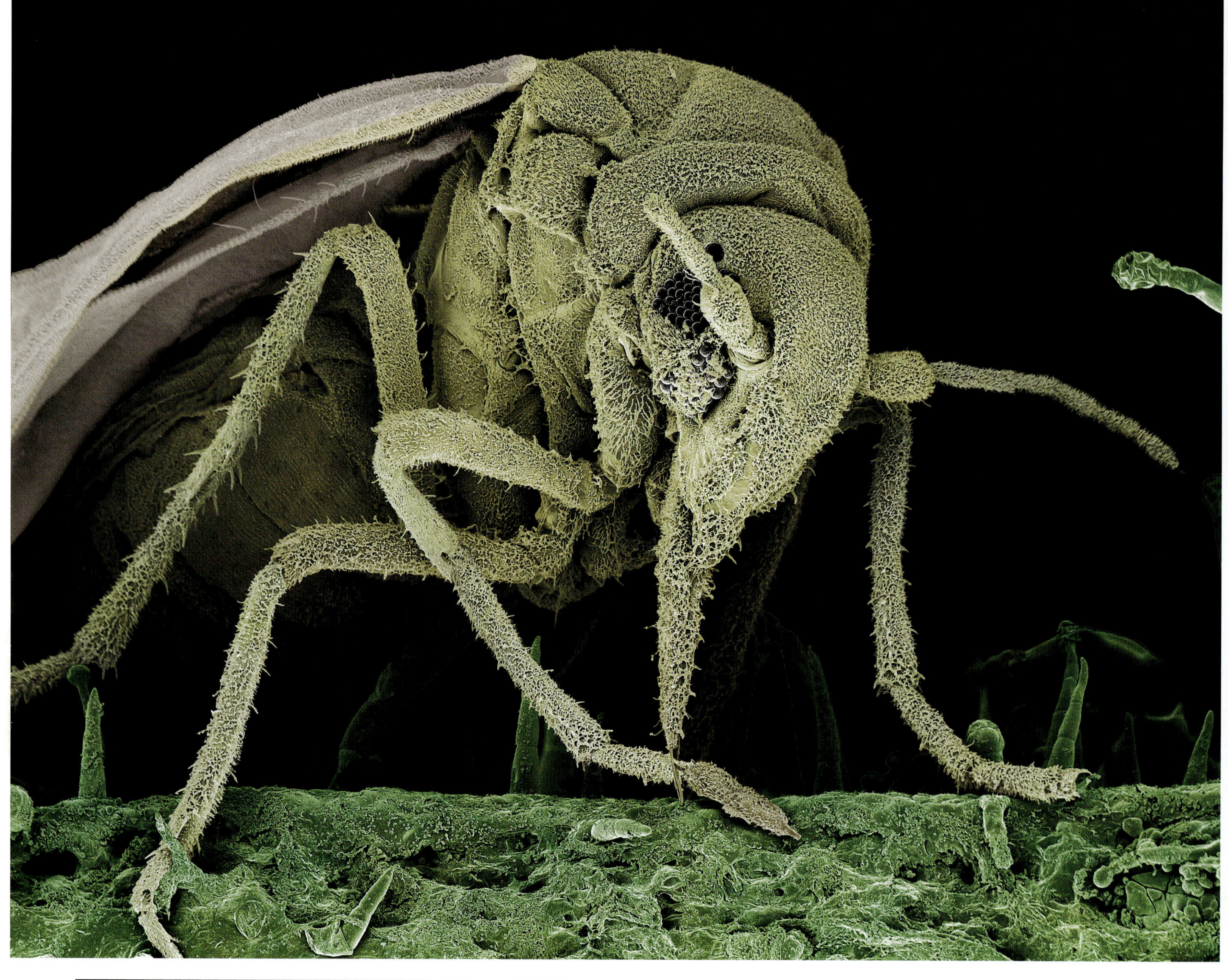

Whitefly, *Trialeurodes vaporariorum* (above and right)

Tiny insects with white papery wings, whiteflies are often seen swarming as clouds of pale flying dust around the host plant when disturbed. They feed on plant sap by piercing the epidermis with a sharp proboscis. Severe infestations can kill the plant.

Magnification: x150   *Scanning Electron Micrograph*
Magnification: x150   *Scanning Electron Micrograph*

Fungus Gnat, *Bradysia paupera* (overleaf and page 4)

Sometimes mistaken for midges, fungus gnats are small flies that run over the surface of the soil or fly slowly around potted or border plants. Harmless as adults, the larvae feed on fungi and decaying plant material but also live on plant roots and burrow into stem bases.

Magnification: x65   *Scanning Electron Micrograph* (overleaf, left)
Magnification: x330   *Scanning Electron Micrograph* (overleaf, right)
Magnification: x60   *Scanning Electron Micrograph* (page 4)

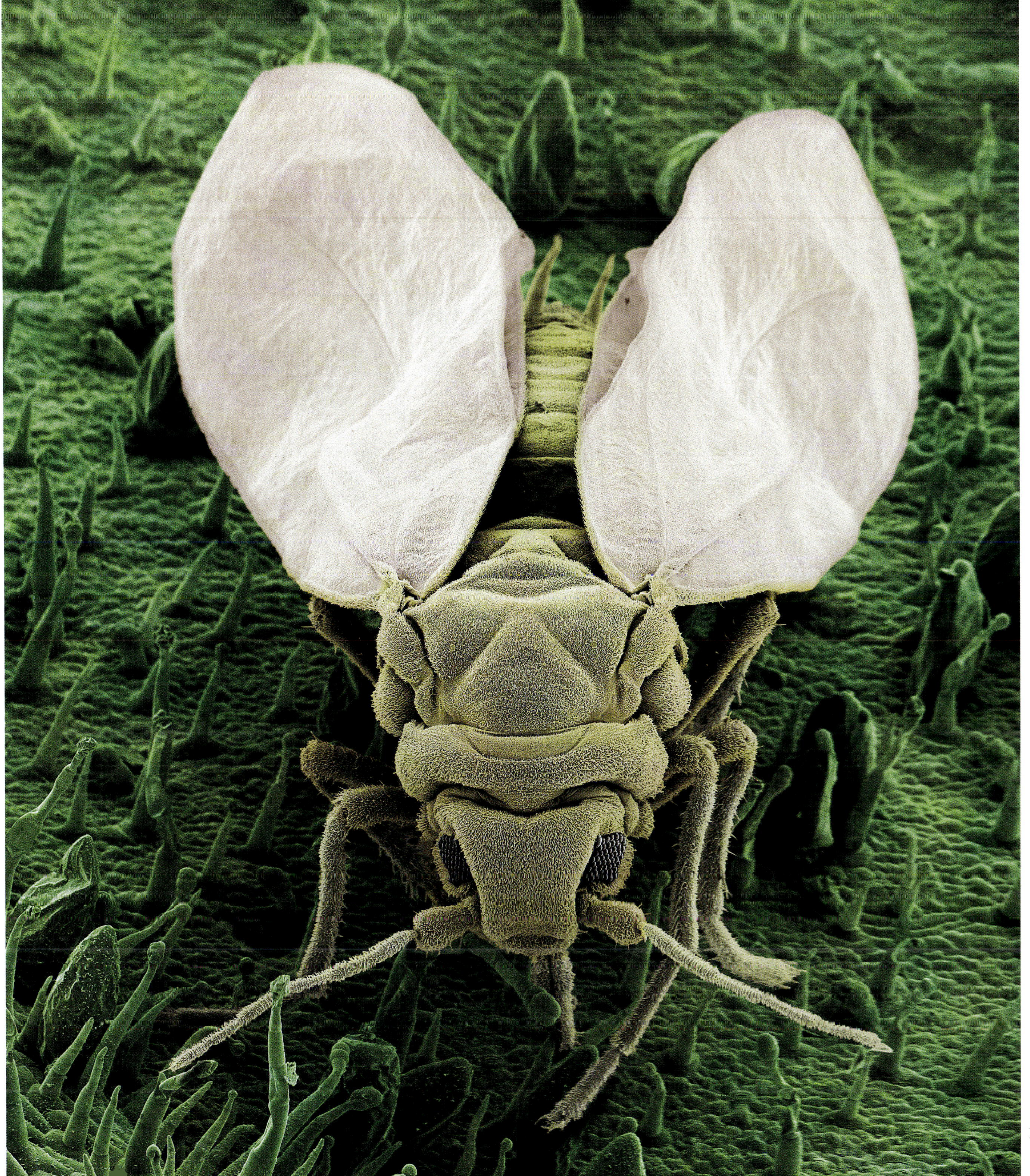

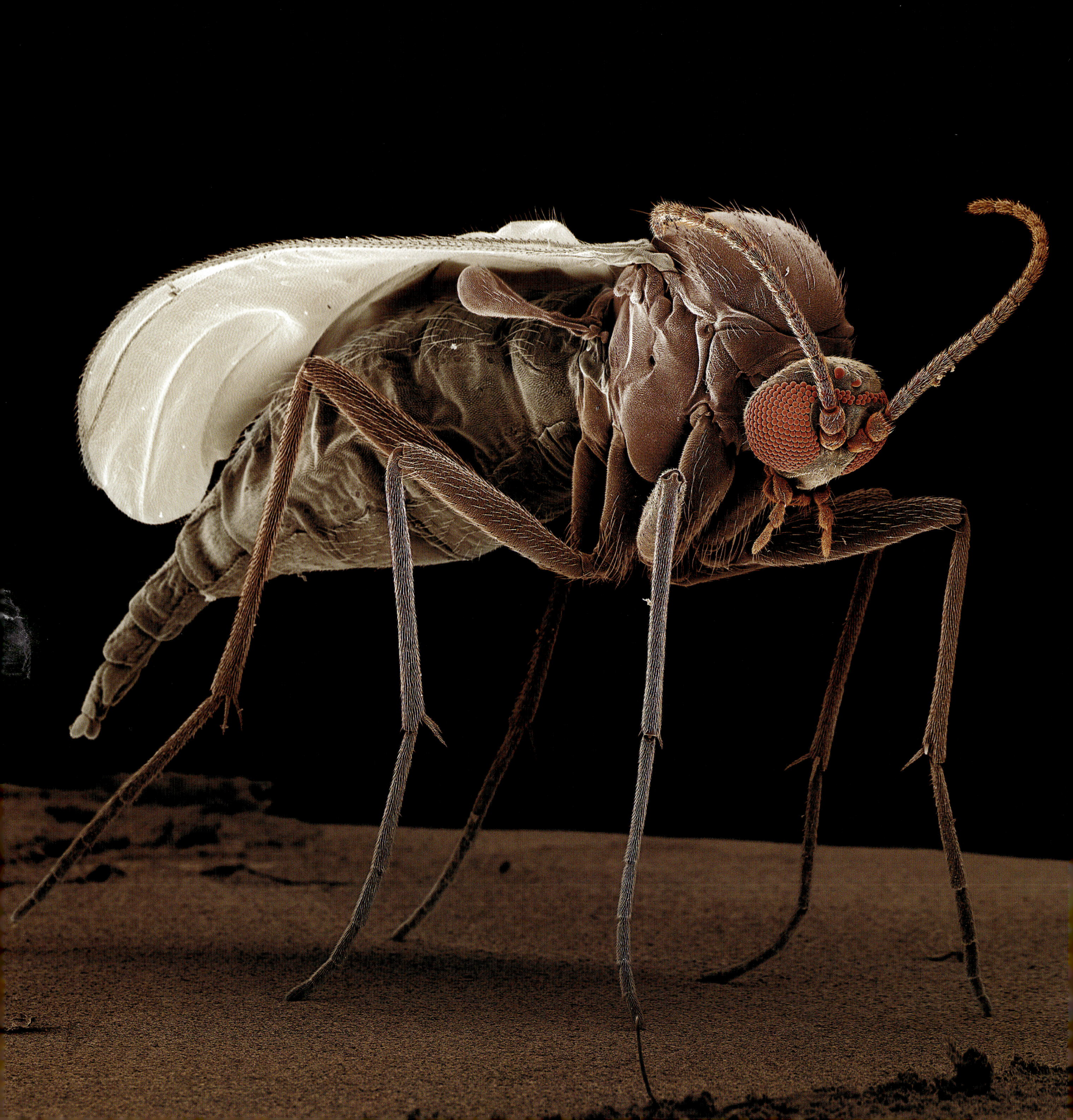

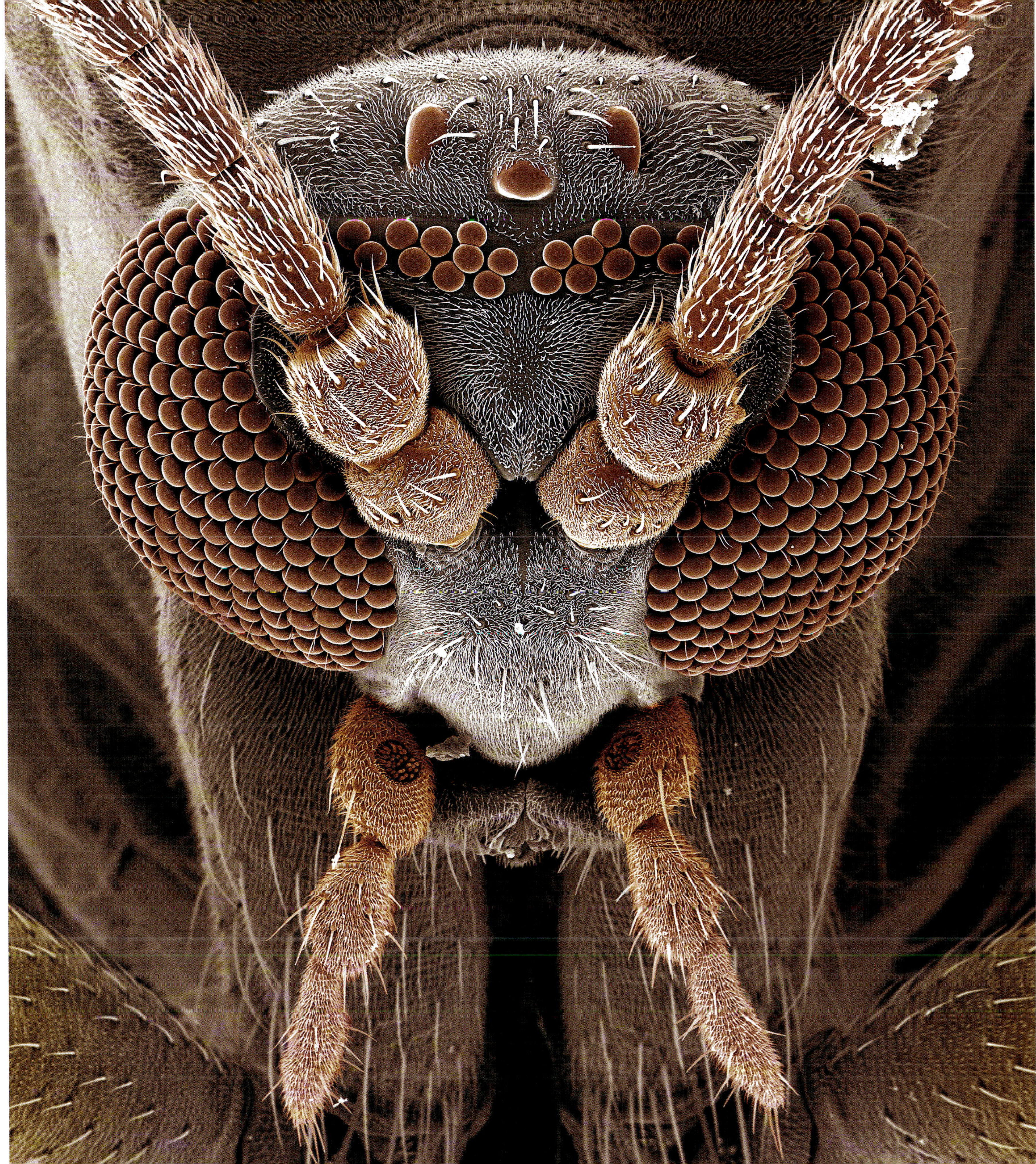

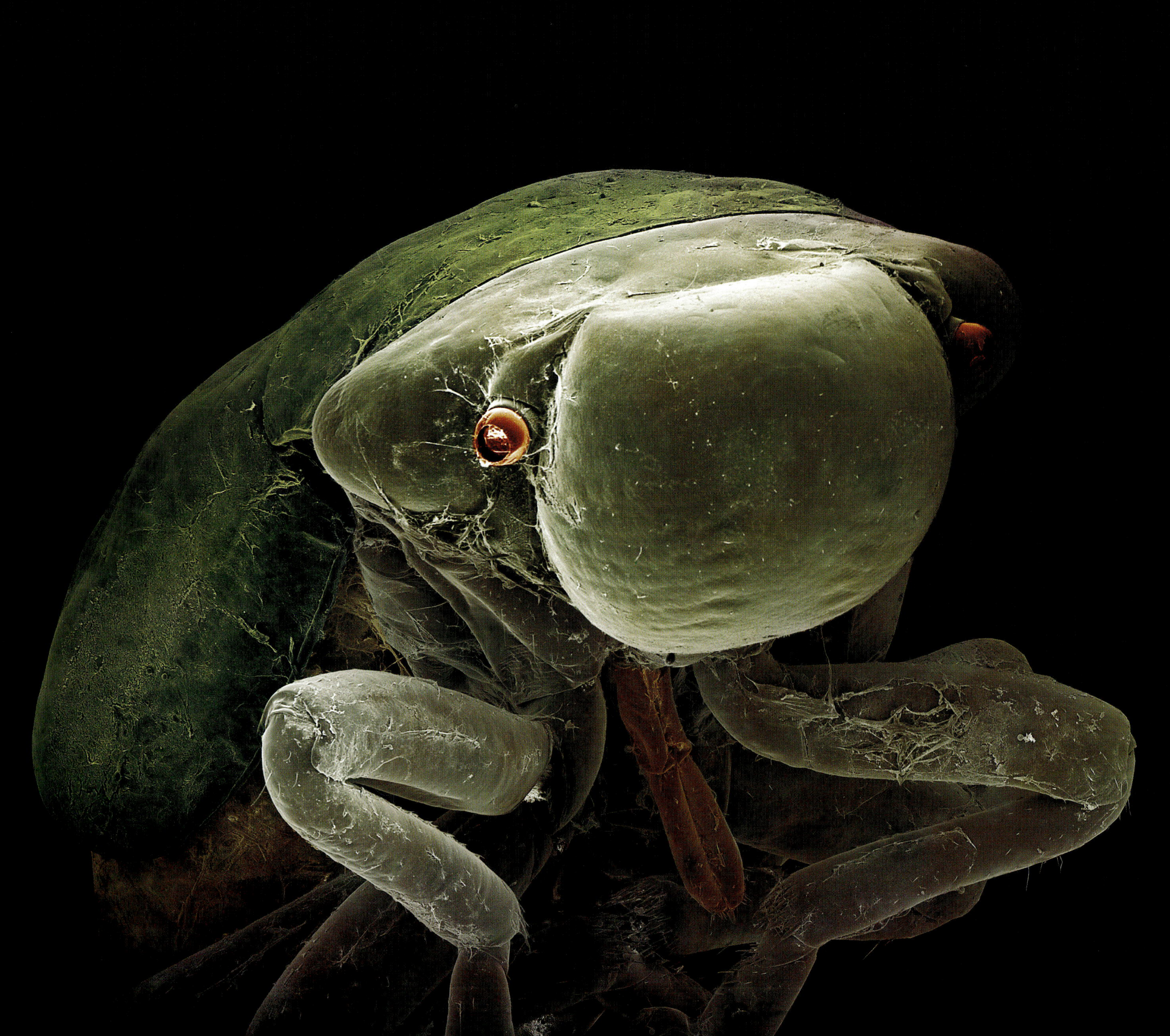

that before pupating. They feed on fungi and decaying plant material, but also feed on live plant roots, and burrow into the soft stem bases of young plant cuttings. Fungus gnats are known pollinators of some orchids, for example the bog orchid *Hammarbya paludosa* and the lesser twayblade *Listera cordata*.

COMPOST

Grass cuttings from the lawn and discarded weeds rot down to form compost. The compost heap is a whole new environment, home to a host of organisms that help to break down what would otherwise be a giant pile of dead material into a form that provides nutrients for plant growth. The breaking down process is aerobic, an oxygen using process performed by bacteria, fungi, nematodes and other animals, such as earthworms, and generates heat as the organic matter decomposes. Just one cubic centimetre of rich garden compost can contain up to a million different species of bacteria. Vegetable peelings, tea bags and food scraps from the kitchen can be added to the pile. Moulds, bacteria and fungi are extremely important in the decay process that releases nutrients such as nitrogen and phosphorous compounds back into the soil for use by plants. The soil contains a mass of tiny threads that make up the mycelium or body of the organism, which appears on the surface as fruiting bodies such as mushrooms and toadstools. These produce millions of spores that drift away on the breeze to grow elsewhere.

FUNGI

While fungi are the main agents for the disintegration of organic matter to increase the fertility of the soil, they also cause the majority of known plant diseases, for example blight, rot, or mildew. The infective fungus grows hyphae over the surface of the plant entering through the stomata of leaves or any convenient aperture in the plant's epidermis to feed on nutrients inside. Whichever part of a plant is attacked, the effect of the fungus either weakens or kills it. The fungus *Fusarium graminearum*, for example, causes head blight in cereal crops, and the famous potato blight, caused by the root-attacking fungus *Phytophthora infestans* caused the Great Irish Famine of 1845-9 during which over a million people died, and continues to be a significant problem. Ironically, some fungus infections are highly beneficial. Nearly all plants rely on their symbiotic associations with root fungi, called mycorrhizae, for their very survival. Fungal hyphae grow around the roots, and while the plant provides the fungus with food, it profits from the huge area of the fungal mycelium, which is better able to absorb mineral nutrients from the soil.

Froghopper (nymph), *Philaenus spumarius* (left)

Froghoppers, such as *Philaenus spumarius*, are small bugs about 6 mm long that can leap a great distance if threatened. They have sharp mouthparts, which they use to pierce plants and suck their sap. The egg and nymph (see left), are protected against dehydration by a blob of white foam, commonly known as "cuckoo spit", formed by blowing air through a viscid fluid expelled from the anus. The blob of white foam, often conspicuous on grasses during the summer, protects the enclosed nymphs from dehydration and from predators.

Magnification: x100   *Scanning Electron Micrograph*

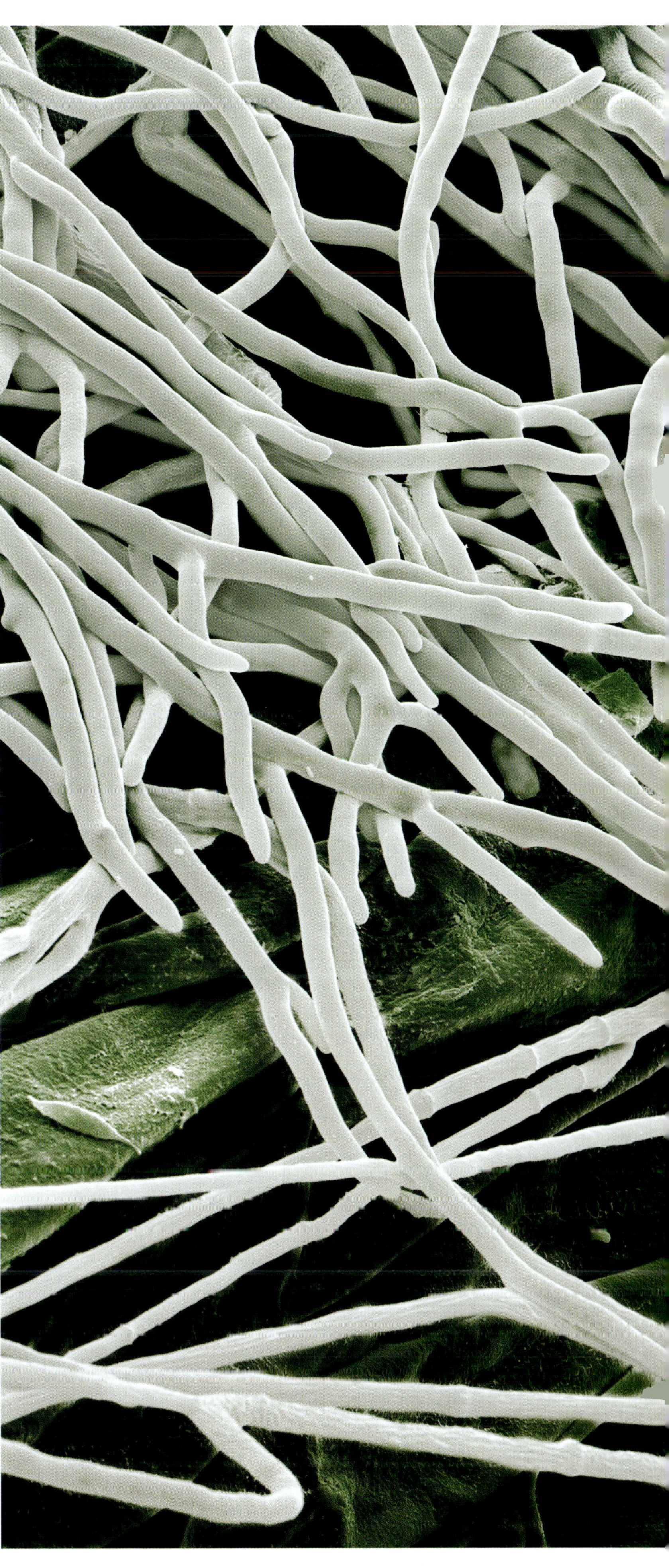

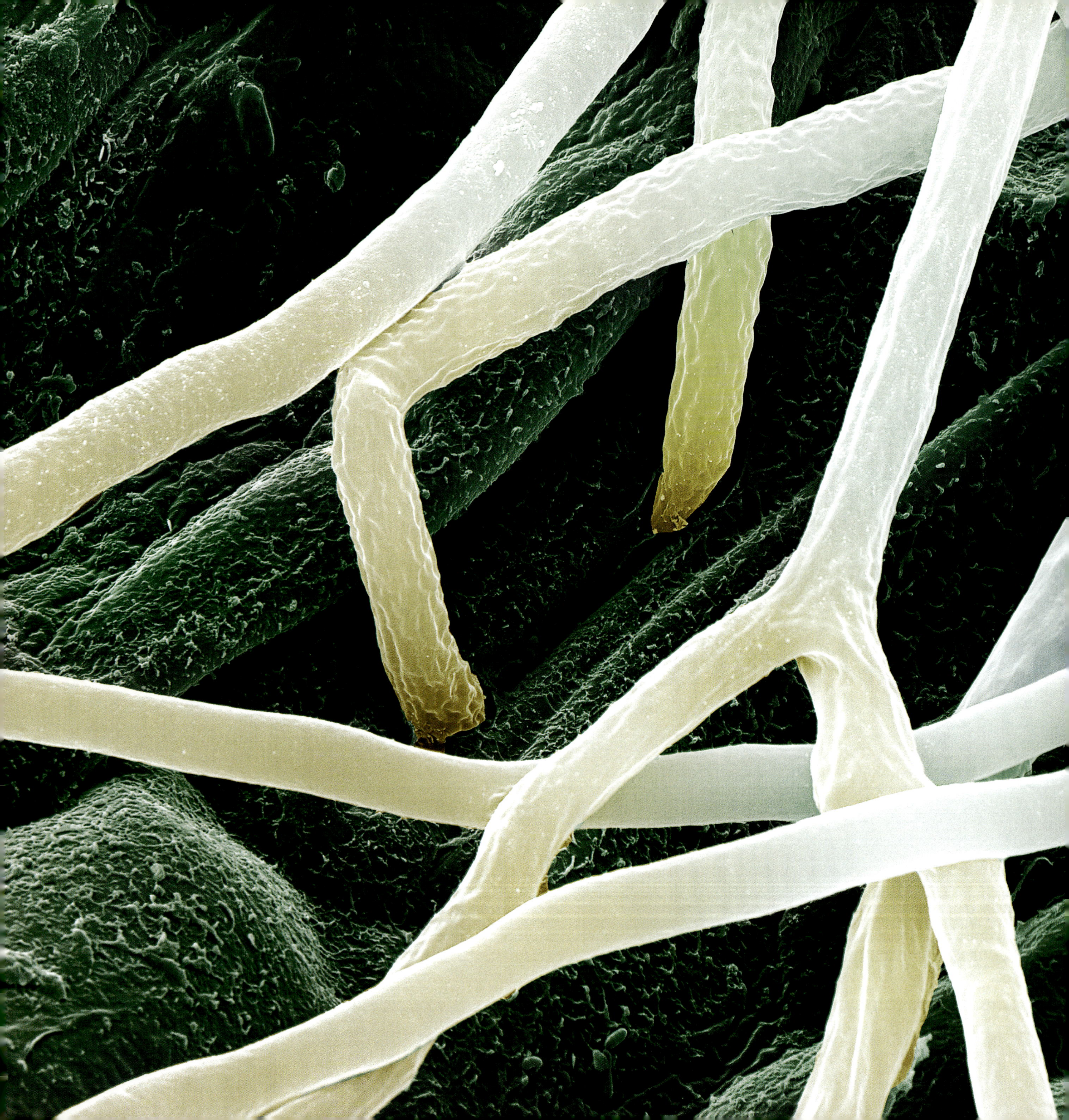

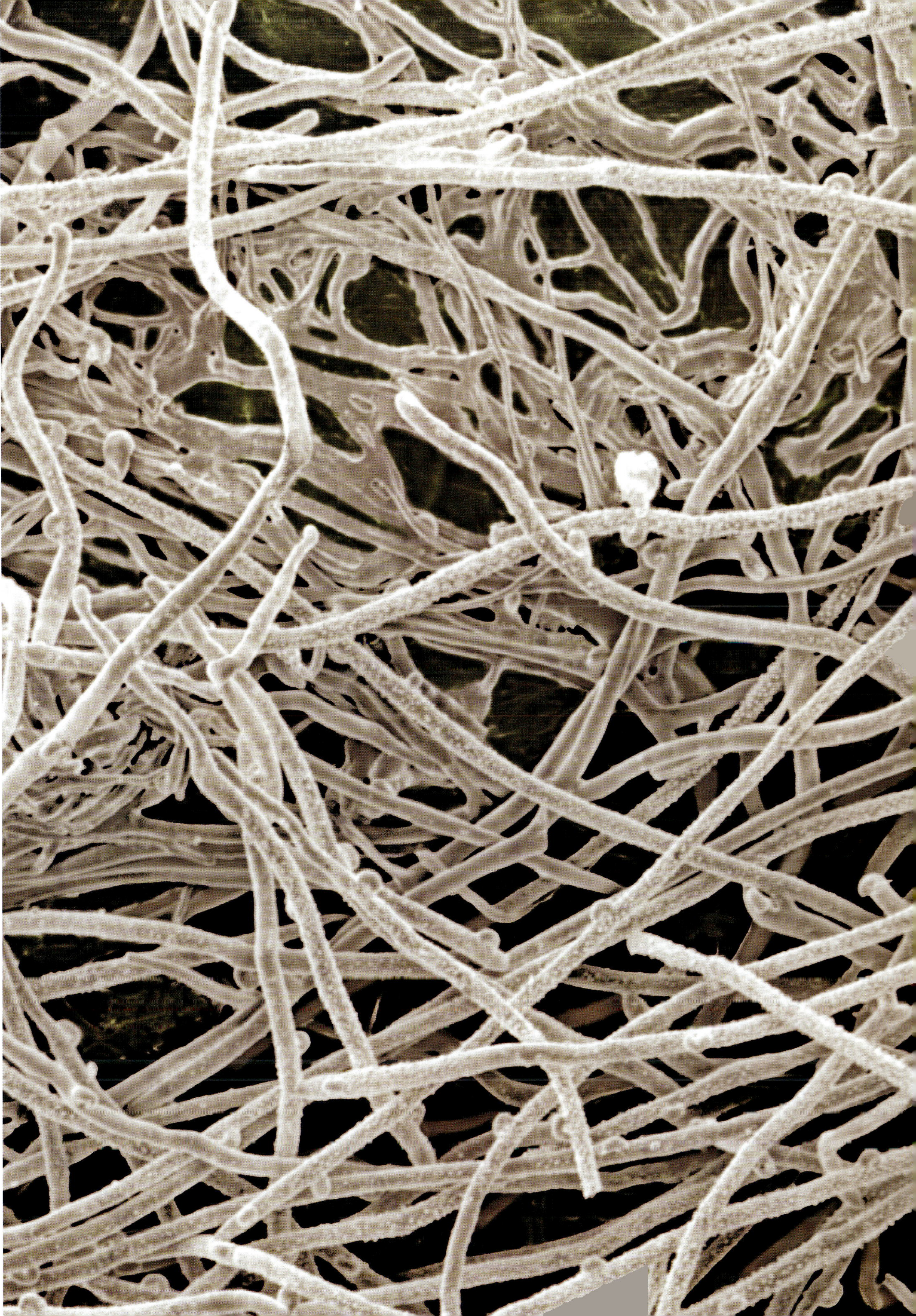

Fusarium,

Fusarium graminearum (left and previous page)

Microscopic hyphae of the fungus *Fusarium* form a branching network, or mycelium, across the surface of a leaf, penetrating the epidermis to absorb nutrients from the plant interior. This fungus causes head blight, a globally important disease of cereal crops, seriously affecting grain quality, and safety due to mycotoxins produced by the fungus during infection. Grown in bulk culture however this fungus is highly nutritious, containing 12% protein, is low in fat and lacks cholesterol. It forms an important ingredient for the meat substitute product 'Quorn'.

Magnification: x2000
Scanning Electron Micrograph
(previous page)
Magnification: x7000
Scanning Electron Micrograph (left)

Mycorrhiza (right)

Mycorrhiza is the mutually beneficial symbiotic relationship between a fungus and a plant. The fungus grows its mycelium of hyphae around the roots of the plant, providing it with food. Meanwhile, the plant profits from the huge surface area of the fungal mycelium, which is better able to absorb mineral nutrients from the soil. Mycorrhizal relationships exist in most plant species; indeed, without them, most plants would be unable to survive.

Magnification: x2280
Scanning Electron Micrograph

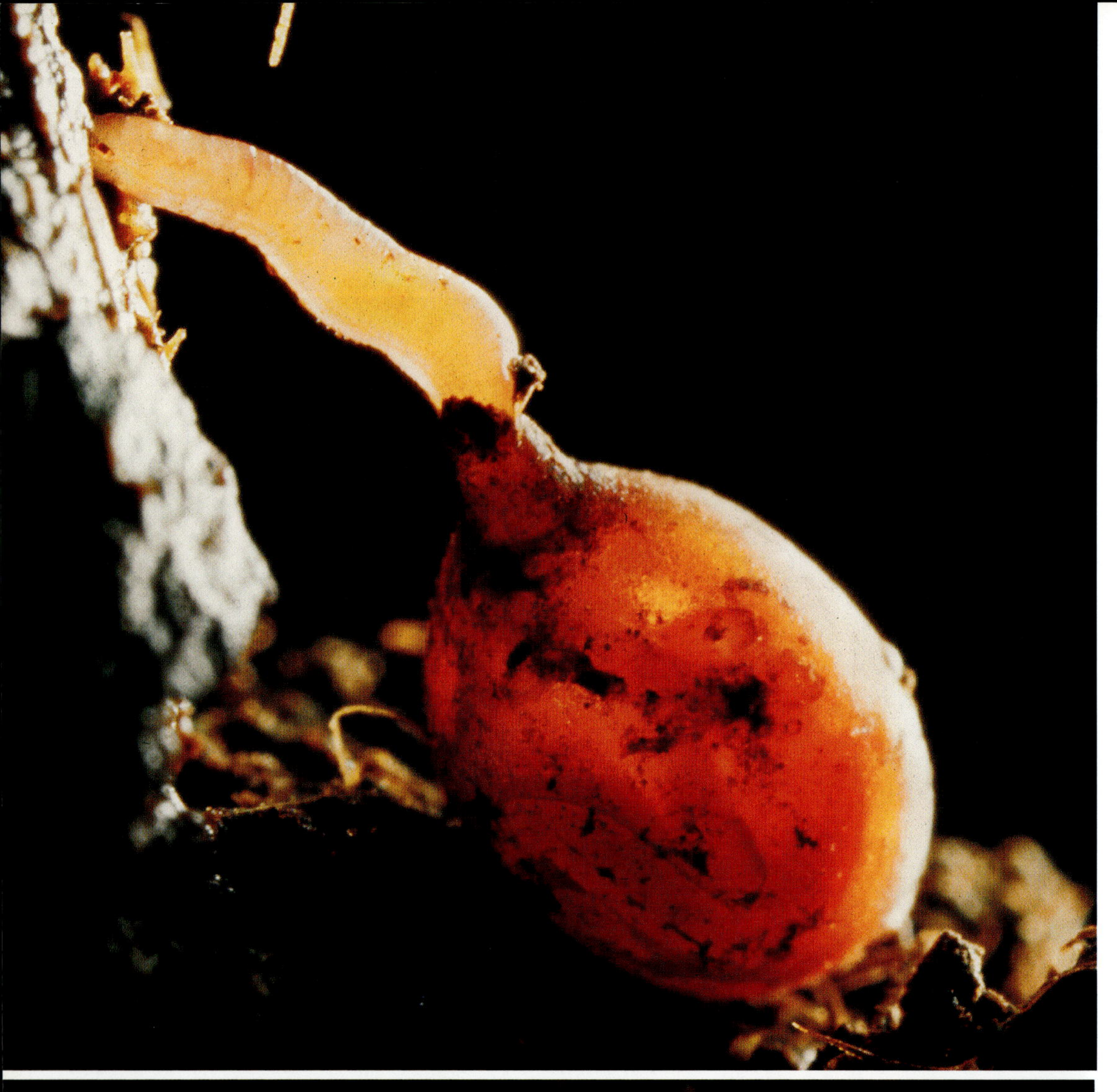

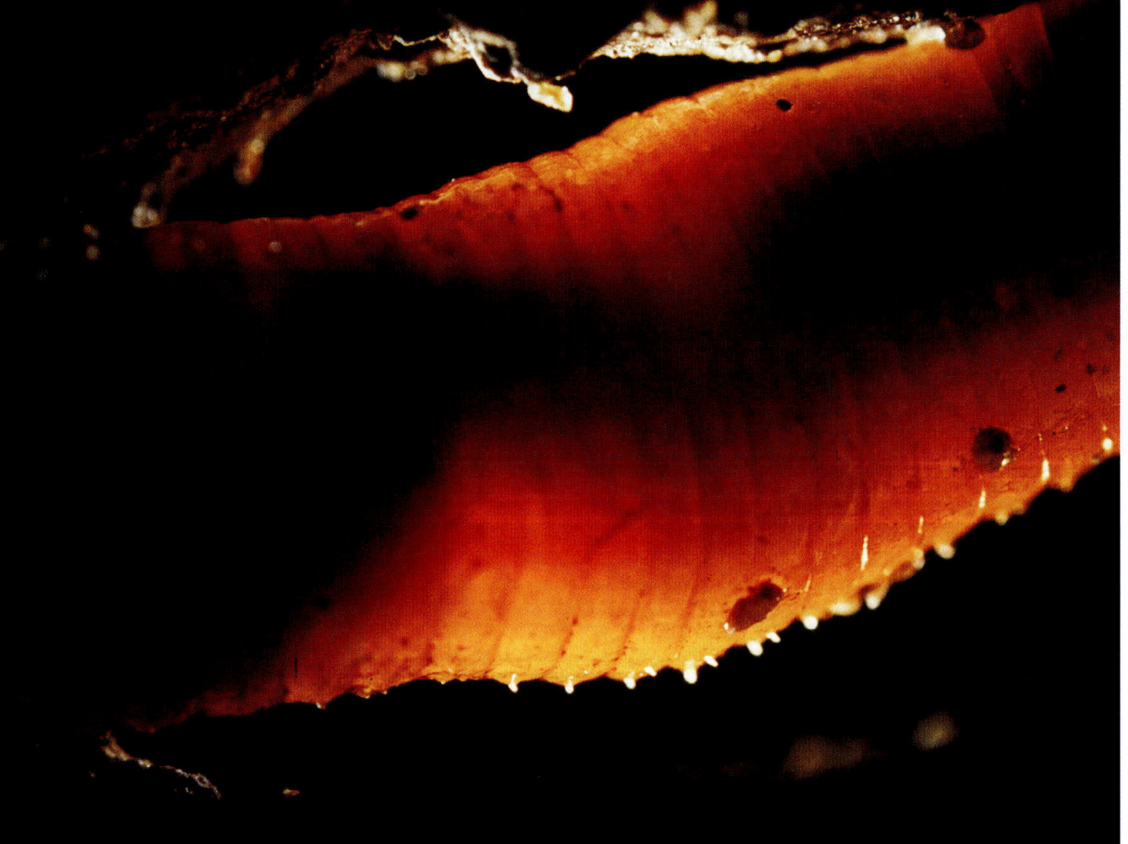

EARTHWORM

Earthworms, *Lumbricus terrestris*, and blood earthworms, *Eisenia foetida*, can be found in huge numbers in compost heaps. They feed on organic matter, which they ingest together with soil particles, excreting it as the familiar, mineral-rich worm casts we find on the surface. They are the natural tillers and aerators of the soil, helping to fertilise it by bringing nutrients closer to the surface. Earthworms are distinctly segmented with annuli, or ring-type sections, which have rows of stiff bristles called chaetae that grip the soil as the worm stretches its body to move.

SLUGS AND SNAILS

Slugs and snails are soft bodied and are scientifically classified as 'gastropods', meaning 'stomach feet', which describes their characteristics quite accurately. They frequent damp places since they are inefficient at controlling water loss, hiding during the day under logs, stones and vegetation and emerging at night to feed on rotting vegetation, garden plants and vegetable crops, especially soft, succulent seedlings. Their mouthparts comprise a rasping organ called a radula, which is used like a file to reduce plant tissue to pulp before being consumed. The radula has neat rows of hundreds of tiny tooth-like denticles, which are continually replaced as they are worn away. Slugs and snails are hermaphrodite, mating in pairs to exchange sperm in a sinuous, slimy slow-motion mating dance. The pearl-like masses of eggs are deposited in crevices in the soil or under rocks and logs. They eat voraciously, and damage to garden plants can be considerable. In damp years, when there are large numbers of snails and slugs, even predators such as hedgehogs and thrushes cannot control the numbers.

Blood Earthworm, *Eisenia foetida* (above left)

A blood earthworm, sometimes known as a 'brandling', hatches from its cocoon (macro-photograph). *Eisenia* lives in rich, rotting vegetation and in dung heaps.

Magnification: x40 — *Macro-photograph*

Earthworm, *Lumbricus sp.* (below left)

A close-up of part of the body of a common earthworm shows the chaetae, or bristles, that grip the sides of its burrow when it stretches its body to move through the soil.

Magnification: x15 — *Macro-photograph*

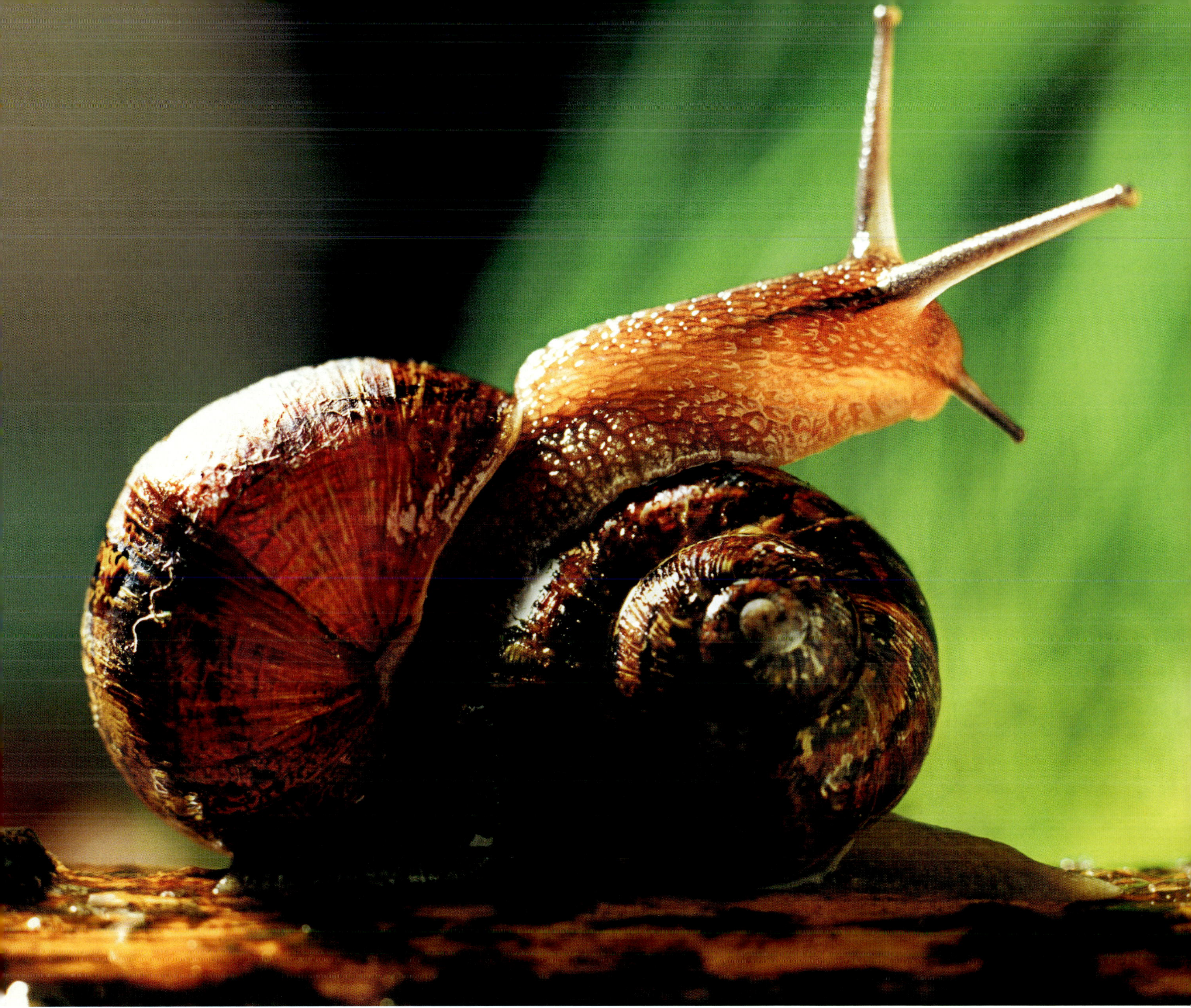

Garden Snail, *Helix aspersa* (above)

A pair of garden snails engaged in courtship behaviour. Snails are hermaphrodite so they will each fertilize the other and lay clutches of pearl-like eggs in the soil, or under rocks and logs. They eat voraciously and cause considerable damage to seedlings and crop plants.

Magnification: x3   *Macro-photograph*

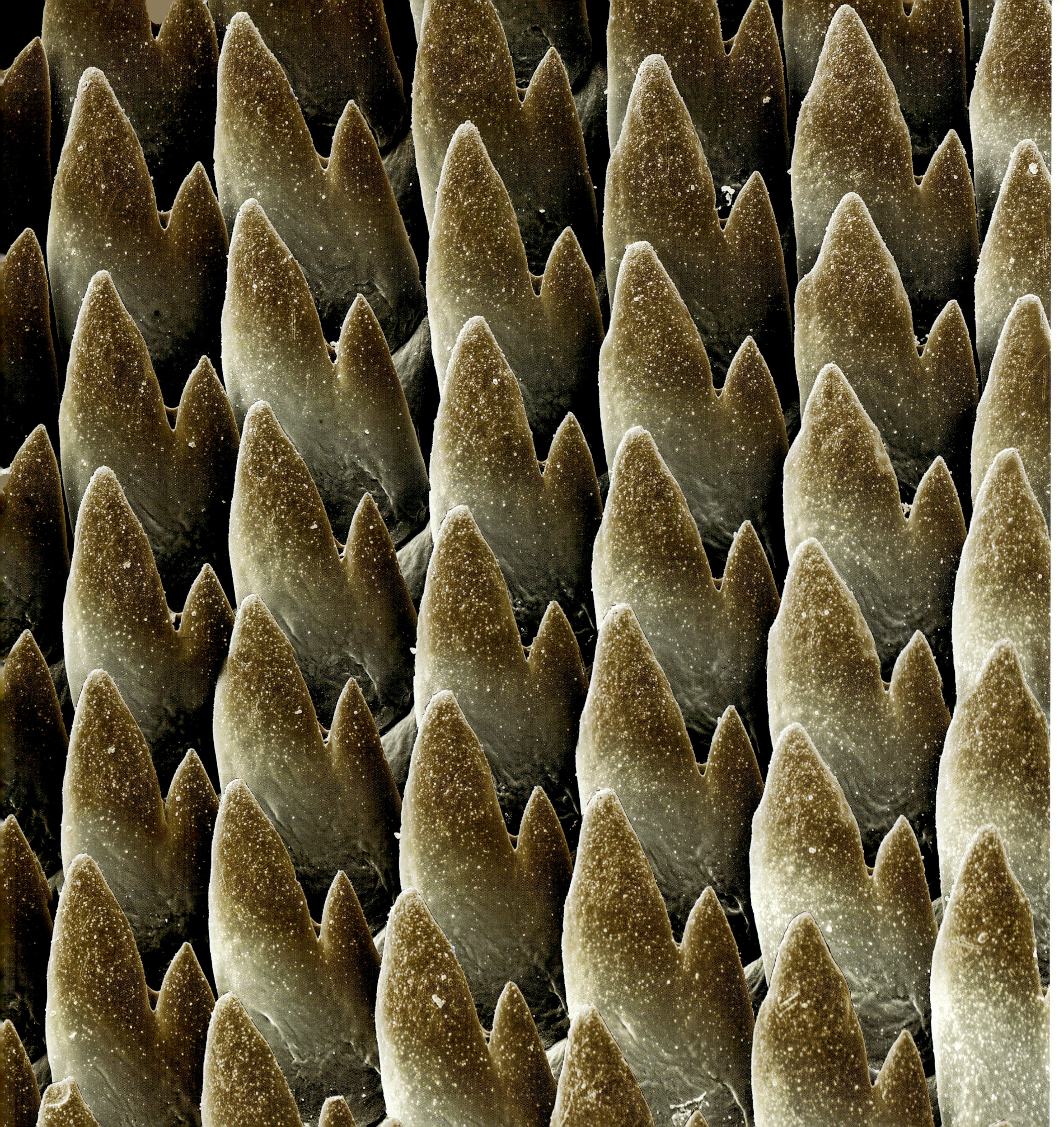

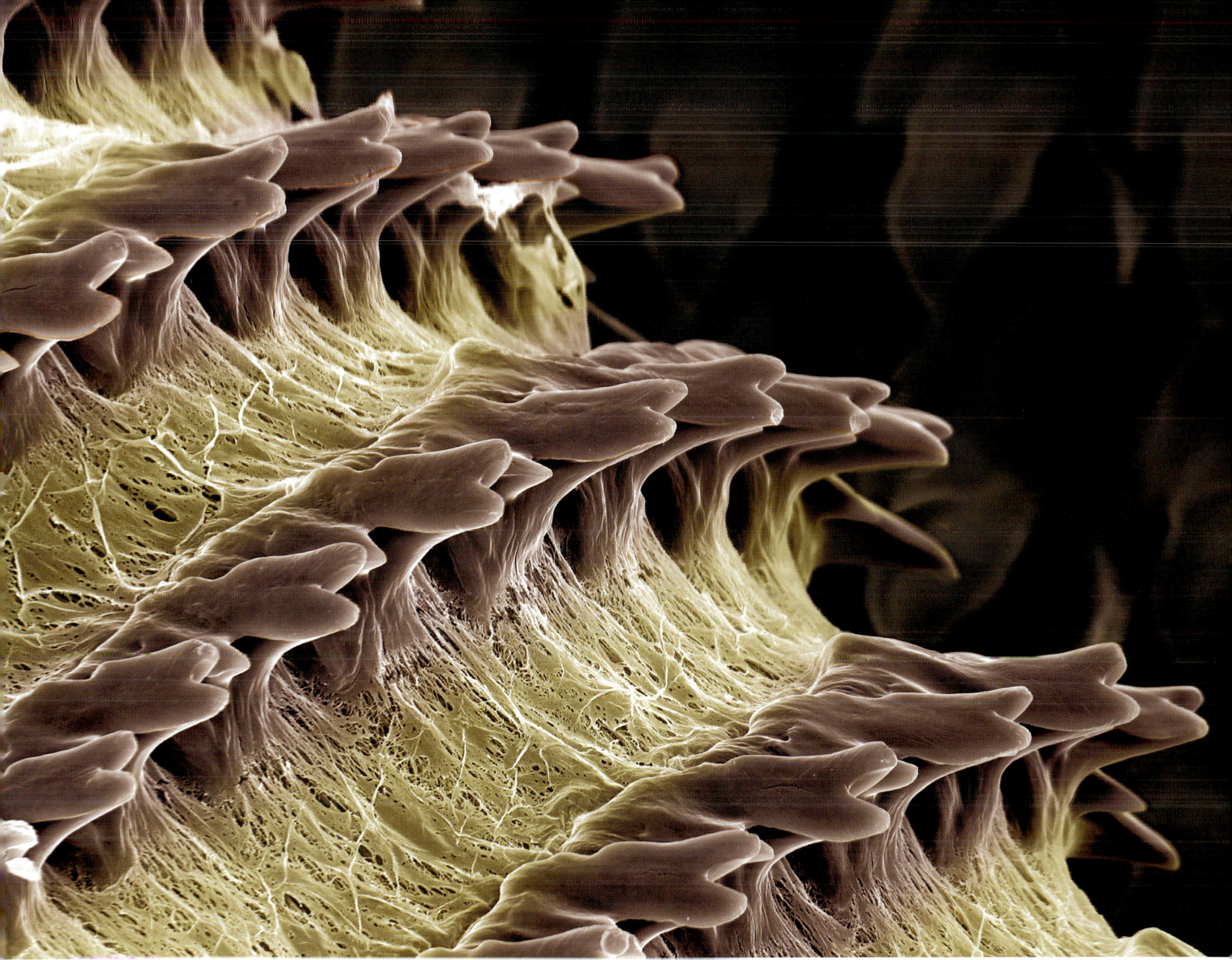

Radula of the Garden Snail, *Helix aspersa*

Garden snails feed on soft, succulent plant material. Their mouthparts comprise a rasping organ called a radula, used like a file to reduce plant tissue to pulp before being consumed. The radula has neat rows of hundreds of tiny tooth-like denticles, which are continually replaced as they are worn away.

Magnification: x2000   *Scanning Electron Micrograph*
Magnification: x2000   *Scanning Electron Micrograph*

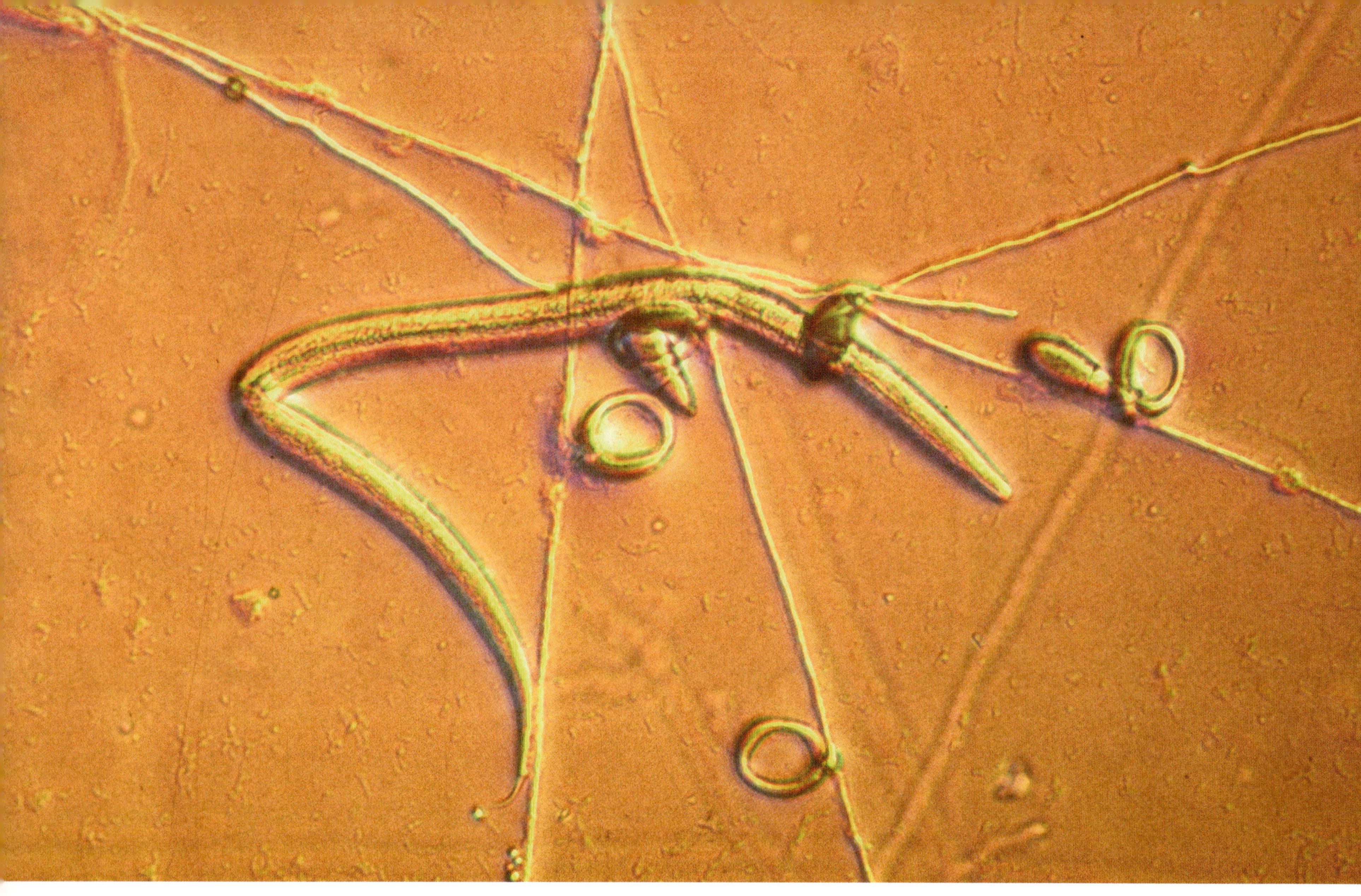

NEMATODES AND NEMATOPHAGOUS FUNGI

Nematodes are the most numerous multi-cellular animals on earth with nearly 20,000 described species. A handful of soil contains thousands of the microscopic worms, many of them parasites of insects, plants or animals. Free-living nematodes are abundant however, including species that feed on bacteria, fungi, and even other nematodes, yet the vast majority of species encountered are poorly understood biologically. Structurally simple organisms comprised of approximately 1,000 somatic cells, nematodes possess digestive, nervous, excretory, and reproductive systems, but lack a discrete circulatory or respiratory system, taking in oxygen from their surroundings through the body wall.

These microscopic worms are usually so small as to be unnoticeable, although they can cause great damage to the roots of plants. They have many enemies, among them a remarkable carnivorous fungus, *Arthrobotrys*, common in soil and decaying plant debris, which traps them in little loops. When a nematode squeezes through one of these loops, the cells of the ring respond by suddenly inflating so it cannot escape. Some species of *Arthrobotrys* have sticky loops that trap the nematode on contact with powerful glue. Either way, the trapped worm dies and as it decays, the fungus grows filaments into its body to digest it.

164

Soil Nematode

Living in the soil, the nematode *Caenorhabditis elegans* (above right) has been caught in a loop trap of the nematophagous fungus *Arthrobotrys sp.* (see above). A loop trap consists of three cells, which rapidly swell if a worm tries to squeeze through it. The trapped worm dies and decays, providing nutrients for the fungus. Most nematodes in the soil (see right) are free living, and are important in releasing nutrients in forms available to plants.

Magnification: x1000
Light Micrograph (above)
Magnification: x1900
Scanning Electron Micrograph (above right)
Magnification: x850
Scanning Electron Micrograph (below right)

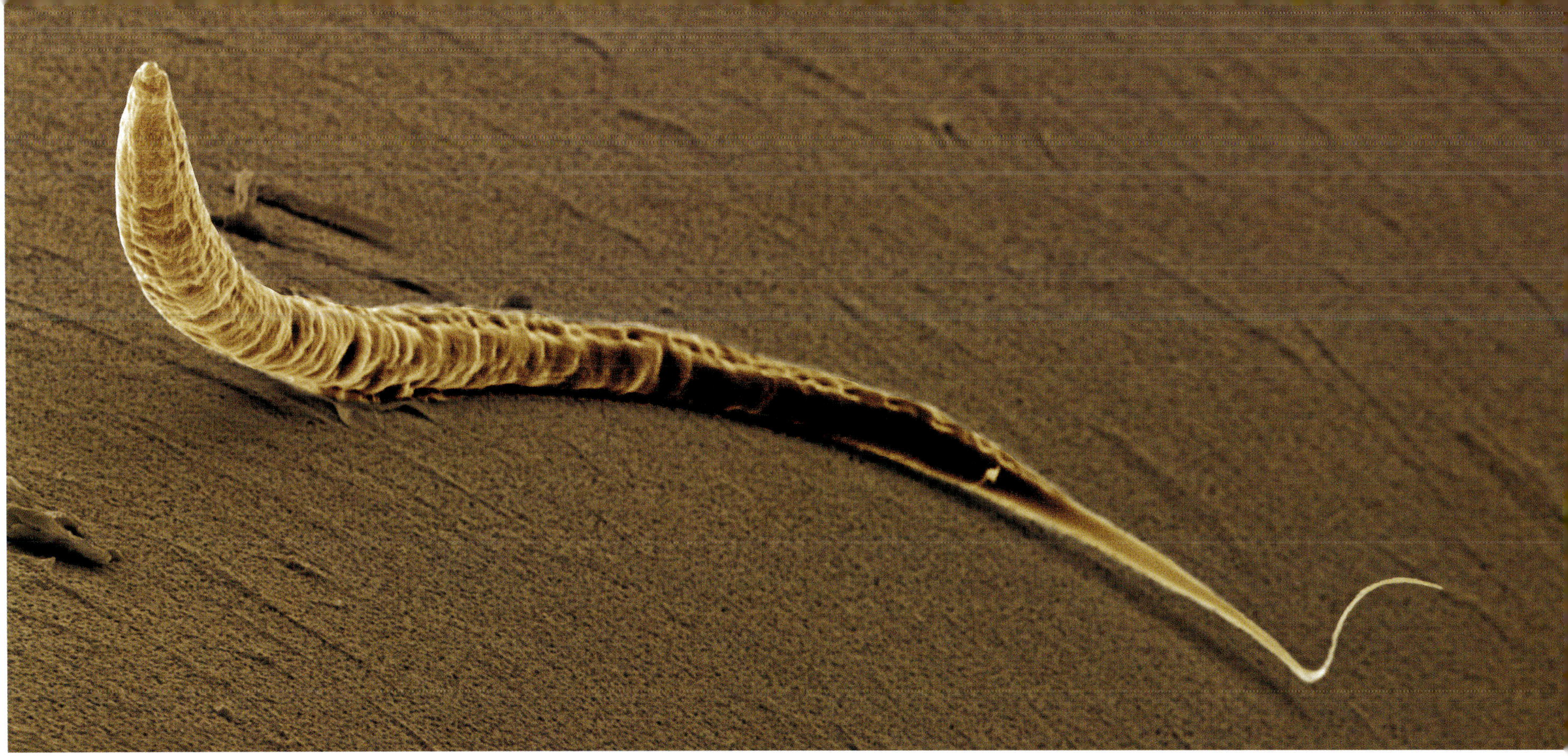

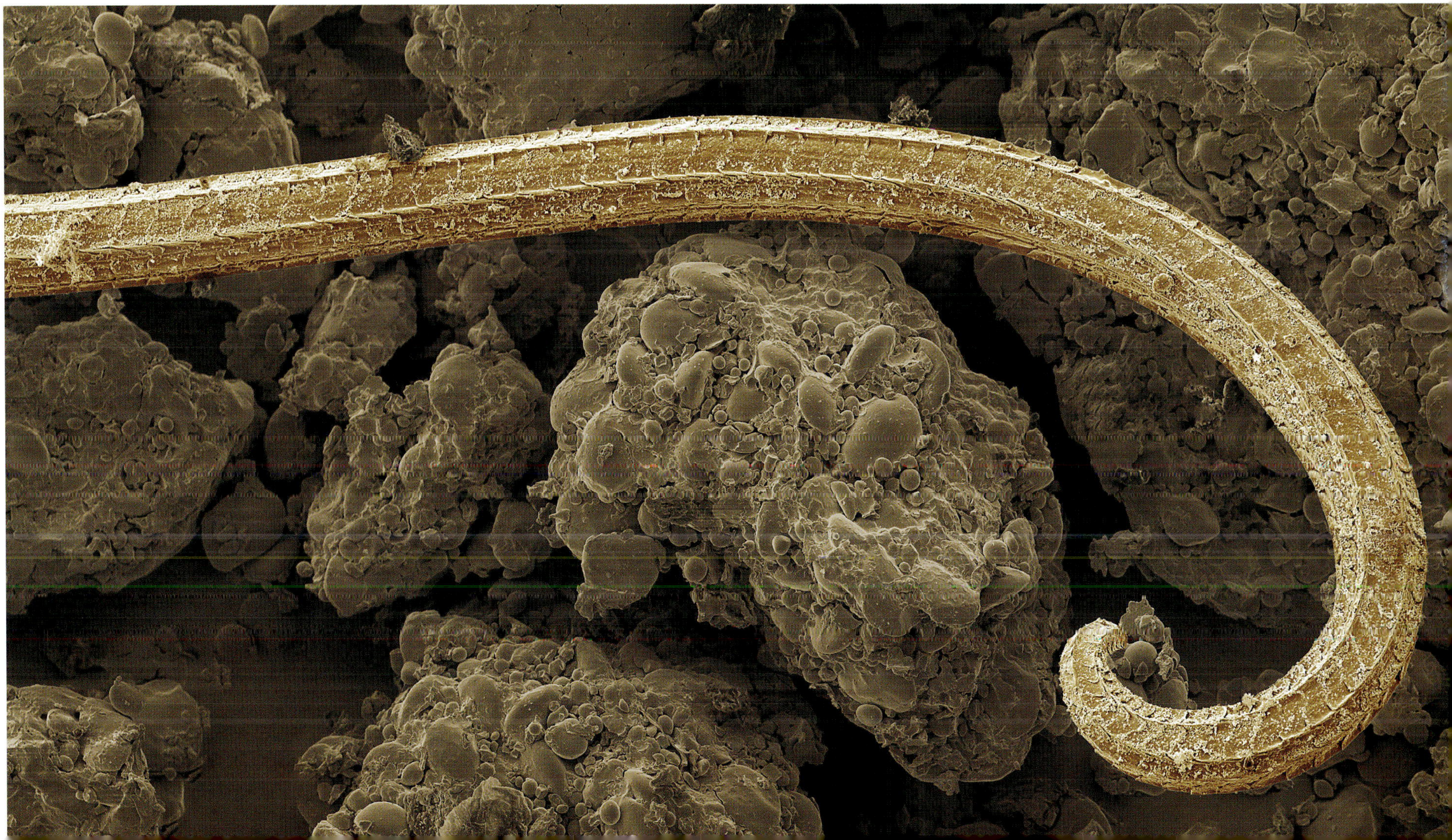

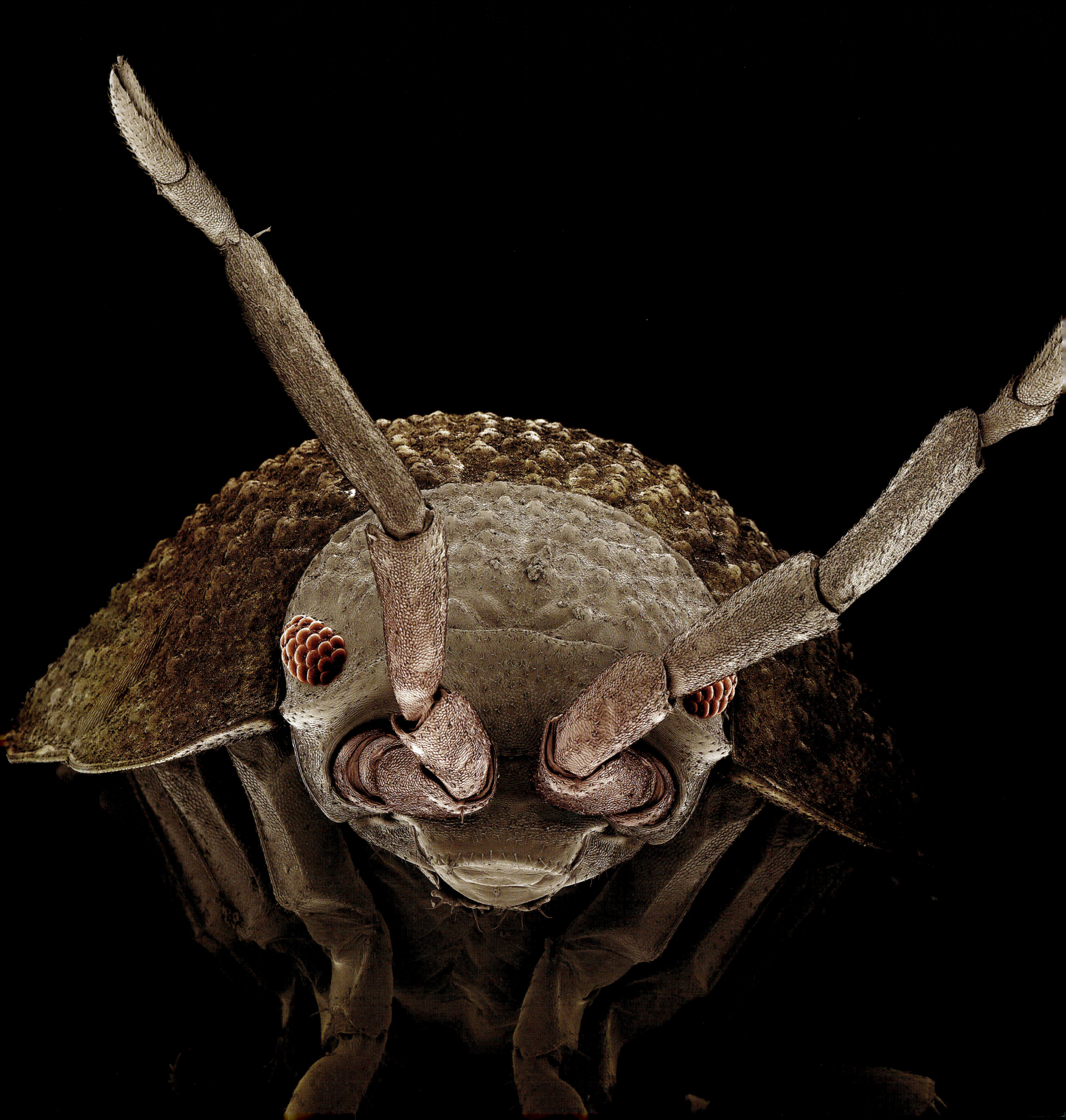

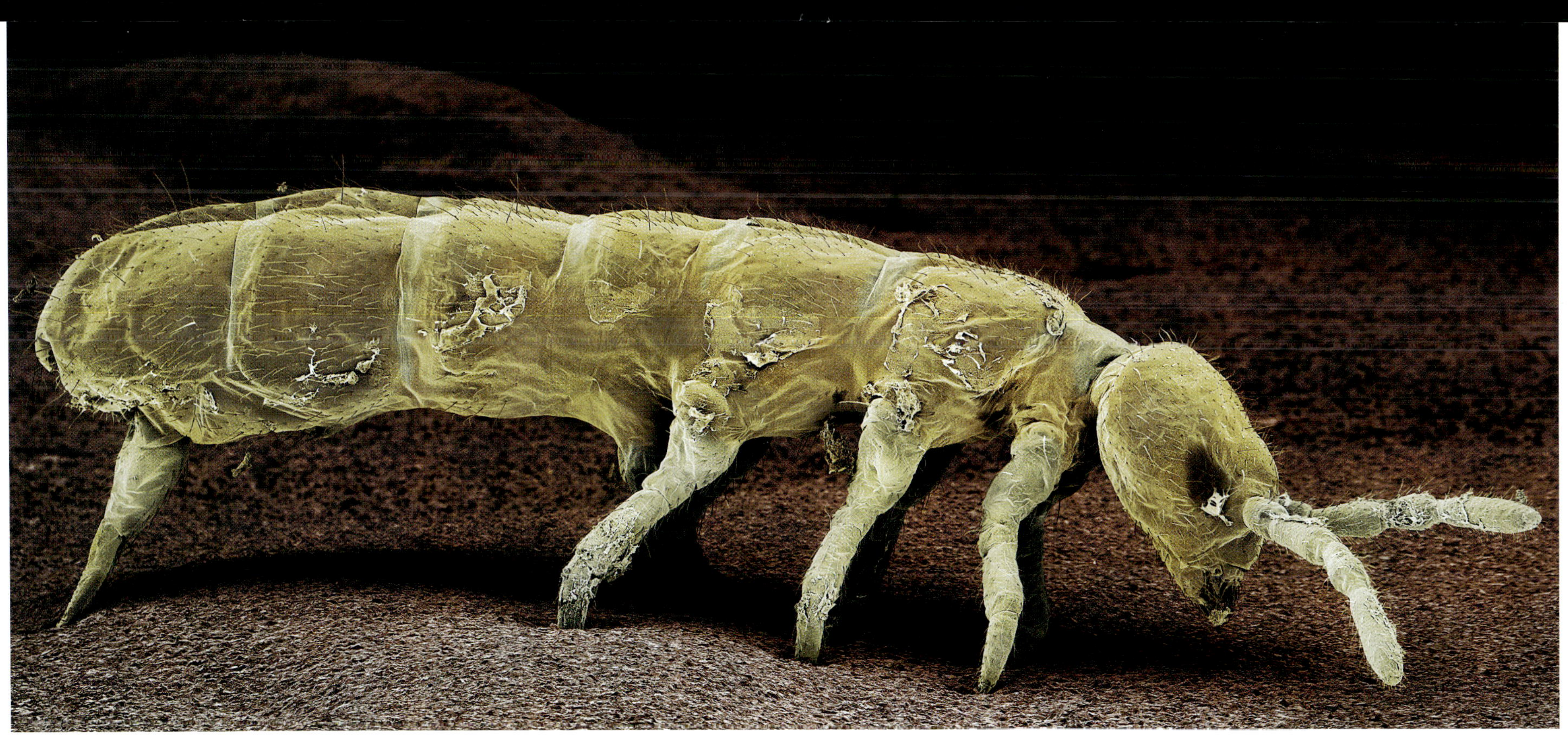

WOODLOUSE

There are 42 species of woodlice recorded from Britain. They are crustaceans, related to crabs and shrimps, but unlike most of their aquatic relatives, they have managed to move onto land, albeit to damp places. This was achieved by evolving protection for their gills to cut down water loss, and a waterproof outer body covering. They occur in moist places in many habitats and, in the garden, are frequently found under stones, bark or in leaf litter, feeding on decaying plant material. The common rough woodlouse *Porcellio scaber* can roll up into a ball if disturbed. It has a grey, segmented body, divided into three main regions, the head, the thorax ('pereion') with seven pairs of legs, and the abdomen ('pleon'). Courtship tends to take place at night. The male climbs on to the back of a receptive female and drums on her with his front legs before moving down, bending his body beneath hers to transfer his sperm to her genital openings. Females develop a 'brood pouch', a fluid-filled chamber underneath the body from which the young emerge when they are fully developed. They undergo a series of moults before reaching maturity, and adult woodlice continue to moult periodically.

SPRINGTAILS

Springtails inhabit garden compost heaps, feeding on rotting vegetation, fungal hyphae and general organic detritus. These tiny creatures, some as little as 0.18 mm long, are among the most abundant insects in the world and are found in vast numbers almost everywhere from the seashore to the tops of mountains. There are around 300 different species of springtails in Britain, an average square metre of soil yielding anywhere between 40,000 and 200,000 individuals. They normally creep about on their six legs, but when alarmed or in danger, they can spring into the air using a forked prong-like appendage, on the last abdominal segment called a furcula, which is normally folded out of the way underneath the animal. Some species are able to run across ponds and pools as they have strongly water repellent surfaces, and they are quite adept at climbing, aided by a ventral tube that protrudes from the body, which acts as an adhesive pad. Some springtails also use these tubes for grooming. *Deuterosminthurus pallipes*, a very common species in the south of England, is about 1 mm in length and exhibits elaborate courtship behaviour in which the male dances and head-butts the female in order to entice her to mate.

Common Rough Woodlouse,
Porcellio scaber (left)

Unlike many species of woodlouse that require moisture to survive, *Porcellio scaber* is quite well adapted to drier conditions because it possesses pseudotrachea, breathing tubes that enable it to breathe dry air. It feeds on rotting vegetation.

Magnification: x85
Scanning Electron Micrograph

Springtail,
Order Collembola (above)

Springtails are tiny, wingless insects commonly found on compost heaps, and any other damp place, where they feed on succulent plant material or decaying organic matter. They normally creep about on their six legs, but when disturbed, they can spring into the air using a forked prong-like appendage on the last abdominal segment, called a furcula, which is normally folded out of the way underneath the animal.

Magnification: x170
Scanning Electron Micrograph

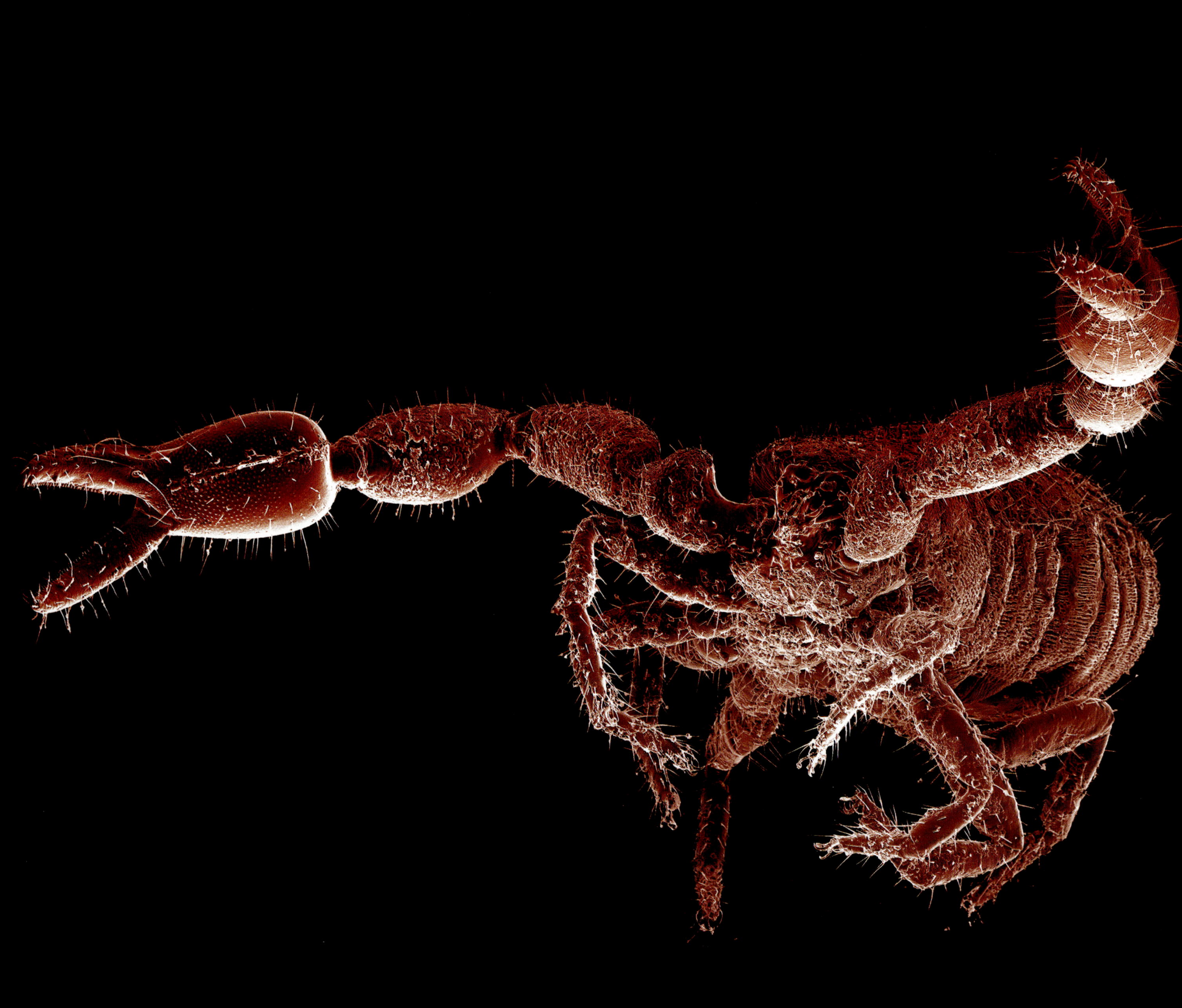

FALSE SCORPION

False scorpions, or pseudoscorpions, are related to true scorpions, spiders, ticks and mites. Widespread in the tropics, there are about 70 species in Europe, of which as many as 20 occur in Britain. They look a little like true scorpions, having large pincers for catching prey, but they do not have a stinging tail and are completely harmless. Most species, like *Chelifer sp.*, 4 mm long, live outside under stones or loose tree bark, and in leaf litter, moss and other debris, but a few, such as the tiny *Chiridium sp.*, just 2 mm long, favour a more sheltered environment and are often found in houses, where they may be seen among undisturbed books or papers. Carnivorous, they feed on tiny insects, worms or other small creatures. They usually crawl slowly, waving their pincers about in front, but they can move swiftly if disturbed. Silk-glands open on the chelicerae, or jaws, that are used, not to spin webs for catching prey like spiders, but to create silk shelters during egg production or hibernation; females remain in these retreats for the entire period of egg development, until the young hatch.

CENTIPEDE

In many parts of the world, people fear centipedes, those long-bodied, fast runners with lots of legs. The bites from some large species can be painful and their venom has been known to kill, but in Britain there are no dangerous ones. 'Centipede' literally means 'hundred-footed', though most have far fewer feet. They are sometimes confused with their many-legged relatives the millipedes, which have two pairs of legs on most body segments whereas centipedes only have one. Millipedes are also rather slow-moving vegetarians, compared to their fast moving, carnivorous cousins. Despite the poisonous bite of some species, centipedes on the whole can be regarded as largely beneficial, particularly around the garden where they often prey on root-feeding insect grubs. Most often seen in Britain is the common centipede, *Lithobius forficatus*, a glossy brown athletic creature with fifteen pairs of legs, which grows up to 3 cm long. Common in gardens, it hides during the day in dark, damp places where there is little disturbance, under logs, stones, in piles of leaves, or inside garden sheds. Its head carries a pair of long, sensitive antennae and, in addition to small chewing mouthparts, has powerful venomous claws for seizing and paralysing prey such as beetle and fly larvae, spiders and worms. The claws are modified legs, which arise from the first body segment and not from the head, like ant or beetle jaws. They can deliver a nip if handled carelessly. Another common garden inhabitant, especially in damp soil and leaf litter on the compost heap, is the garden, or yellow snake, centipede, *Geophilus sp*, which has a very thin, elongated body up to 7 cm long, with around 60 pairs of short legs.

There is one more species that occasionally enters our homes. *Scutigera coleoptrarta* is really a native of southern Europe and, although established in the Channel Islands, it occurs only spasmodically in Britain, probably arriving from warmer climes hidden in imported fruit and vegetables and carried into our homes with the groceries. Up to 4 cm long, *Scutigera* has 15 pairs of legs, some of which are very long and slender, in fact, the hind legs of the female are more than twice the length of her body. This centipede is a very fast runner and darts after flies and other insect prey with amazing speed. It is venomous and there are a few records of this species biting people, an experience equivalent to a bee sting.

EARWIG

The common earwig in Britain is *Forficula auricularia*, easily recognised by its long, reddish brown body up about 1.5 cm in length, and a large pair of pincers on the tip of the abdomen, used to deliver a nip if threatened. The belief that they enter the ear to burrow into the brain is folklore, but they do seek dark places to hide in. They are most active at night, emerging from their refuges under stones or wood piles to feed on rotting vegetation, small insects, fruit and flower petals. The female is unusual for a non-social insect in caring for its young, defending the eggs against predators through the winter, keeping them clean and free of mould, then feeding and protecting the nymphs after hatching until they are large enough to fend for themselves.

False Scorpion or Pseudoscorpion,
Lamprochernes nodosus (left)

Related to true scorpions, these small arachnids are harmless to man, lacking a stinging tail. They feed on tiny insects, worms and other small soil creatures.

Magnification: x100   *Scanning Electron Micrograph*

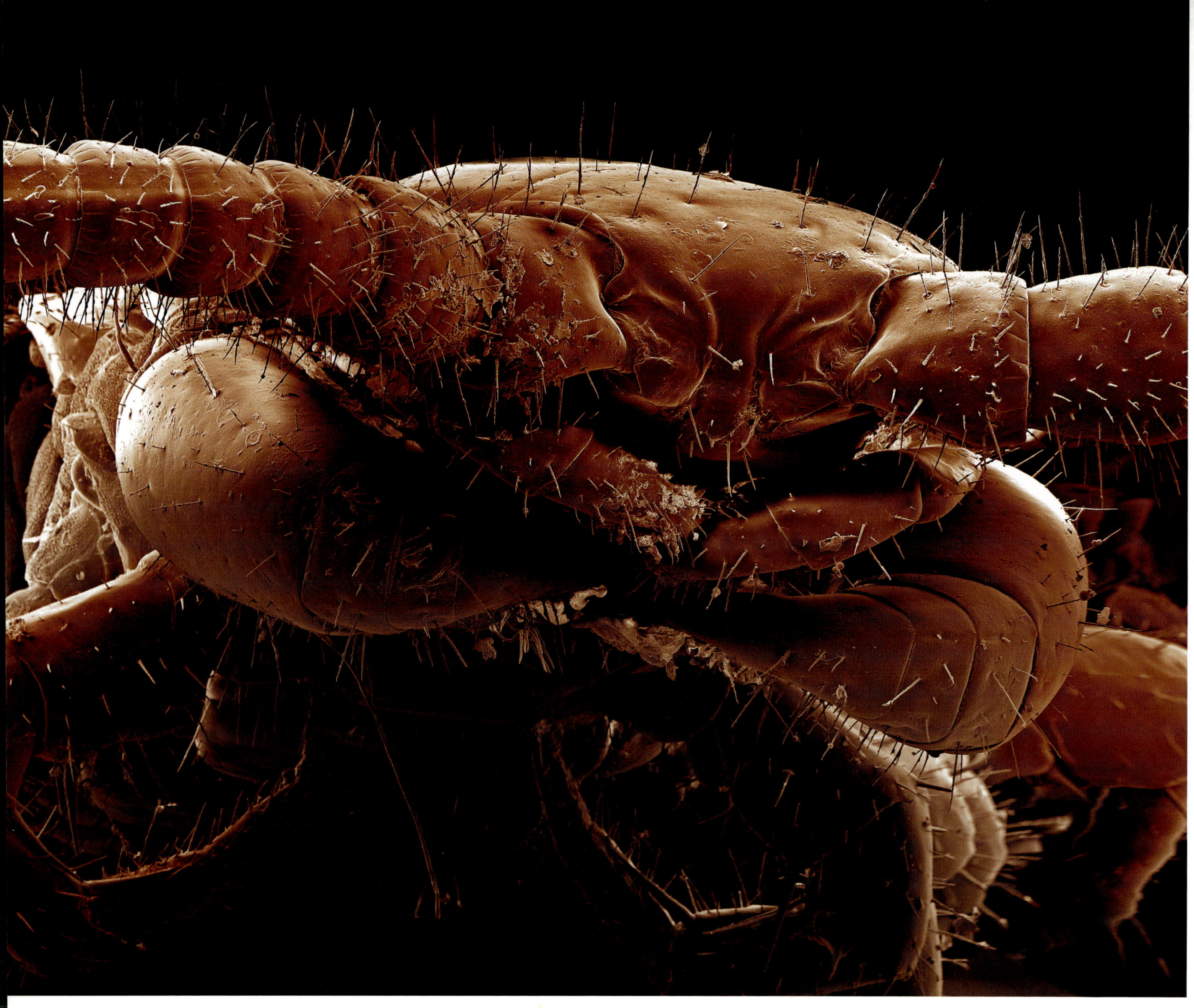

Common Centipede, *Lithobius forficatus*

'Centipede' literally means 'hundred-footed', though most have far fewer feet. Most often seen in Britain is the common centipede, *Lithobius forficatus*, a glossy brown athletic creature with fifteen pairs of legs. Its head carries a pair of long, sensitive antennae and, in addition to small chewing mouthparts, has powerful venomous claws, which are modified legs, for seizing and paralysing prey such as beetle and fly larvae, spiders and worms.

Magnification: x70   *Scanning Electron Micrograph* (above)
Magnification: x25   *Scanning Electron Micrograph* (right)

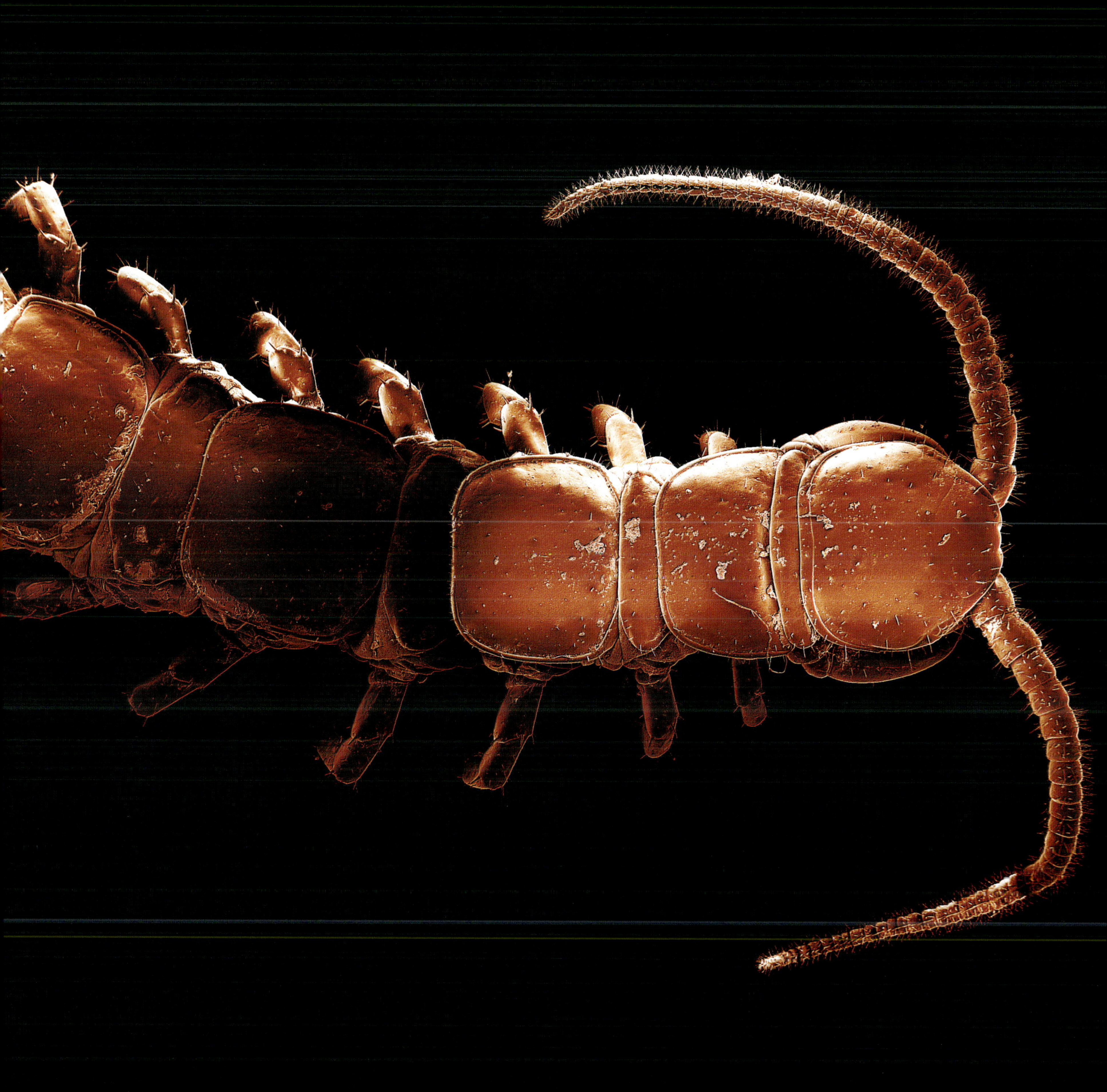

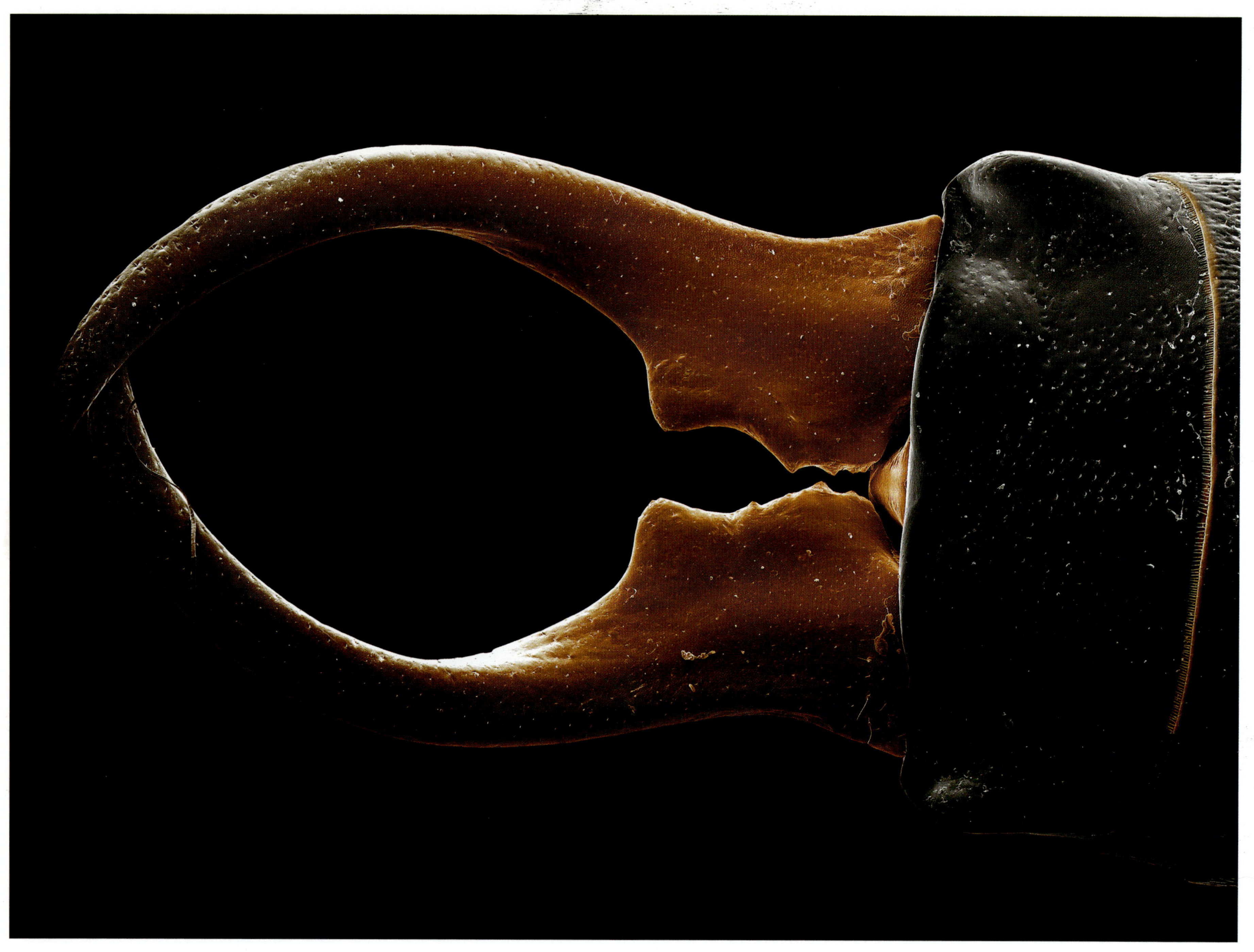

Earwig, *Forficula auricularia* (above, right and page 5)

The curved pincers on the tip of the abdomen belong to a male earwig; those of the female are almost straight. They frequent damp dark places, feeding on decaying organic matter and occasionally small insects.

Magnification: x75   *Scanning Electron Micrograph* (above)
Magnification: x100   *Scanning Electron Micrograph* (right)
Magnification: x20   *Scanning Electron Micrograph* (page 5)

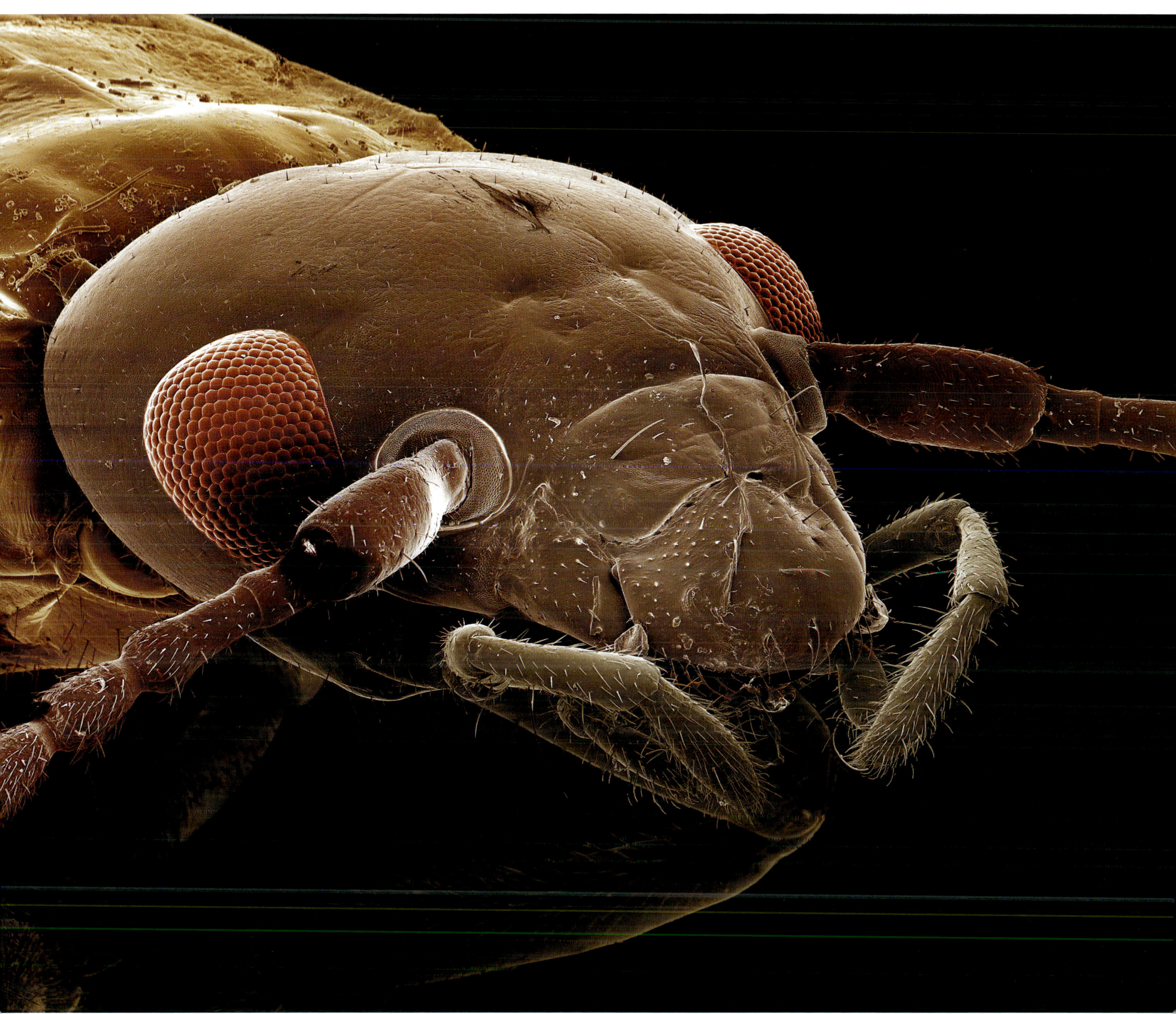

THE POND

A pond adds a new dimension to a garden, changing the landscape and providing a soothing vista, a place to relax and ease the spirit. It is also an important resource for wildlife, home to a wide variety of plants, invertebrates, amphibians, birds and mammals, with an entire ecology of its own. Natural ponds and lakes become established over a long period of time, but in a garden, a pond starts its life as an essentially artificial site designed to mimic nature in a controlled way, with plants and animals that we introduce. Over time, however, a huge variety of plants and animals will miraculously arrive on their own by natural dispersal methods, such as flight, wind, or on the feathers of birds, helping gradually to establish a balanced ecosystem. Every organism that lives and dies in the pond will contribute to the delicate balance that is necessary to keep it healthy.

If something is out of balance though, a chain reaction occurs causing conditions to deteriorate quickly. Fish in high numbers, for example, can significantly undermine a developing community of smaller organisms, so their introduction early in a pond's life is not a good idea. Later, the presence of a few fish such as rudd or roach, which are surface feeders, helps to prevent floating plants like duckweed from choking the pond; and sticklebacks, which eat all sorts of invertebrates, are small enough to fit into the developing ecosystem without doing any damage. Larger fish, however, might deplete the invertebrate population too much, stirring up the mud, making the water cloudy, preventing sunlight from reaching underwater plants and leading to unwanted growth of algae.

A sudden increase in the density of algae can be catastrophic. It can be caused by external factors too, for example by the leakage of garden fertiliser into the pond. This leads to increased levels of nitrogen and phosphate, which will trigger an explosion of plant growth, with algae reproducing so fast they form a mass of green slime on the surface called an algal bloom. Unfortunately, this prevents sunlight from penetrating below the surface, without which underwater plants cannot produce oxygen; and without oxygen, animals can no longer respire. As a result, the pond quickly becomes stagnant, and effectively dies. This process of decline is called eutrophication; anaerobic bacteria that do not need oxygen take over, producing by-products such as hydrogen sulphide, making the water uninhabitable for most organisms.

ALGAE AND CYANOBACTERIA

At the lower end of the food chain, algae provide essential food for many species of invertebrates. Planktonic algae drift about in the water, although some species are mobile in a limited way, moving by means of whip-like flagellae. Mostly microscopic, many forms are single cells but some are colonial, or groups of single cells joined together, for example the spherical *Volvox*, which grows up to 2 mm across so is visible to the naked eye. Many, such as diatoms, display beautiful symmetry in form and possess delicately sculpted silica cases. Less than quarter of a millimetre in length, some of them can glide about slowly, but there is debate about how they move, one theory suggesting that a mucous-like substance is excreted at one end of the central slot and sucked in at the other to create a current. Desmids, a type of green alga, are also symmetrical, and consist of two joined semi-cells, which are mirror images of each other. *Micrasterias* is a desmid that is very common in ponds. Colonial algae sometimes grow as long, thin filaments, for example *Spirogyra,* whose identical cells are joined end to end. Cyanobacteria, or blue-green algae, are a mysterious group, originally considered to be algae but in fact are more closely related to bacteria. The cells of *Anabaena* are barrel-shaped, linked together like a string of beads and like some other cyanobacteria, produces neurotoxins that are harmful to animals or humans drinking contaminated water.

Cyanobacterium, *Anabaena sp.*

Formerly known as blue-green algae, cyanobacteria are a mysterious group of organisms closely related to bacteria. *Anabaena*, which is found in ponds, lakes and reservoirs, can suddenly multiply in hot weather, causing 'algal blooms'. It fixes nitrogen, converting atmospheric nitrogen to ammonia, and produces neurotoxins that are harmful to animals or humans drinking contaminated water.

Magnification: x3000   *Light Micrograph*

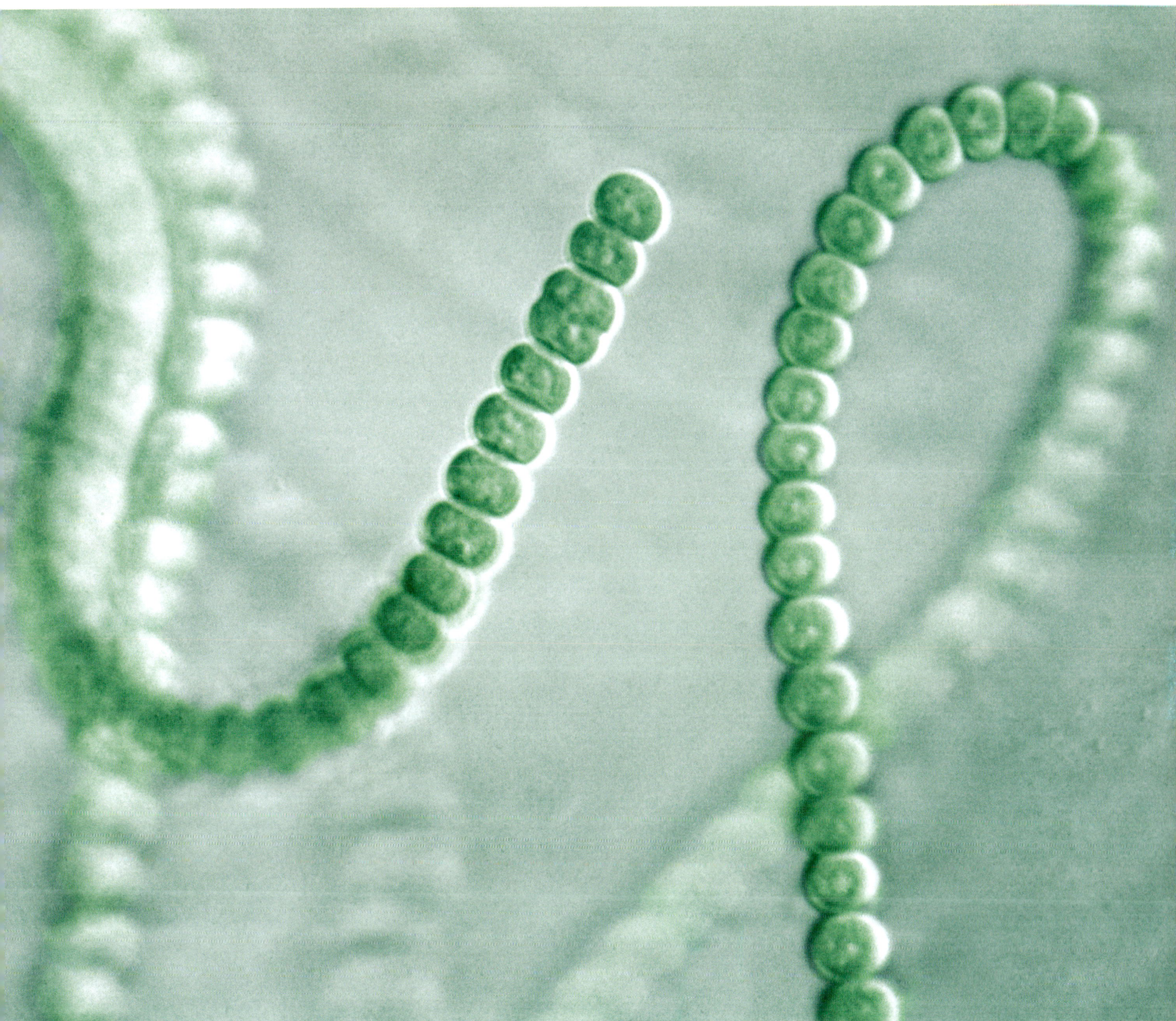

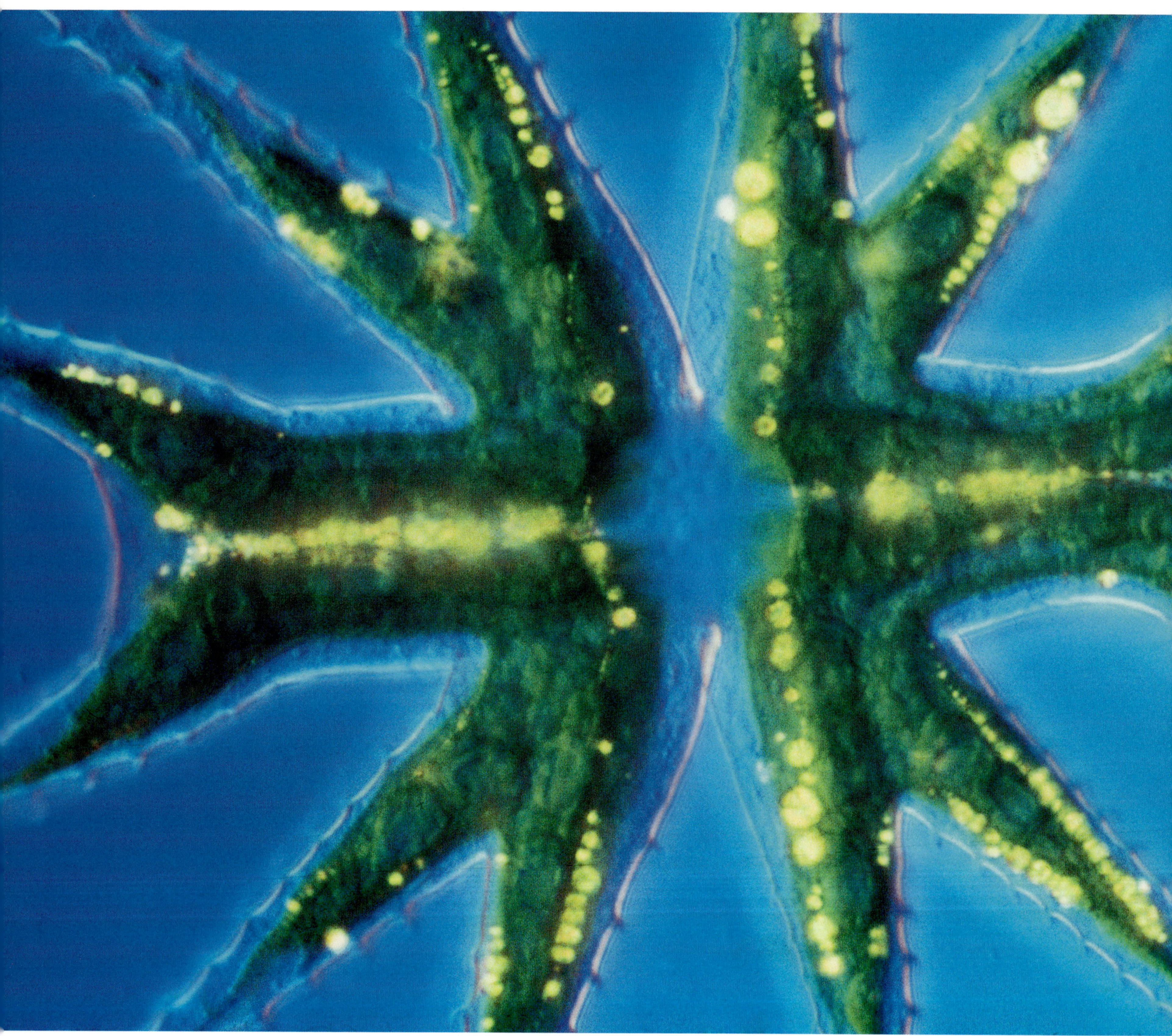

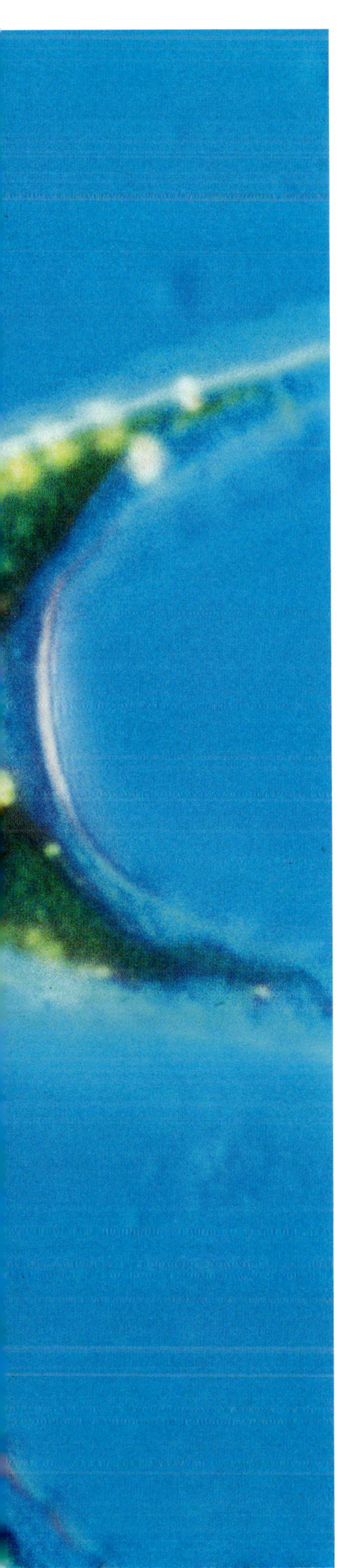

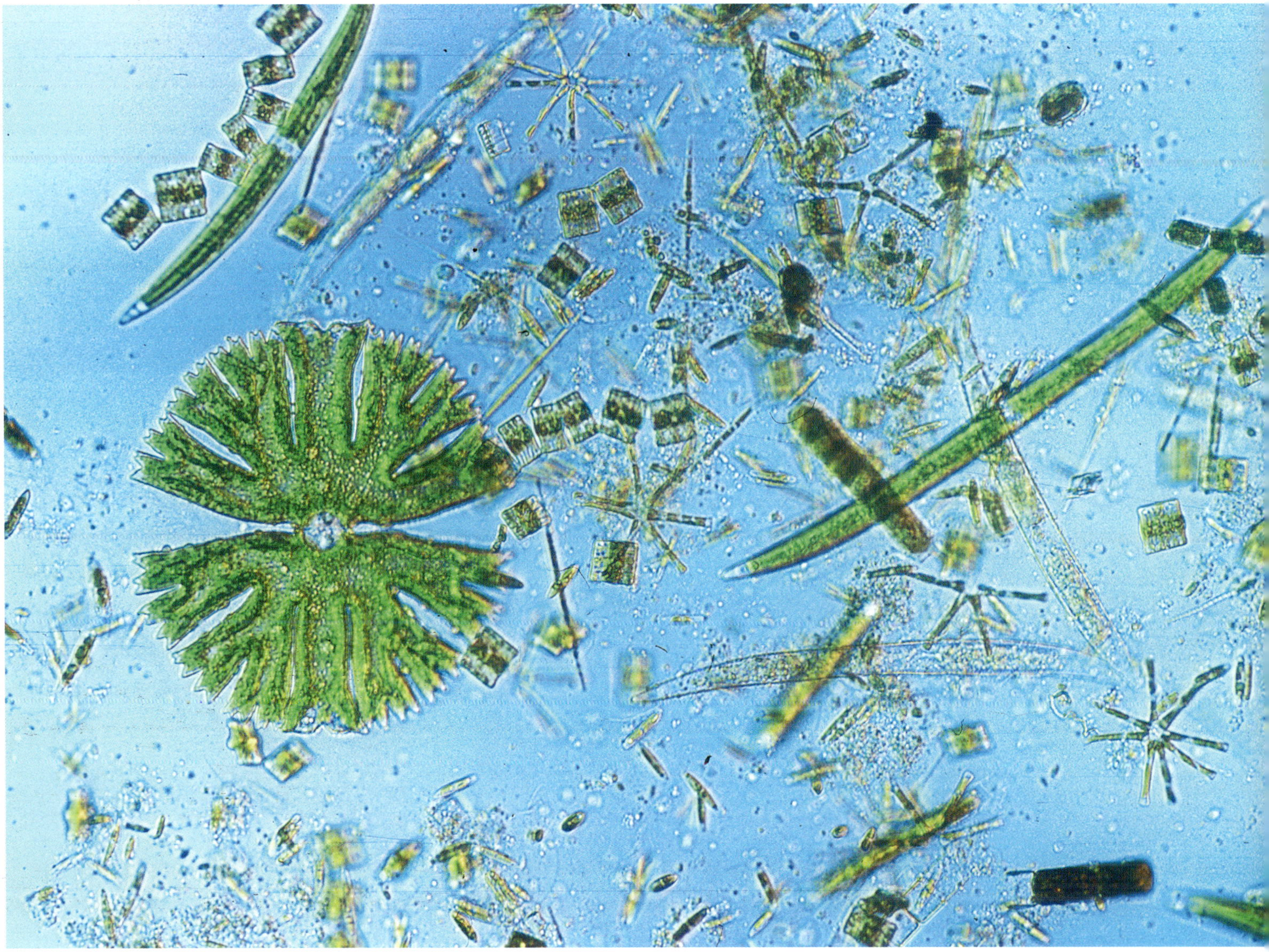

A selection of freshwater micro-organisms (left, above and overleaf)

Desmids, for example *Micrasterias sp.* (see left), are symmetrical green algae, consisting of two joined semi-cells that are mirror images of each other. Even a tiny sample of pond water contains a rich selection of micro-organisms (see above). Along with the desmid are several different species of algae and diatoms. (Overleaf) Left page: (above left) a filamentous blue-green alga on a bed of spiky diatoms; (centre left) a cylindrical diatom; (lower left) a dumbbell-shaped diatom; (above right) Desmid, *Micrasterias sp.*; (centre right) a magnified view of the same specimen; (lower right) a colony of diatoms, cells that stick together like ladders, on a bed of spiky diatoms. Right page: a diatom at higher magnification. Diatoms, like Desmids, show beautiful symmetry in form, and they are unusual in that they have a silica component in their rigid cell walls. They photosynthesise, possessing chlorophyll and can move slowly, supposedly by excreting a mucous-like substance at one end and drawing it in at the other.

Magnification: x3000   *Light Micrograph* (left)
Magnification: x160   *Light Micrograph* (above)

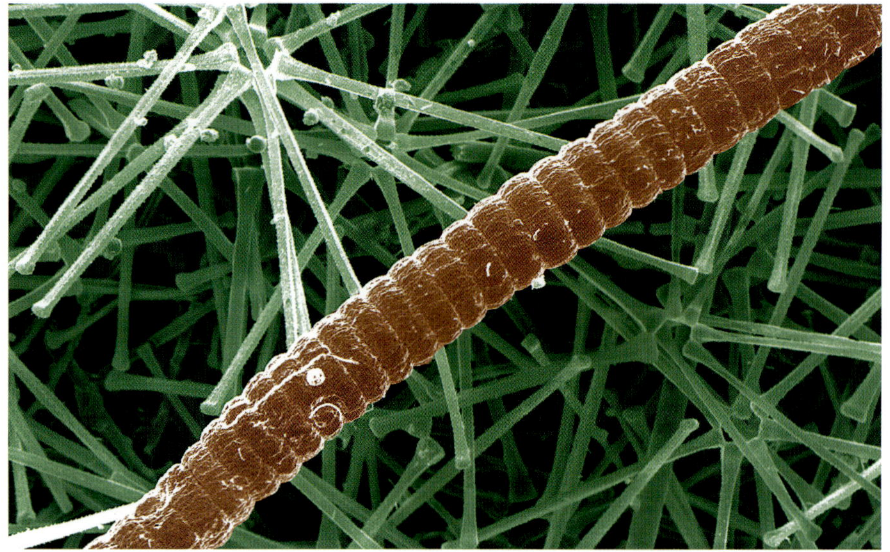

Magnification: x652   *Scanning Electron Micrograph*

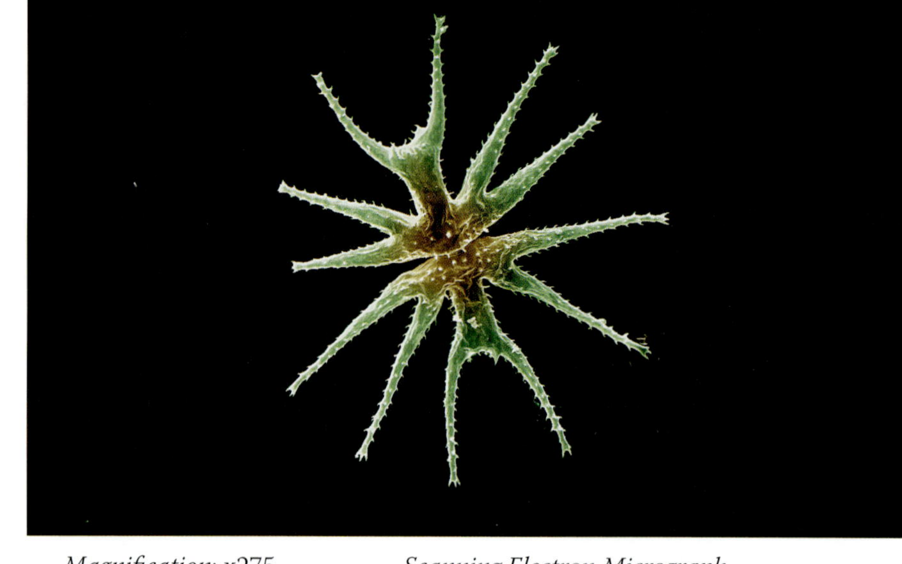

Magnification: x275   *Scanning Electron Micrograph*

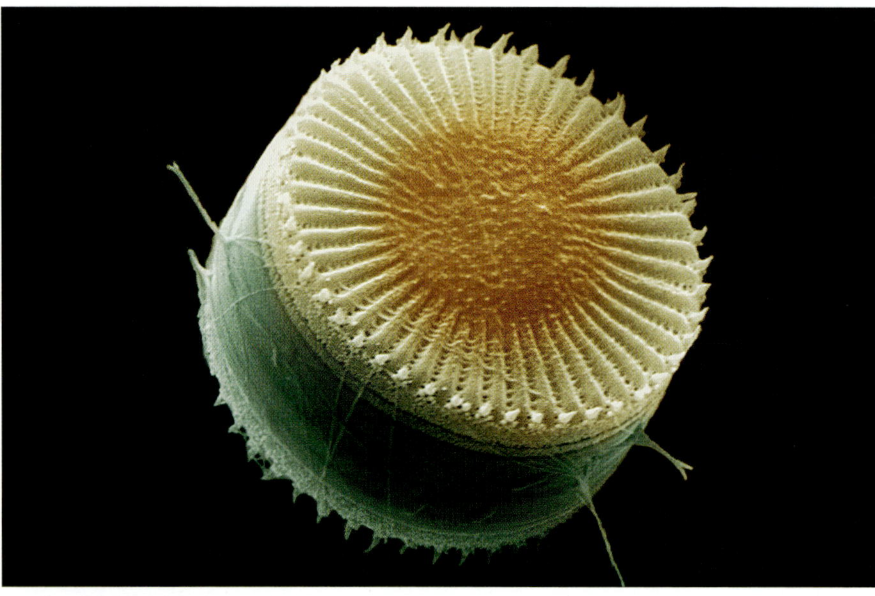

Magnification: x3140   *Scanning Electron Micrograph*

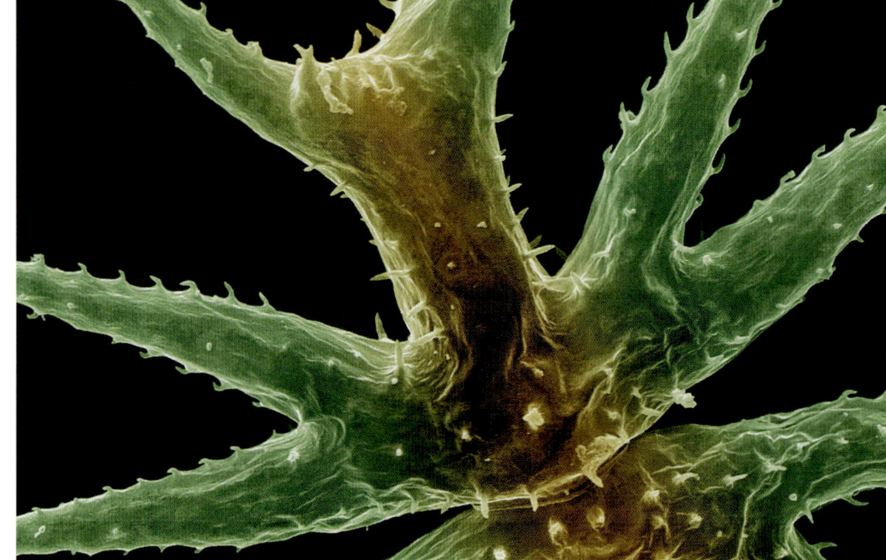

Magnification: x1120   *Scanning Electron Micrograph*

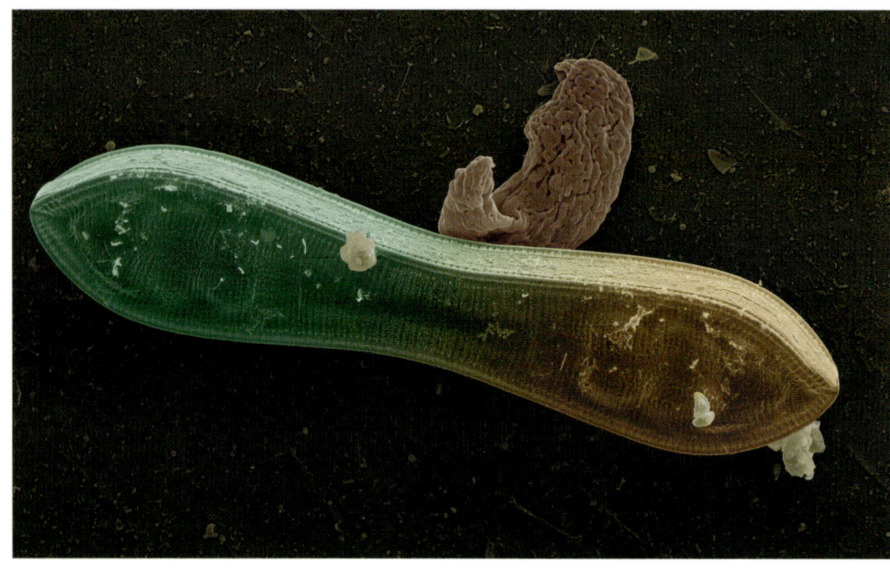

Magnification: x800   *Scanning Electron Micrograph*

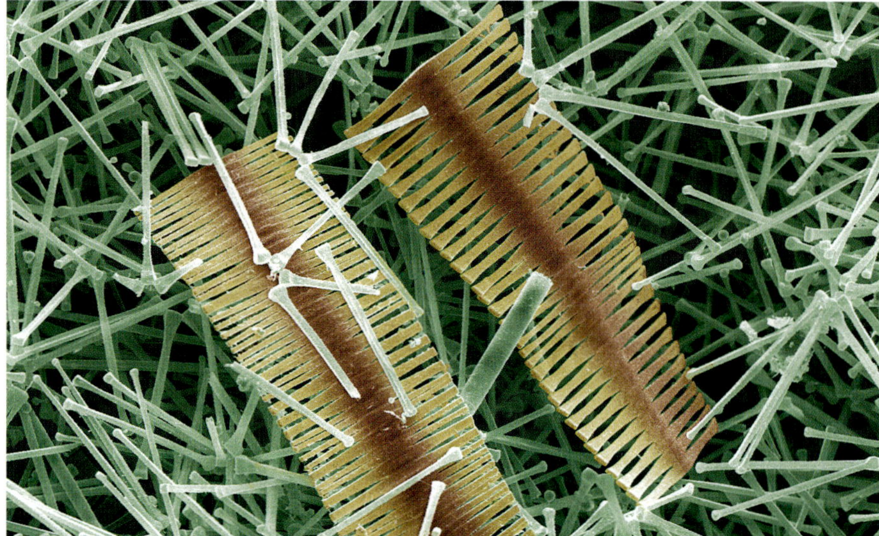

Magnification: x385   *Scanning Electron Micrograph*

178

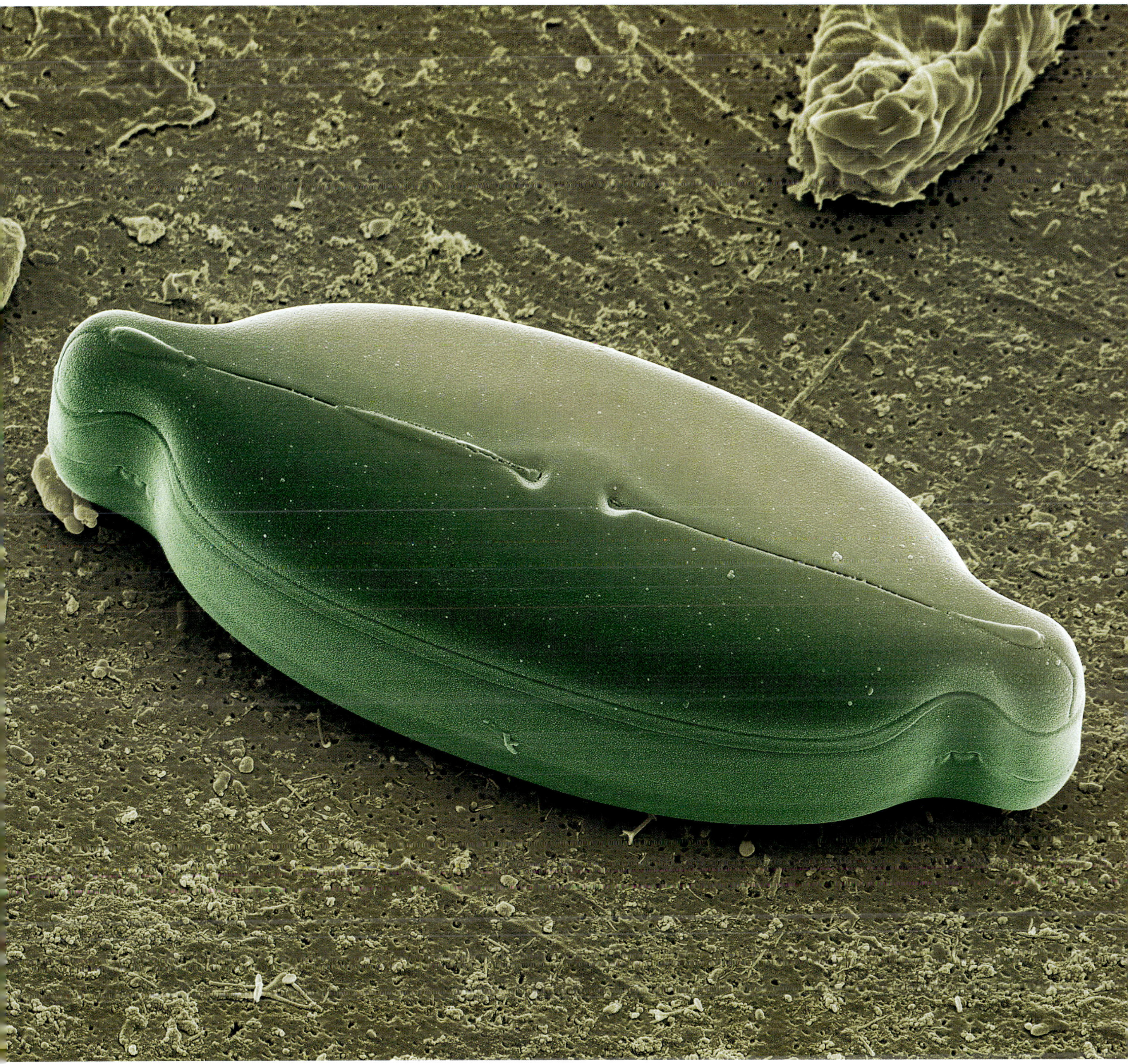

Magnification: x4500   *Scanning Electron Micrograph*

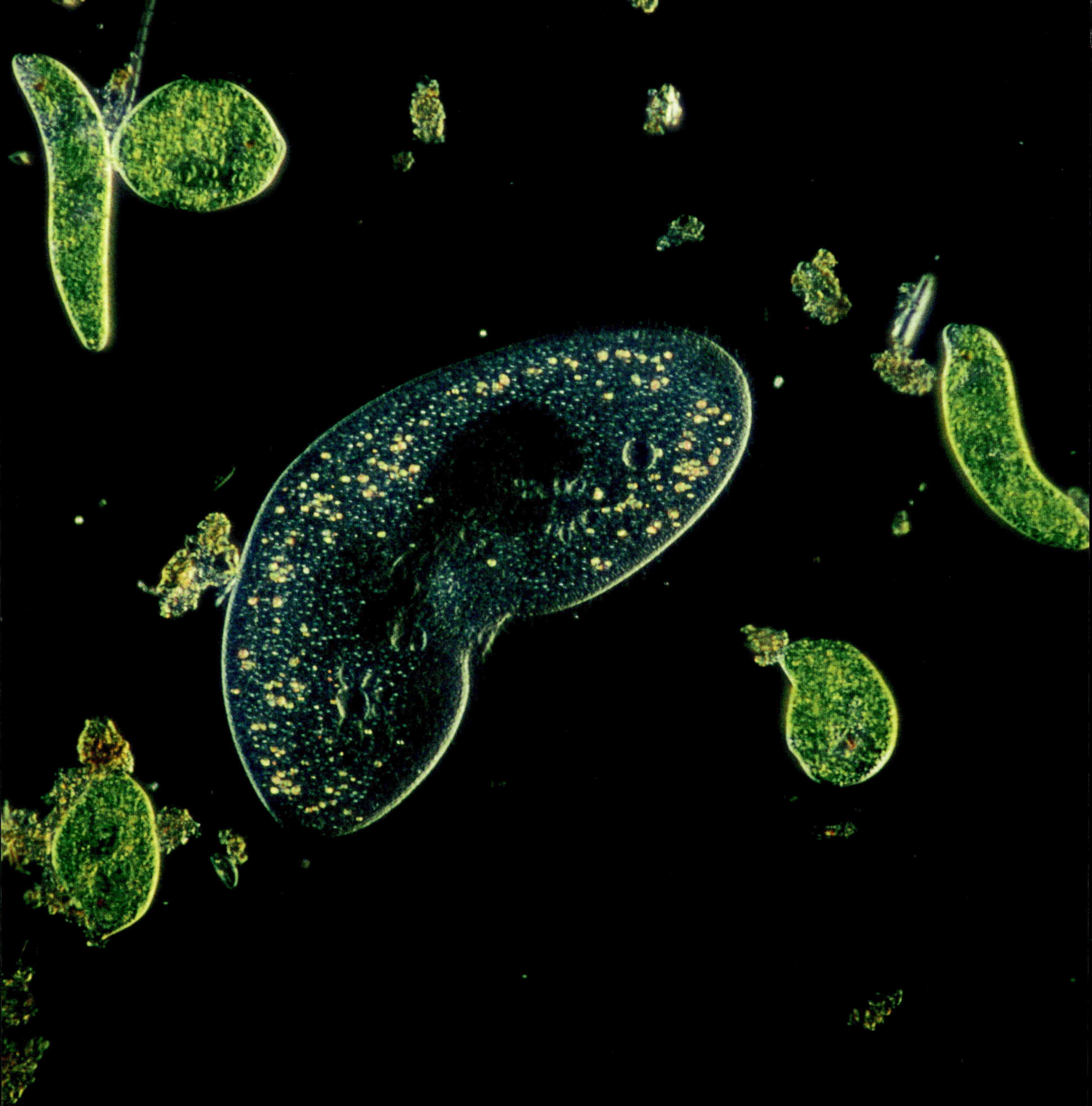

PROTOZOA: AMOEBA and CILIATES

Under healthy conditions, algae support a whole community of planktonic life, tiny creatures that drift about in the water, using a variety of techniques for swimming and gathering food. One of the simplest is *Amoeba*, a single-celled organism that moves by changing its shape with flowing body extensions called pseudopodia. It feeds by moving over and engulfing tiny water creatures, usually other single-celled animals and algae. While many amoebae spend their lives as unprotected blobs, some species called testate Amoebae, such as *Euglypha*, make a protective case for themselves out of tiny disks, probably of silica compounds, pushing pseudopodia out of a small opening at one end to move and for capturing food. Ciliates, such as *Paramecium*, *Frontonia*, *Colpoda* and *Spirostomum* are also single-celled protozoa that are more rigid in form and able to swim by means of spiral rows of small beating hairs, called cilia, that also draw food particles into a large gullet on one side of the cell wall.

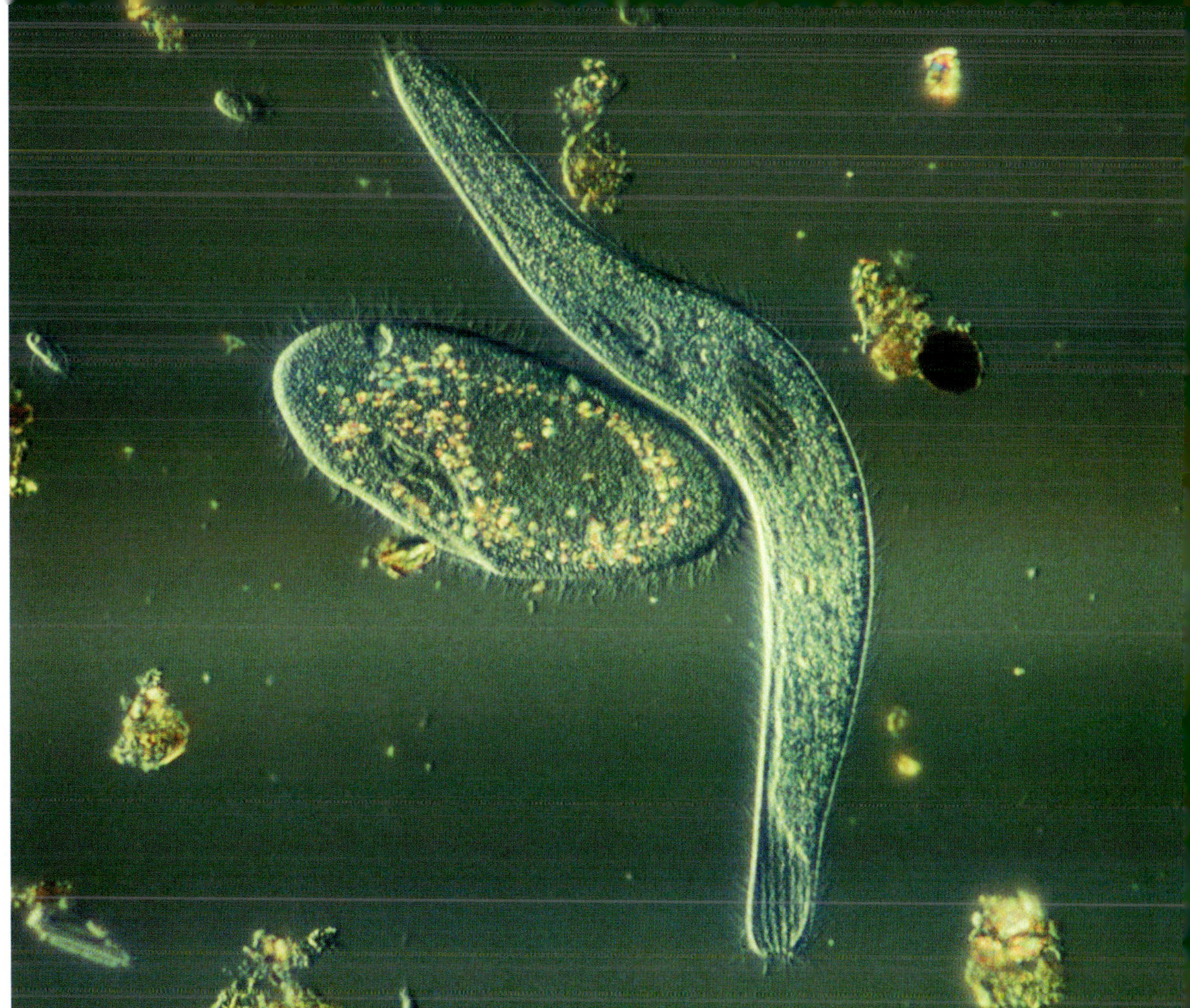

Ciliate Protozoa

Protozoans are single-celled organisms that have evolved a level of complexity in cell structure through the specialisation of different parts, or organelles, and a cytoskeleton, or inner proteinaceous layer that gives shape and support. Ciliates swim by beating hundreds of tiny surface hairs, or cilia, and use them for feeding by wafting small organisms into a gullet.

A kidney shaped ciliate, *Tillina sp.* (see left) is surrounded by several *Euglena sp.*, protozoans that swim by means of thrashing whip like flagellae and possess chlorophyll for photosynthesis. The elongate ciliate *Spirostomum teres* (above right) is accompanied by a *Paramecium bursaria* ciliate. A very large oval shaped ciliate, *Frontonia sp.* (below right), shows numerous green food vacuoles, containing the remains of the ciliate's recent meals.

Magnification: x850   *Light Micrograph* (left)
Magnification: x780   *Light Micrograph* (above right)
Magnification: x380   *Light Micrograph* (below right)

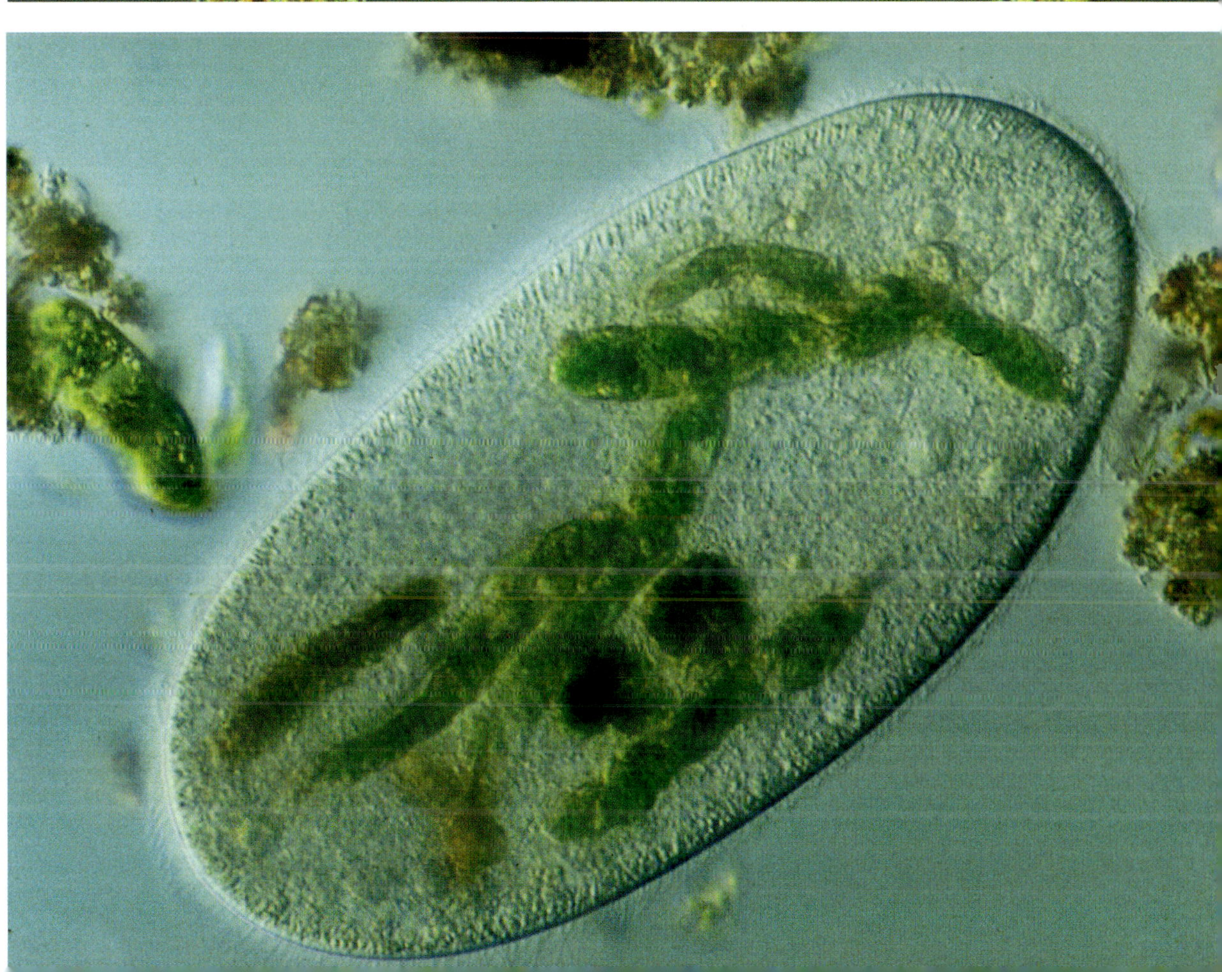

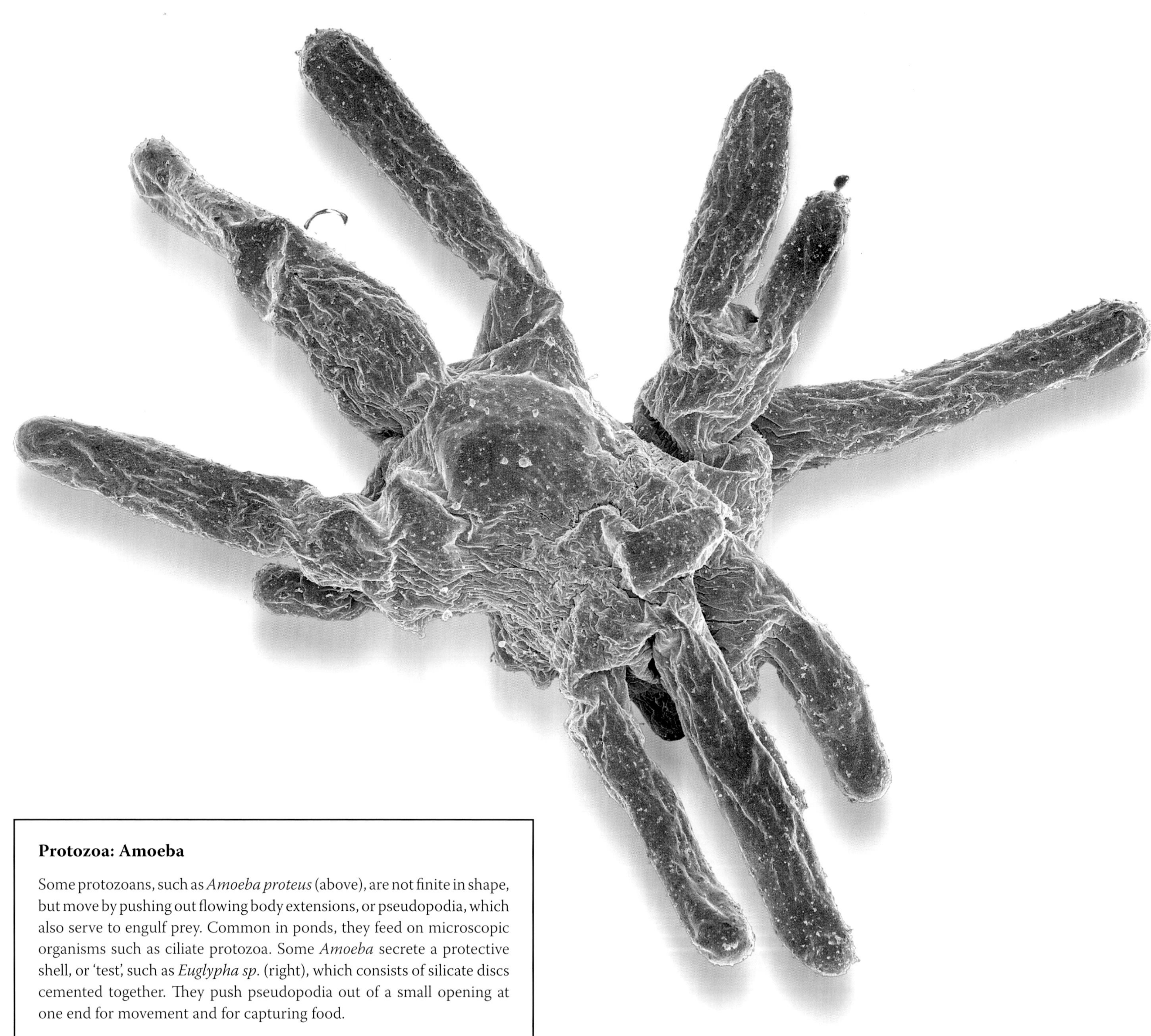

Protozoa: Amoeba

Some protozoans, such as *Amoeba proteus* (above), are not finite in shape, but move by pushing out flowing body extensions, or pseudopodia, which also serve to engulf prey. Common in ponds, they feed on microscopic organisms such as ciliate protozoa. Some *Amoeba* secrete a protective shell, or 'test', such as *Euglypha sp.* (right), which consists of silicate discs cemented together. They push pseudopodia out of a small opening at one end for movement and for capturing food.

Magnification: x1275   *Scanning Electron Micrograph* (above)
Magnification: x600   *Scanning Electron Micrograph* (right)

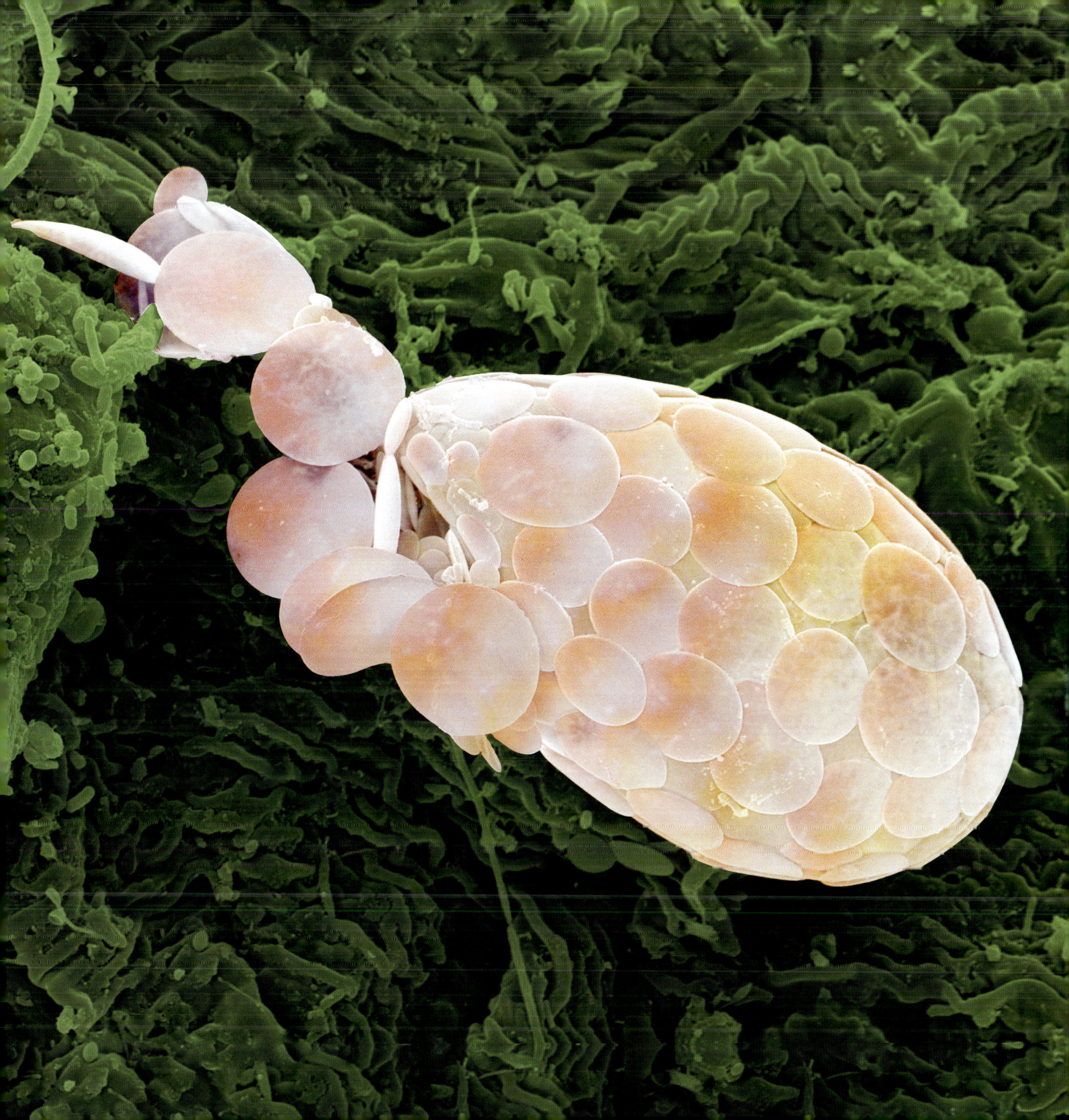

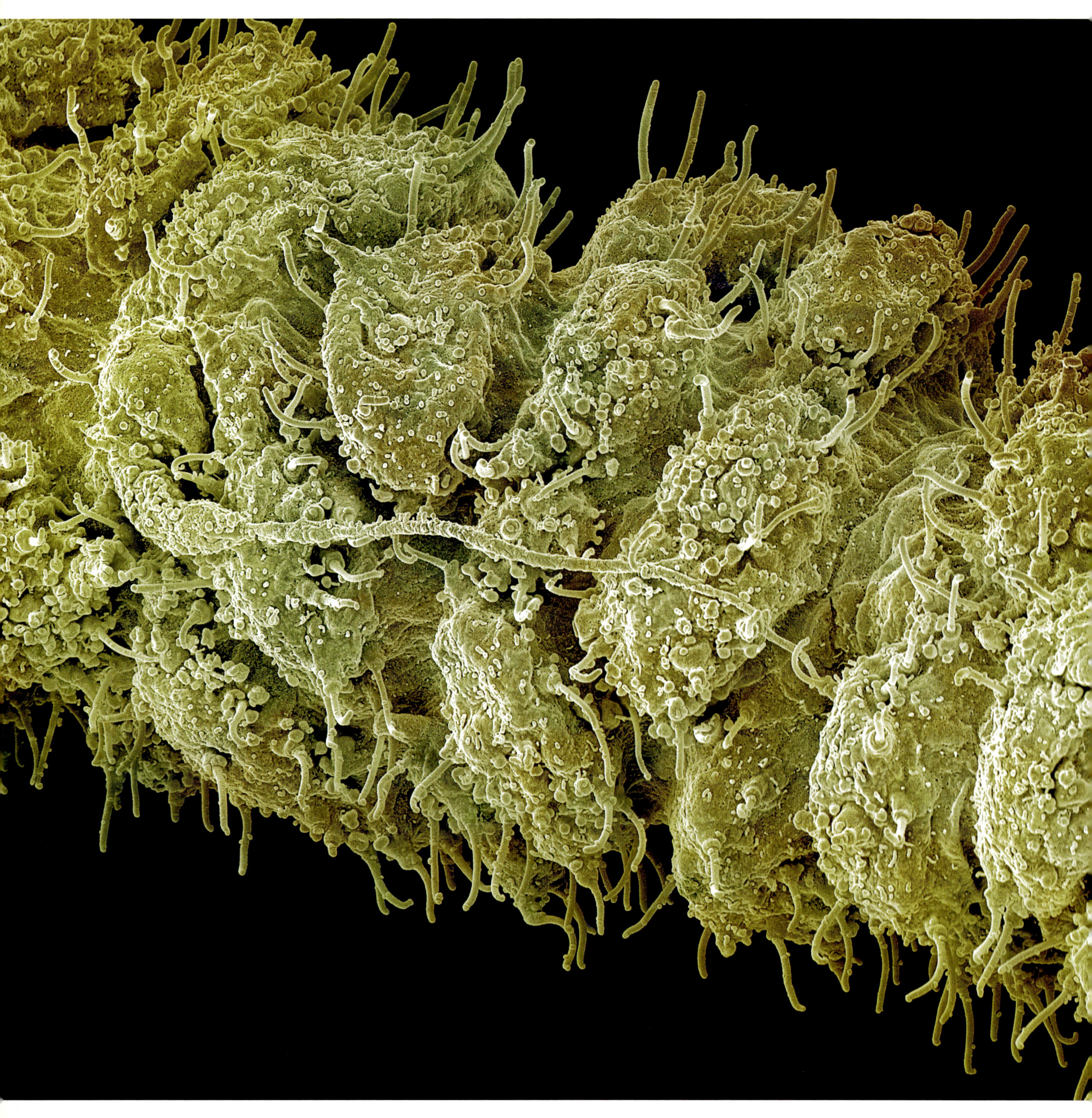

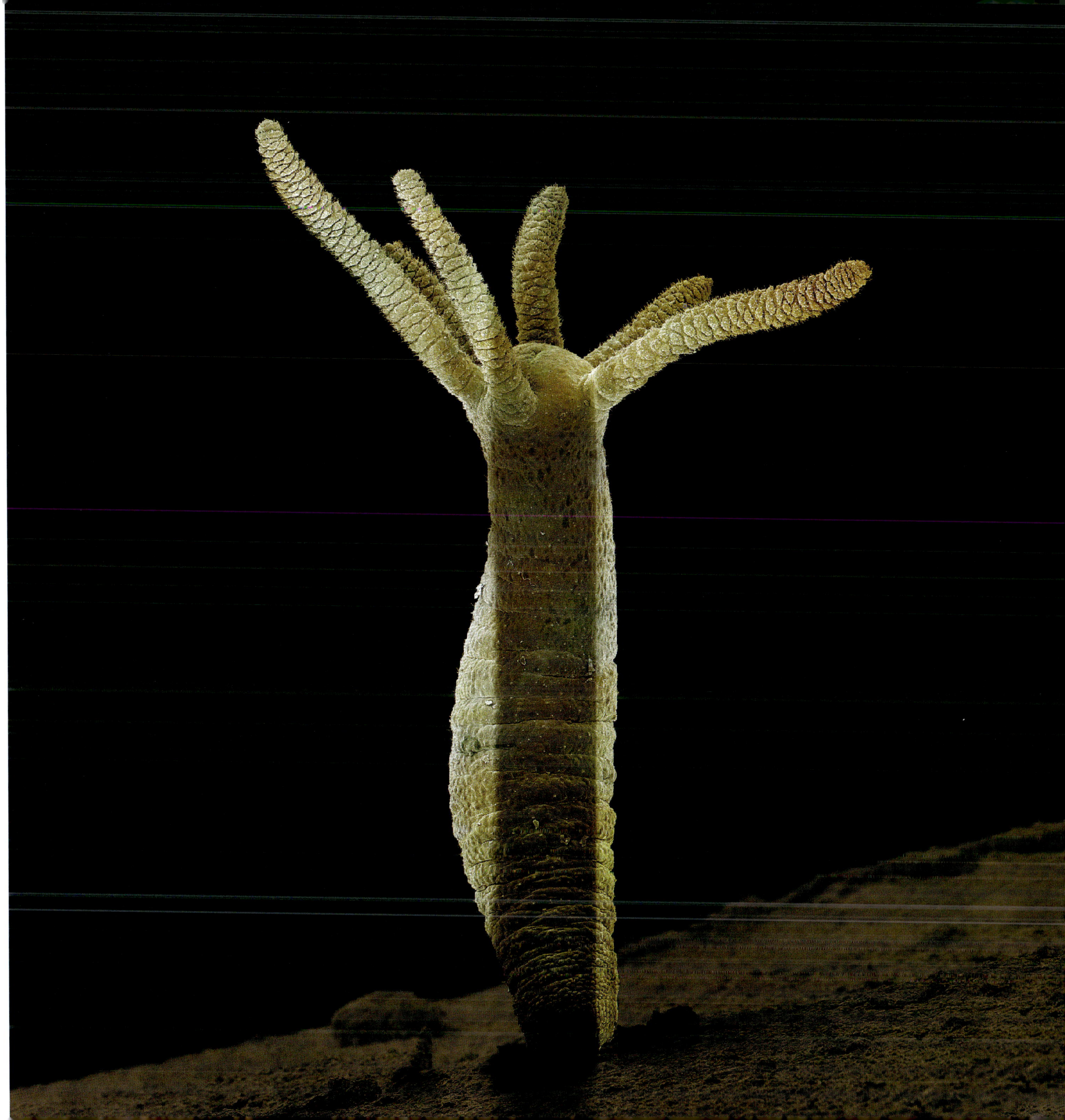

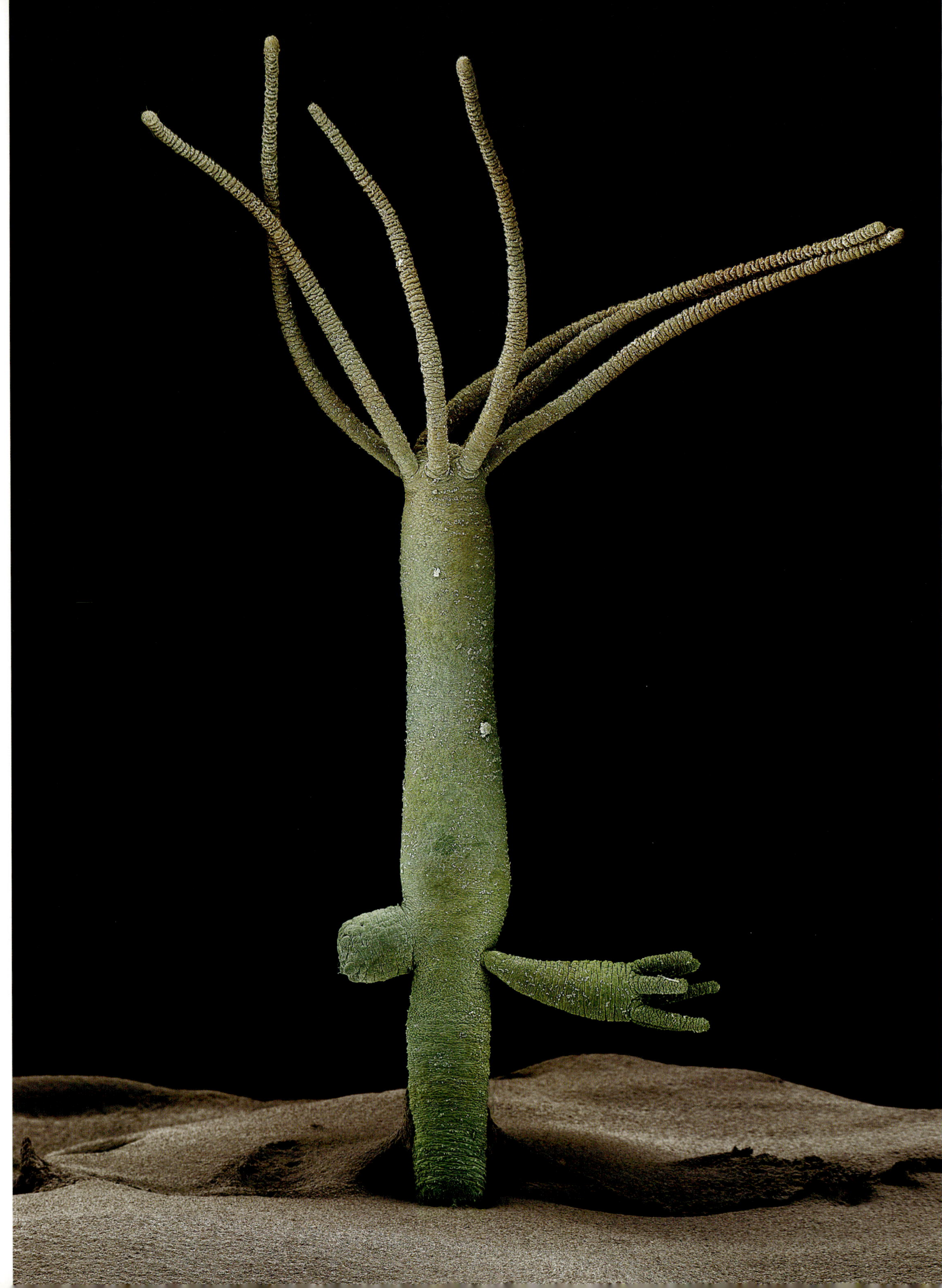

Hydra, *Hydra sp.*

Hydras are strange animals that may be found anchored to plant stems or rocks. They belong to the same group as jellyfish, corals and sea anemones, which evolved in the sea. Very few members of this group are found in freshwater since not many succeeded in evolving away from the marine environment. The genus *Hydra*, however, is quite a common inhabitant of ponds, and also occurs in slow-flowing streams and rivers. The body looks rather like a miniature tree (see previous page, right), a narrow trunk, varying between 1 mm and 2 cm in length depending on the species or its age, topped by a ring of tentacles, which are used to catch small organisms, preventing their escape by using stinging organs or nematocysts, which cover their surface (see previous page, left). Some of the nematocysts discharge a hollow thread that penetrates the prey, injecting a paralysing toxin, while some others are sticky, used to capture prey and for anchorage during locomotion. The tentacles transfer food to the mouth, which opens wide to swallow it. Movement is limited but is achieved by shuffling along or by cartwheeling, bending over and using the tentacles to hold on until the base is attached again in a new position. Hydras are primitive, their bodies made up of only two layers, each of them one cell thick. There are only a few different types of cells, which, if separated, are able to reorganise themselves. This ability was demonstrated in an early experiment when a number of Hydra were pushed through layers of muslin fine enough to fragment the bodies; they were then watched, as the loose cells reorganised themselves into *Hydra* again. Reproduction is usually by a process of 'budding' (see left and right), when an infant hydra simply grows from the side of the trunk of the adult and separates when mature. Sexual reproduction can occur however when conditions are poor and there is a shortage of food. Sex cells, both male and female, are released from the same adult into the water, where fertilization takes place, producing small ciliated larvae, which can either develop into new Hydra immediately or encyst until better conditions prevail.

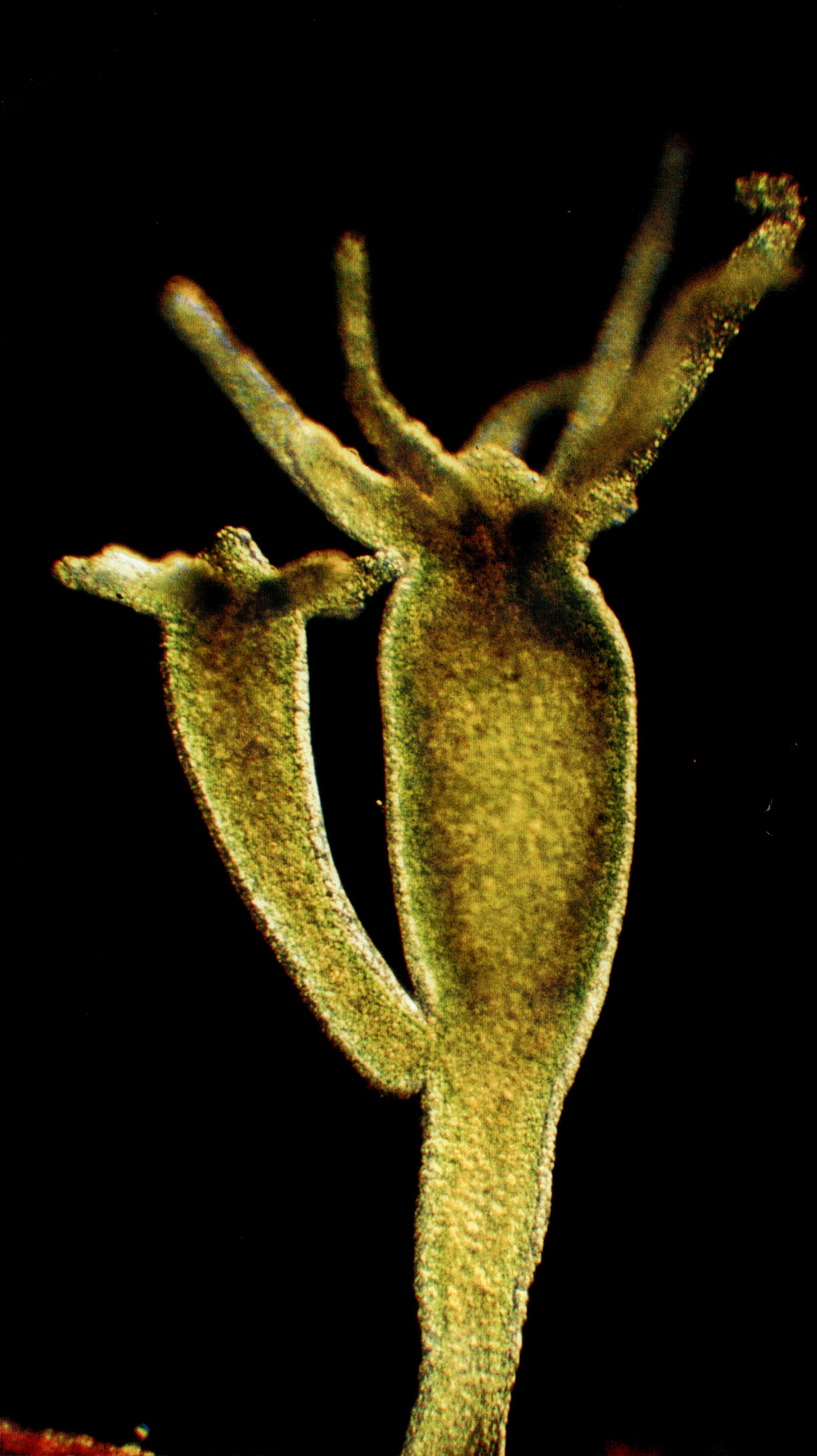

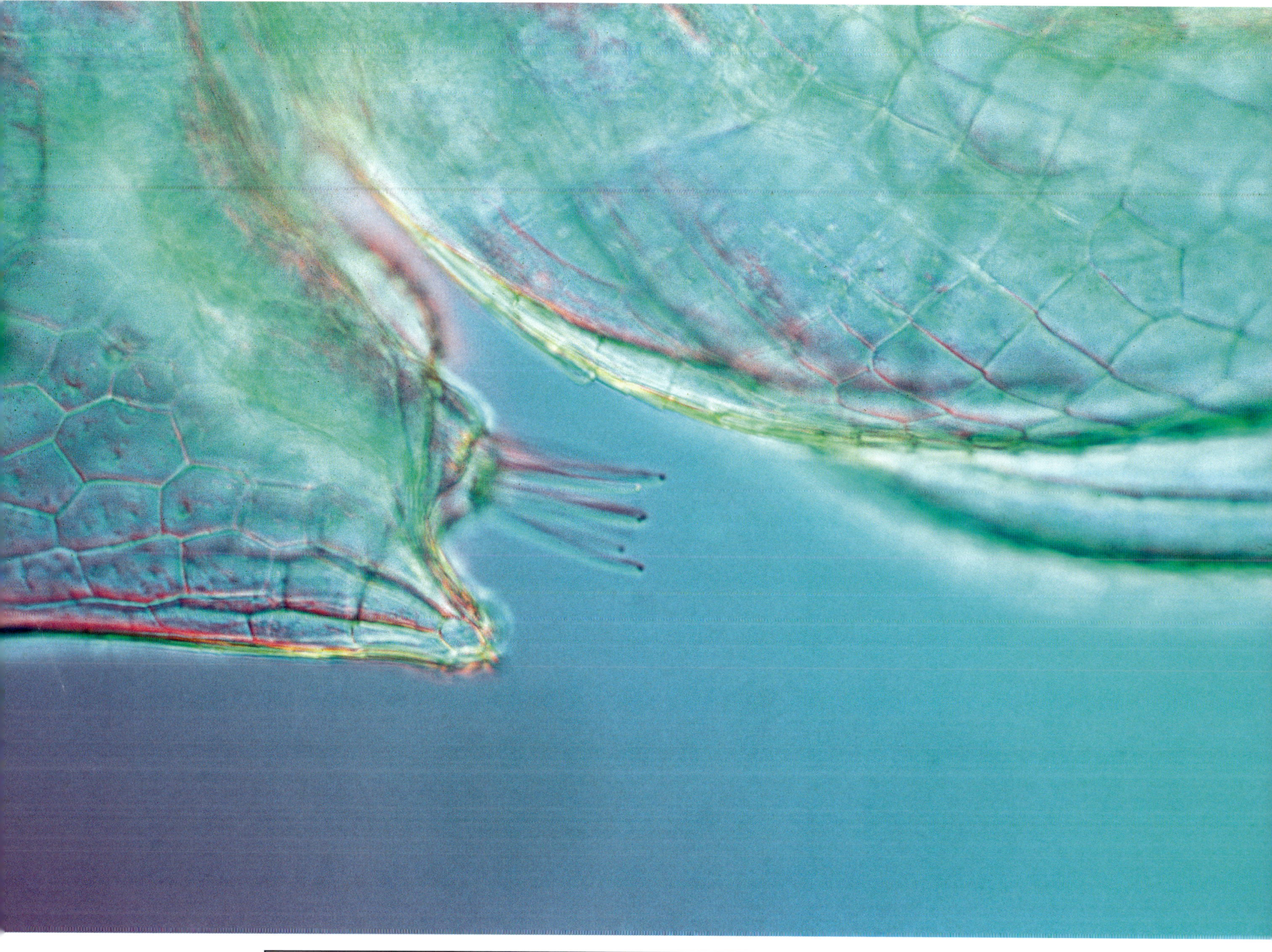

Water Flea, *Daphnia sp.*

A tiny crustacean that filter feeds in the plankton, *Daphnia sp.* uses its branched, paddle-like legs to enable a jerky swimming motion. During spring and summer, *Daphnia* reproduce parthenogenetically, the female carrying her young in a brood pouch (see right) until their release. The photograph above, of part of the exoskeleton of *Daphnia,* was taken using the Nomarski technique to show the individual cell margins.

Magnification: x1500   *Light Micrograph* (above)
Magnification: x100   *Light Micrograph* (right)

The water flea, *Daphnia sp.*, is a miniature crustacean, with a single eye, a little over 1 mm across that filter feeds in the plankton, and is very common in fresh water. Its paddle-like legs beat back and forth to enable a jerky swimming motion and aid in the collection of small algae, bacteria and fine detritus on which it feeds. The body case is essentially transparent so it is possible to see the beating heart and blood coursing around the body, as well as any eggs being carried by the females. During the summer, females reproduce parthenogenetically, the eggs developing without any need for fertilisation by males. Towards the end of the summer however, some of the eggs produce male *Daphnia*, which reproduce sexually with the females. The fertilised eggs survive the winter and hatch in the spring. In Britain, there are some 80 species of water fleas varying in size from less than 1 mm to about 5 mm across.

EXTRACTING OXYGEN

Before animals can exist in a pond, plants must be established to produce oxygen, the basic requirement for life. Aquatic animals have evolved a broad range of designs and methods to tap this precious resource. Some tiny animals, like *Daphnia* the water flea, have a huge surface area compared to their body bulk so they can absorb dissolved oxygen across their entire surface. Some insects have adapted their trachea, or breathing tubes, to act like gills, with extensions of the blood system that enable gaseous exchange. Dragonfly larvae do it by pumping water in and out of a highly vascularised rectum, a mechanism that also provides for emergency jet-propelled escape when needed. Fish have gills, although carp also visit the surface to swallow air, and then extract oxygen from it through their gut wall. Pond snails possess lungs, and they can exchange gases through their skin, as can frogs and newts, a useful adaptation making it possible to remain underwater for long periods before returning to the surface to gulp air. In their developmental stages, amphibians employ a variety of methods. As young tadpoles, they have external gills at first that look like feathery extensions, which become internal as they grow older, then, as they metamorphose into adults, they also develop lungs, making it possible to live indefinitely out of the water, only returning in order to breed.

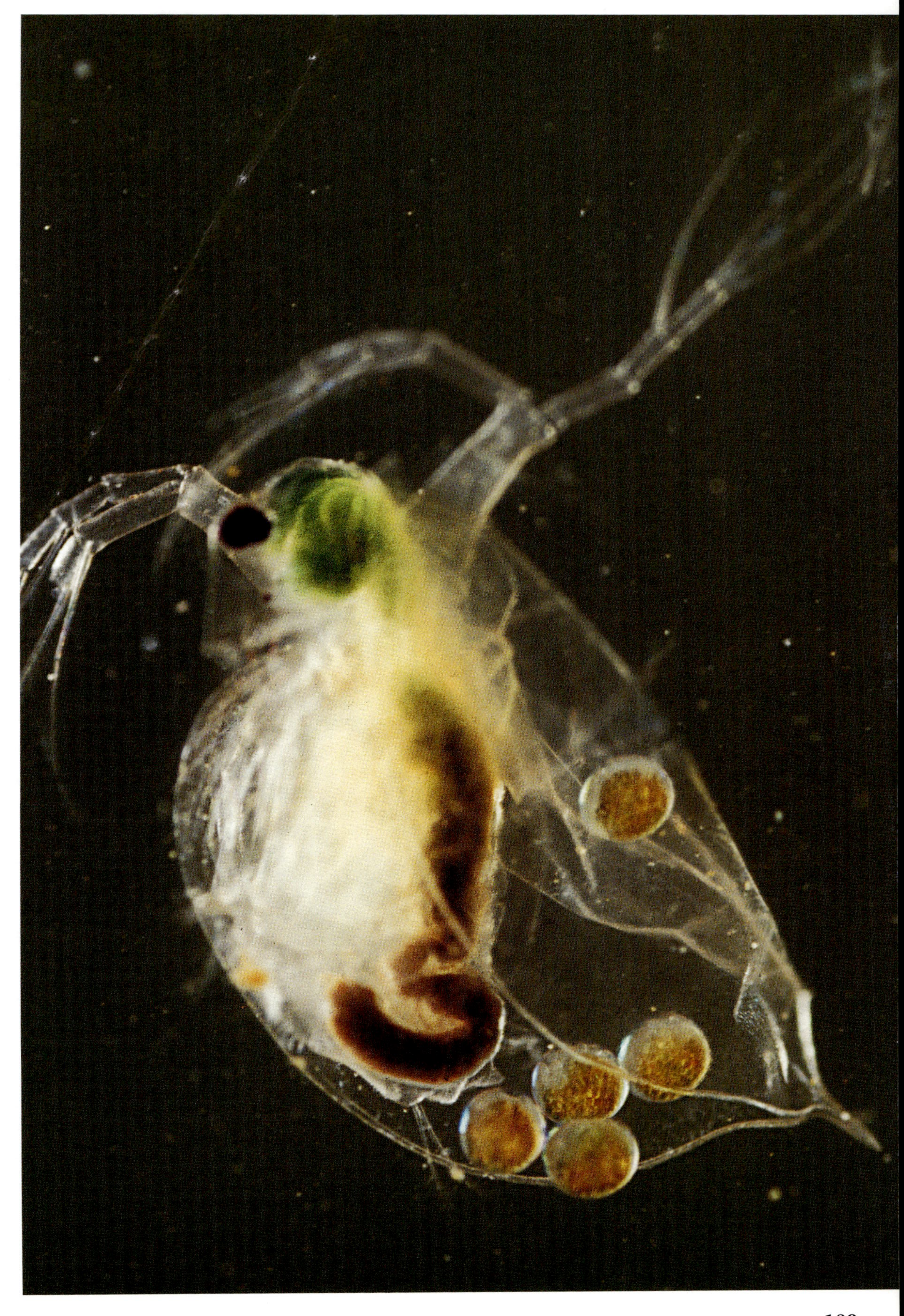

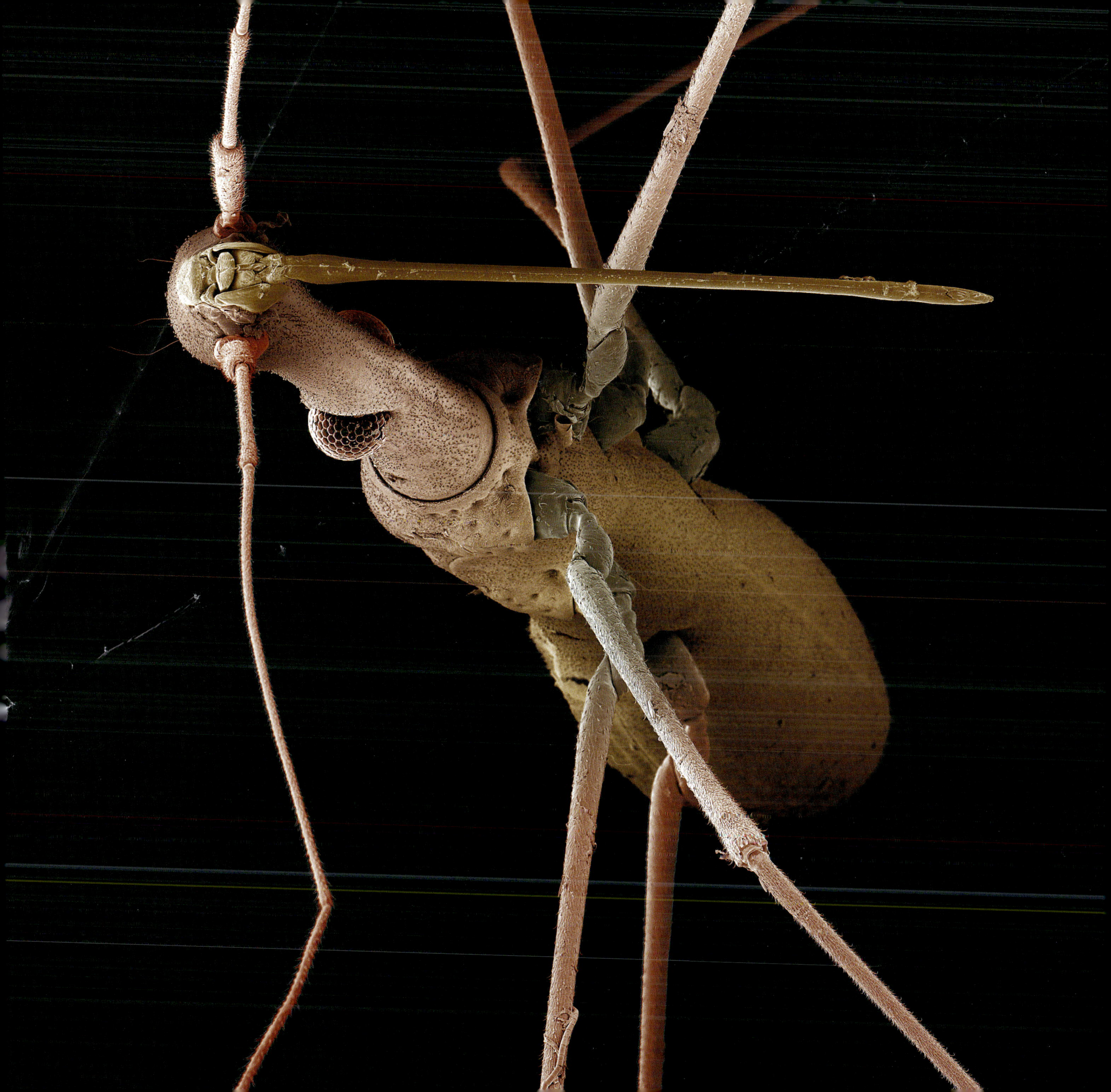

BREATHING

Some aquatic insects need to breathe air so they must periodically visit the surface of the pond to replenish their supply. Insects normally possess a network of trachea, or breathing tubes, in their bodies that open to the air through holes called spiracles. Some underwater species link these to breathing tubes or snorkels, which are pushed through the water surface, as in the larvae of the mosquito, *Culex*, and the water scorpion, *Nepa cinerea*, which, despite the intimidating name, is a relatively harmless bug. The front legs of *Nepa*, modified as pincers to grab prey, look like those of a scorpion, and the long 'tail' is the breathing tube, but it can remain submerged for a long time creeping slowly over the mud and relying on camouflage to approach its prey unseen. They grow up to 3.5 cm long but 330 million years ago, giant water scorpions - *Hibbertopterus*, up to 2 metres in length, were the largest arthropods that ever lived, and, judging from fossilised tracks discovered recently in Scotland, may have occasionally emerged from lakes to wander overland to seek new bodies of water.

SURFACE DWELLERS

There are a quite a few organisms that dart about on the surface of a pond. Small, whirligig beetles, *Gyrinus substriatus*, derive their common name because they swim rapidly in circles when alarmed, and for their grouping behaviour, a survival mechanism that helps them to avoid predation. They are also notable for their divided eyes, which can see both above and below water. A bubble of air trapped underneath their abdomen aids flotation, and allows them to dive and breathe underwater for quite a long time. Perhaps the most familiar of the surface dwellers is the pond skater *Gerris lacustris*, a bug with long legs that runs around on the water surface, its weight supported by the surface tension, assisted by a dense covering of velvety hairs anointed by an oily secretion on the legs and body that effectively repels the water. A slender, dark brown or grey insect, about 2 cm long, the pond skater has eyes that project from the sides of its head giving it good all-round vision. As it floats on the water surface, it senses vibrations and ripples caused by any disturbance. If a small invertebrate accidentally falls into the water, radiating ripples on the surface send information about its direction, prompting the pond skater to dart across the surface to catch the stricken prey. Like all bugs, it has sharp tubular mouthparts, used to puncture and suck body fluids from its prey. A pond skater will also hunt other surface-dwelling insects, even nymphs of its own species. The water measurer, *Hydrometra stagnorum*, is another bug that lives on the water surface. Like the pond skater, it has water repellent feet, but it is smaller and its way of life is more relaxed, slowly patrolling the pond surface hunting mosquito larvae or water fleas, which it spears through the surface with its long proboscis.

CADDIS FLIES

Caddis larvae are well known for many of them protect their soft bodies by building cases out of various materials from their surroundings, fastened together with a sticky secretion produced by a gland near the mouth. Each species favours particular case building materials - grains of sand, plant fragments or empty snail shells - and makes them to a specific design. Not all caddis larvae, however, make cases - some species spin silken nets attached to underwater plants, and a few are free-living. Most of the case-bearing and net-living forms are omnivorous, feeding on scraps of plant or animal matter, but free-living species are largely carnivorous. The adults are moth-like insects, with wings densely covered with tiny hairs. Normally found near water, some species feed on nectar, but most probably do not feed at all.

WATER BOATMEN AND BACKSWIMMERS

Common and distinctive inhabitants of ponds, water boatmen, *Corixa punctata*, and backswimmers, *Notonecta sp.*, trap bubbles of air with the fine hairs on their bodies and carry it down with them when they dive to allow them to continue breathing as they swim. They use their long legs like oars to swim around, and have wings that can transport them from pond to pond. They are both bugs, the water boatman about 12 mm long, feeding on algae or other tiny morsels of plant material. Water boatmen are often confused with backswimmers, which are much larger, swimming on their back, and hunting prey up to the size of tadpoles and small fish. Backswimmers can also bite if handled carelessly.

Water Measurer, *Hydrometra stagnorum*

A bug with piercing, sucking mouthparts, the water measurer walks slowly on the surface of a pond stalking prey such as mosquito larvae, which it spears with its long proboscis. It maintains its oily hydrophobic coating that enables it to walk on the water by frequent preening.

Magnification: x30   *Scanning Electron Micrograph*

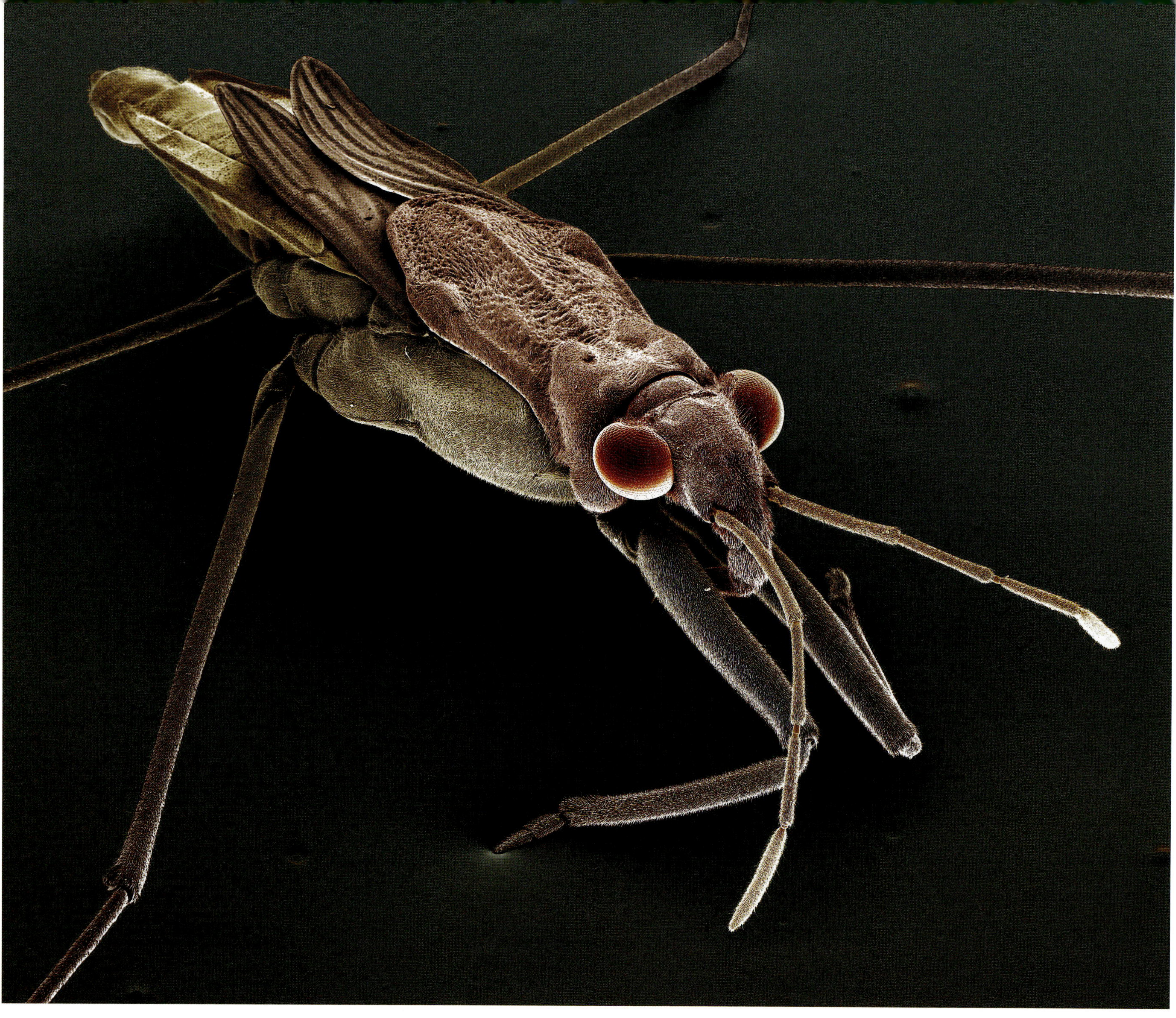

Pond Skater, *Gerris lacustris* (above)

The pond skater is perhaps the most familiar of the surface dwellers, a bug with long legs that runs around on the water surface, its weight supported by the surface tension, assisted by a dense covering of velvety hairs anointed by an oily secretion on the legs and body. If a suitable prey is nearby, radiating ripples on the surface send information about its direction, prompting the pond skater to dart across to catch it.

Magnification: x20   *Scanning Electron Micrograph*

Caddis Fly, *Order Trichoptera* (right)

Adult caddis flies resemble moths, but their wings are almost transparent and they possess very long antennae. There are around 200 species in Britain. The larvae are aquatic, some of them well-known for protecting their soft bodies by building cases out of various materials from their surroundings, and feeding by scavenging on the bottoms of streams and ponds. The adults often have reduced mouthparts and rarely feed, living only for a short time to court and to mate.

Magnification: x30   *Scanning Electron Micrograph*

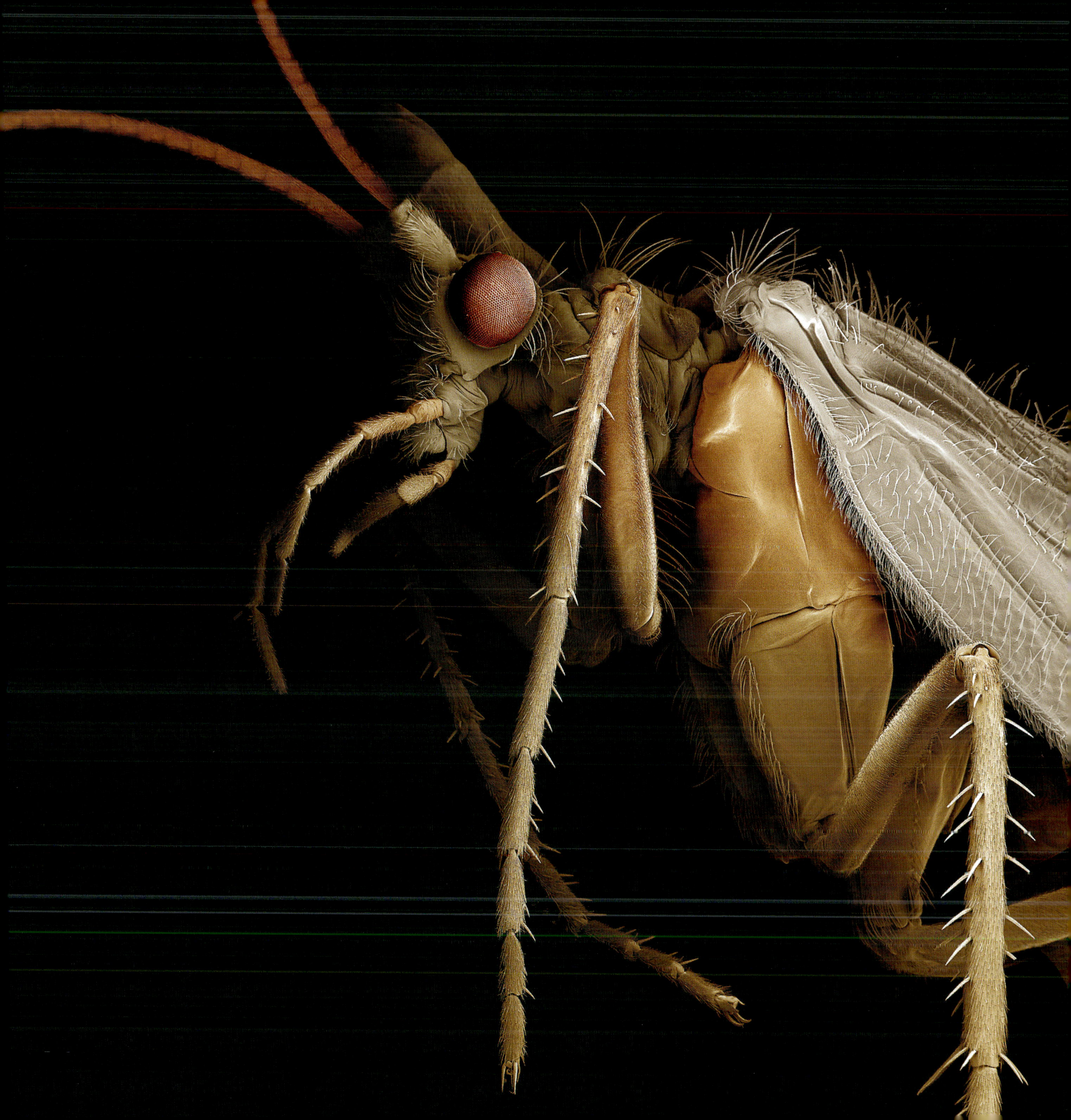

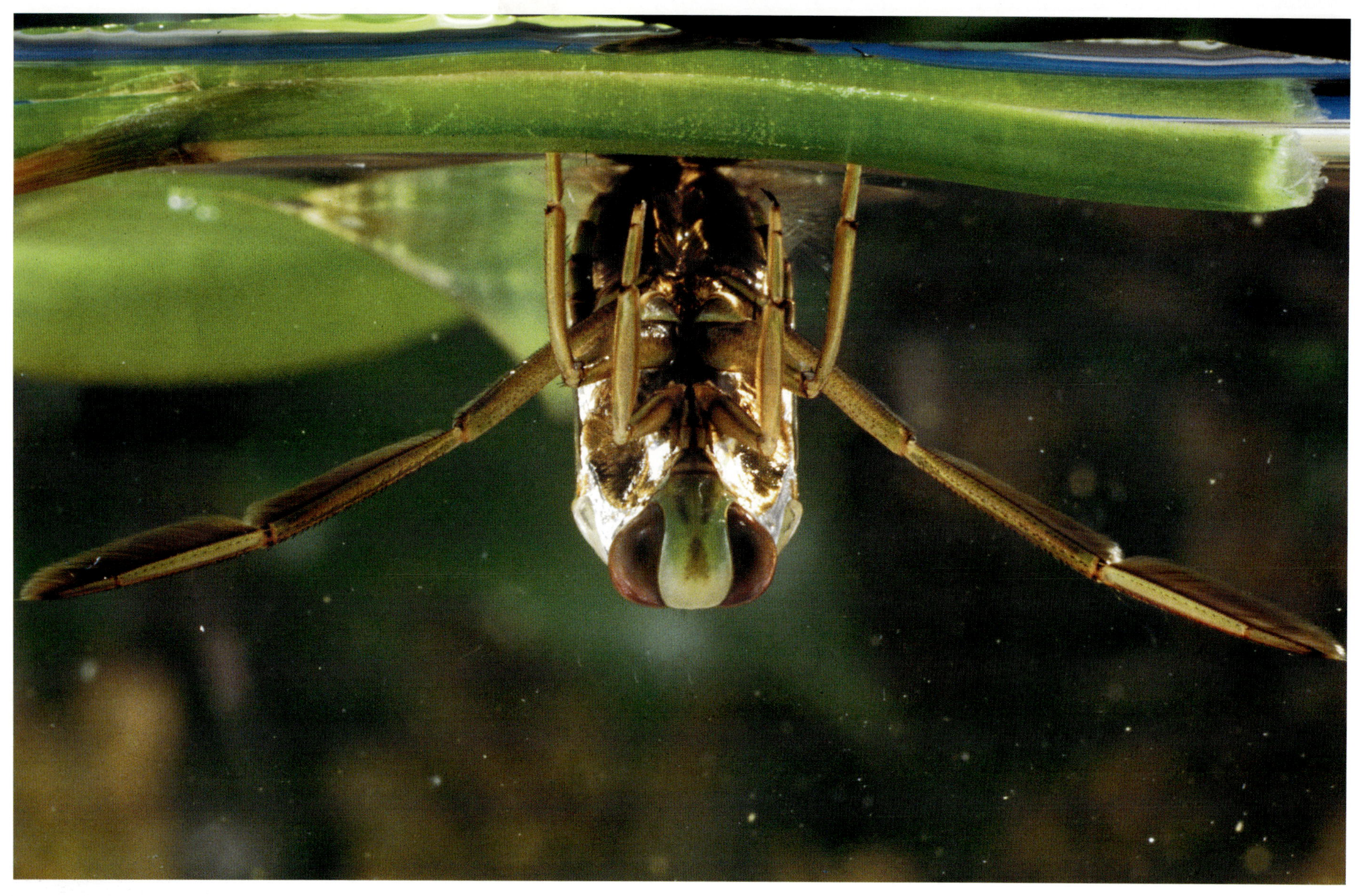

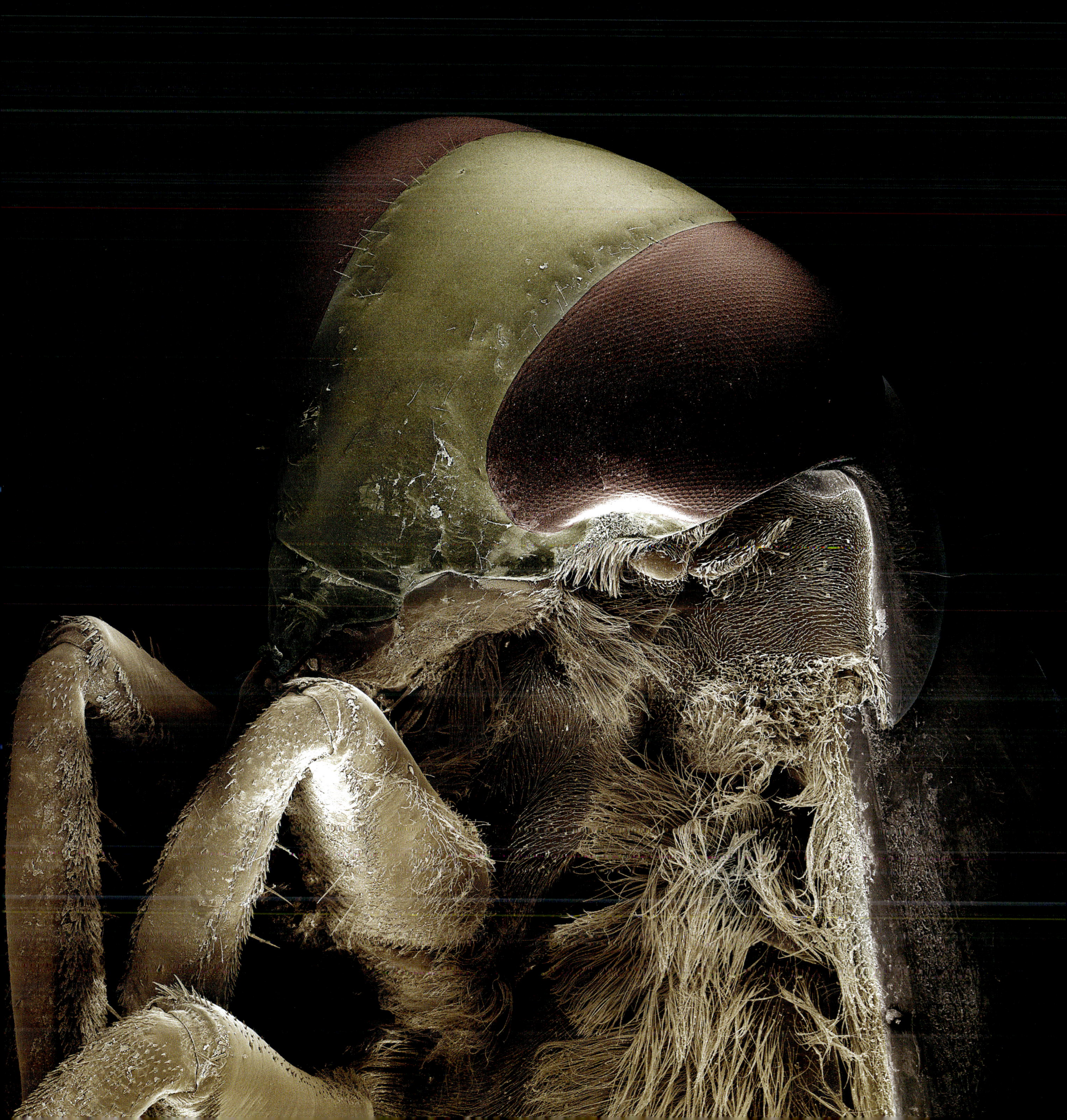

DRAGONFLY, MAYFLY

Dragonflies, damselflies, mayflies and stoneflies are amongst the most primitive groups of freshwater insects. They were alive 300 million years ago and some early dragonflies had wingspans up to one metre. Although not as spectacular, modern dragonflies and damselflies are noted for their striking colours and aerial agility, always catching attention as they dart back and forth over the water.

The eggs of many dragonflies and damselflies are laid inside a plant stem, which the female slits open with her scimitar-shaped ovipositor, or egg tube, to protect the eggs from predators or desiccation. During oviposition, the female climbs down the stem and may completely submerge herself underwater. Dragonfly larvae, for example *Aeshna sp.*, can jet-propel themselves if they feel threatened by predators or are chasing fast-moving prey by pumping water in and out of the rectum. They are fierce hunters, and to feed, dragonfly larvae use a modified lower lip, or labium, which has a pair of sharp jaws at the tip and is hinged at the base so it can be withdrawn under the head. When the prey is within range, the labium is shot out at high speed to grab it with the jaws. Most dragonfly larvae complete their development within a year, but larger species may take longer. An adult emerges after the fully developed larva has climbed up a stem or rock near the water's edge. Adult dragonflies may survive four to eight weeks while damselflies live for less than a month.

Mayflies, *Order Ephemeroptera*, are aquatic insects whose nymphs, or naiads, live in clean, well-oxygenated freshwater, favouring streams where they cling to stones or vegetation. The nymphs breathe by absorbing oxygen through external gills on the abdomen. The adult flies only live for a few hours, often swarming in large numbers to mate, the females laying eggs a few at a time by dipping their abdomens in the water whilst in flight. Once their work is done, the adult mayflies die, providing a feeding bonanza for fish.

Stoneflies are distantly related to cockroaches. The larvae are flat in shape and creep around on stones or burrow into the mud at the bottom of the pond. The smaller species feed on algae, detritus and debris, while the larger ones are carnivorous. Most spend the winter as larvae, completing their development in one year. When ready, the larvae crawl to the water's edge and the adults quickly emerge, immediately flying to find shelter. On warm still days the adults will fly, otherwise they run around on foliage or stones near the water's edge. Male stoneflies drum their abdomens on the ground to attract females. After mating, females fly just above the surface, dipping their ovipositor, or egg tube, into the water to wash off the eggs.

SCAVENGERS AND TUBIFEX

Rotting plant material, uneaten animal body parts and excreta collects at the bottom of the pond as detritus. Bacteria and fungi, together with a whole community of scavengers, help to break this down into nutrients that can be used by plants. One of the most common scavengers in the pond is the pond slater or water louse, *Asellus aquaticus*, a small crustacean about 1 cm long that is related to wood lice. This little creature is very tolerant of low oxygen levels, and can survive periods of pond stagnation. Another small crustacean scavenger, the freshwater shrimp *Gammarus pulex*, is less tolerant of low oxygen and more likely to occur if a stream runs through the pond. These active scavengers are helped in their work by flatworms, *Planaria sp.*, which glide about on the surface of the mud, and by sludgeworms, *Tubifex*, sometimes called 'bloodworms', because they possess haemoglobin giving them a red colour. *Tubifex* makes a tube in the mud, waving its body about to create ventilating currents in order to help it survive in conditions of low oxygen.

Flattened Mayfly Nymph, *Ecdyonurus venosus*

The nymphs of mayflies are varied in form; some burrow, some crawl on vegetation, while others, such as *Ecdyonurus venosus* (see right) are streamlined for swimming or clinging to rocks in fast-flowing streams. All of them, however, have three long, spiky tails. They feed, for the most part, on algae and plant matter but may also eat small creatures.

Magnification: x10   *Light Micrograph*

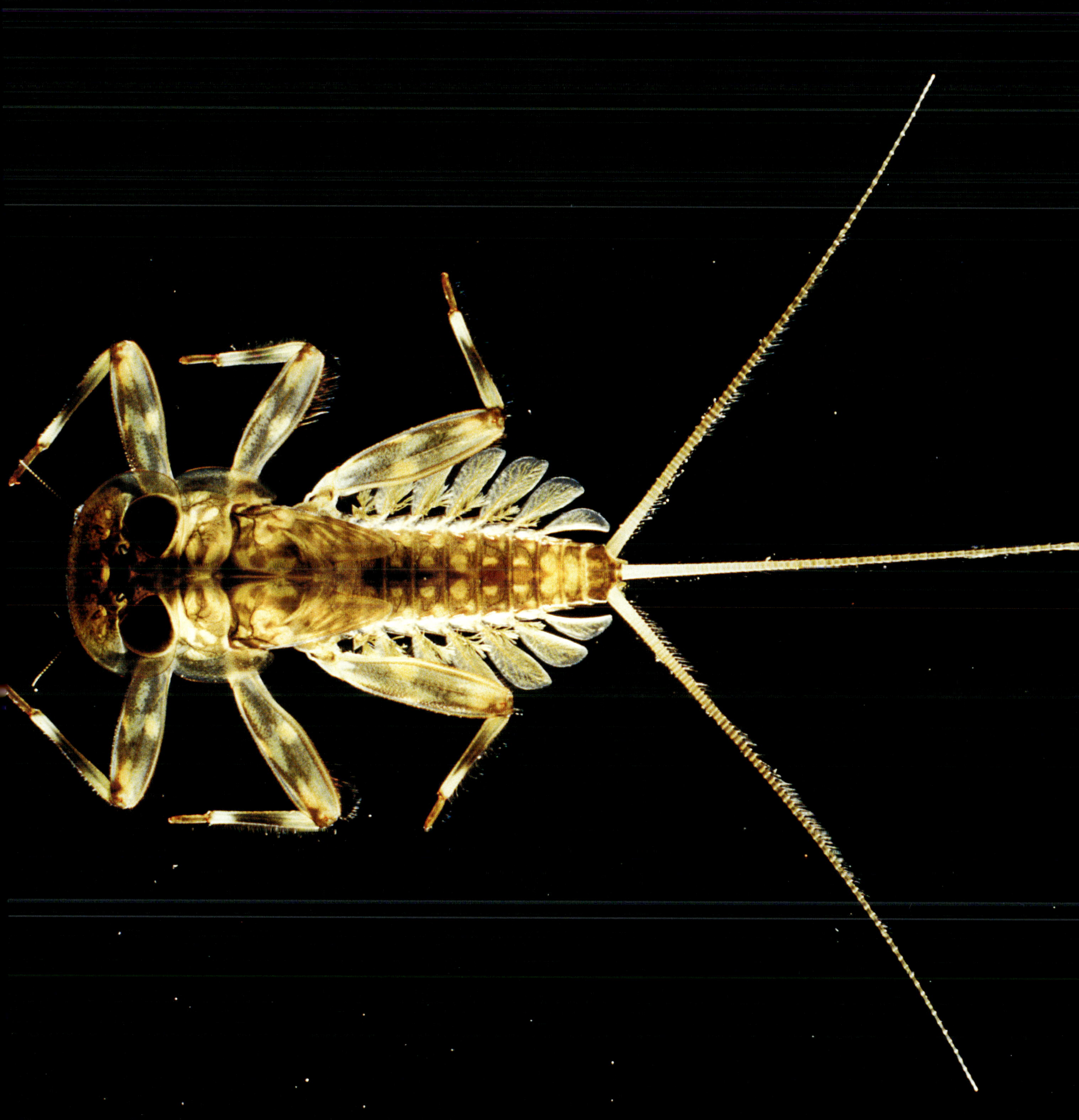

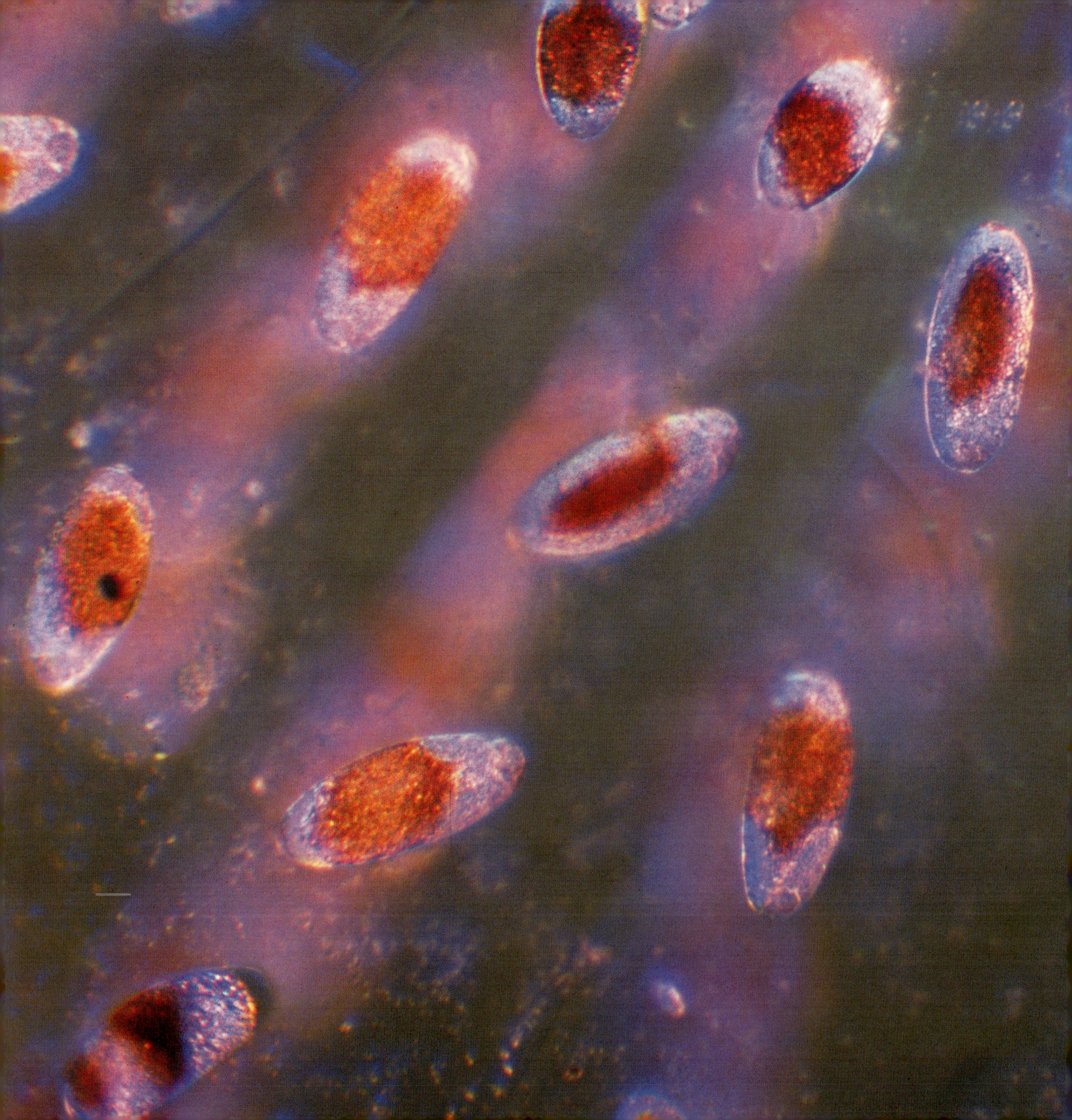

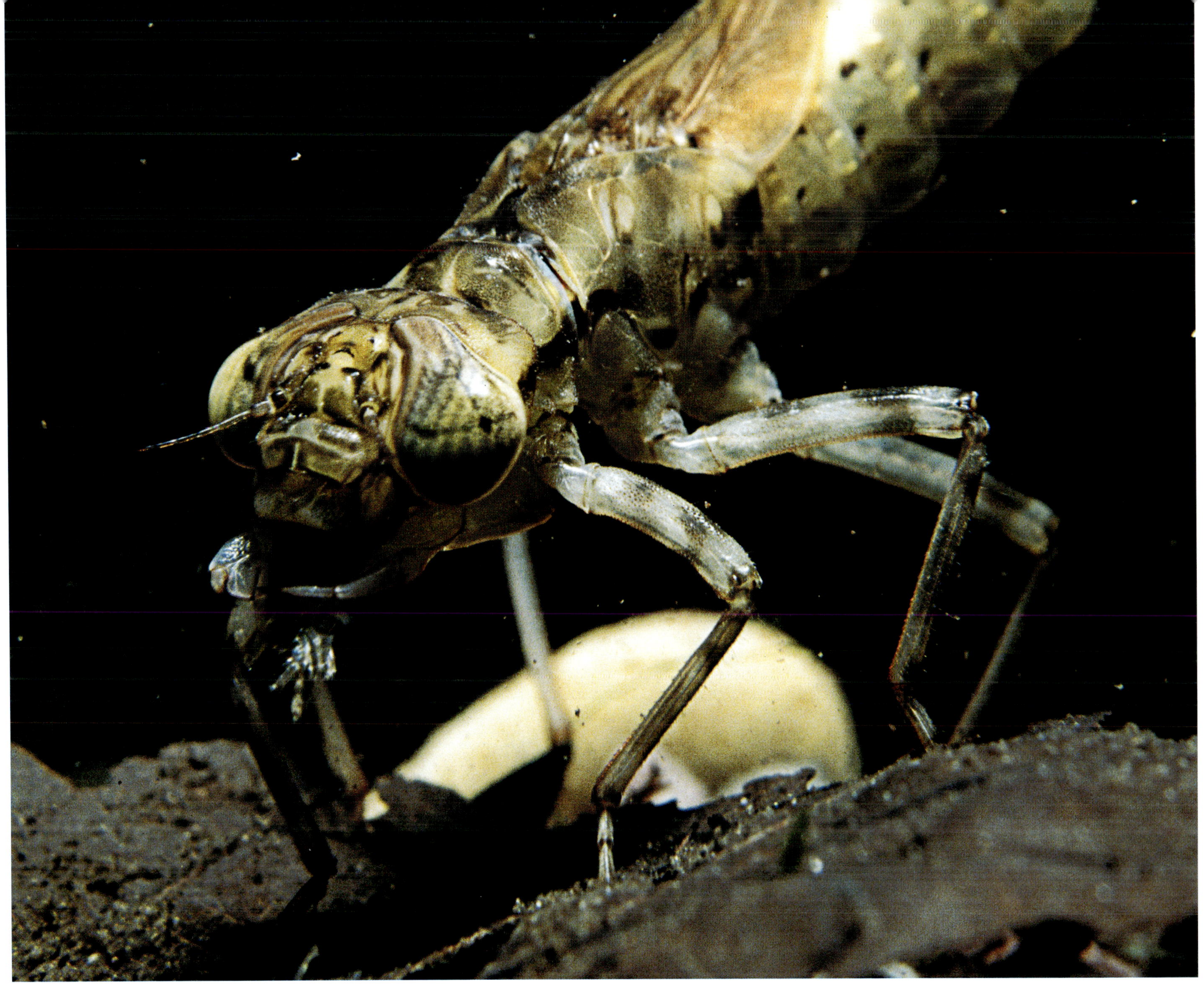

Eggs of a Chironomid Midge, *Family Chioronomidae* (left)

Chironomid midges resemble mosquitoes but do not bite. There are about 400 species known in Britain and over 5,000 species worldwide. They can be recognised by the thorax which is conspicuously humped, and the male possesses feathery antennae. As adults, they do not live long and may not feed at all, emerging in swarms in the summer months to mate. The eggs are laid in strings under water and hatch out into small worm-like larvae, which in some species are red due to the presence of haemoglobin, and, like the *Tubifex* confusingly called 'bloodworms'. They live in the detritus at the bottom of a pond feeding on organic matter, the haemoglobin enabling them to obtain sufficient oxygen.

Magnification: x500   *Light Micrograph*

Dragonfly Nymph, *Order Odonata* (above)

The spectacular dragonflies that gracefully fly back and forth over ponds have aquatic larvae that are fierce hunters. The nymph lies in wait for unsuspecting prey to venture within range, then strikes, grabbing it with jaws at the tip of a hinged lower lip, or labium, which shoots out from beneath the head.

Magnification: x8   *Macro-photograph*

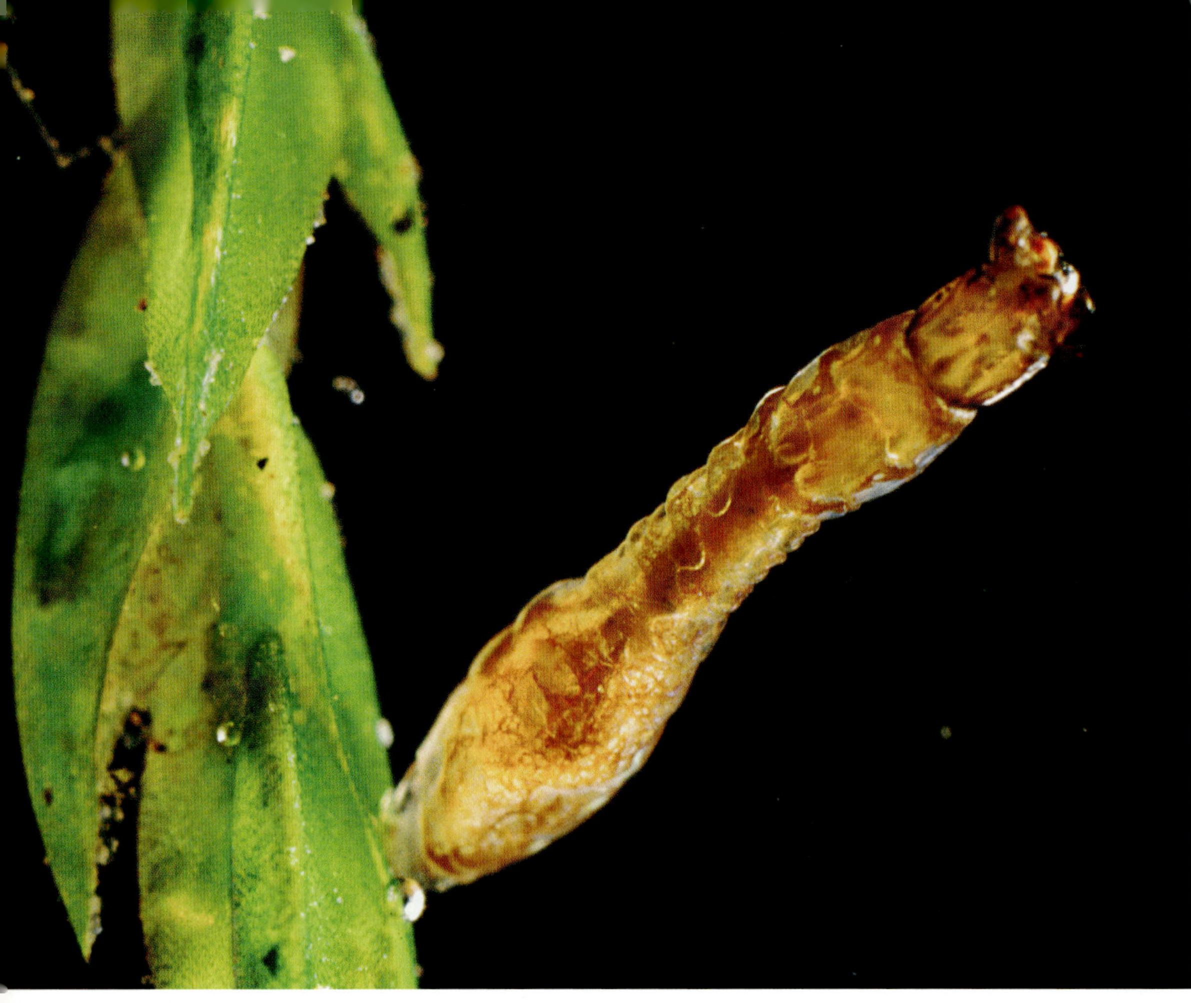

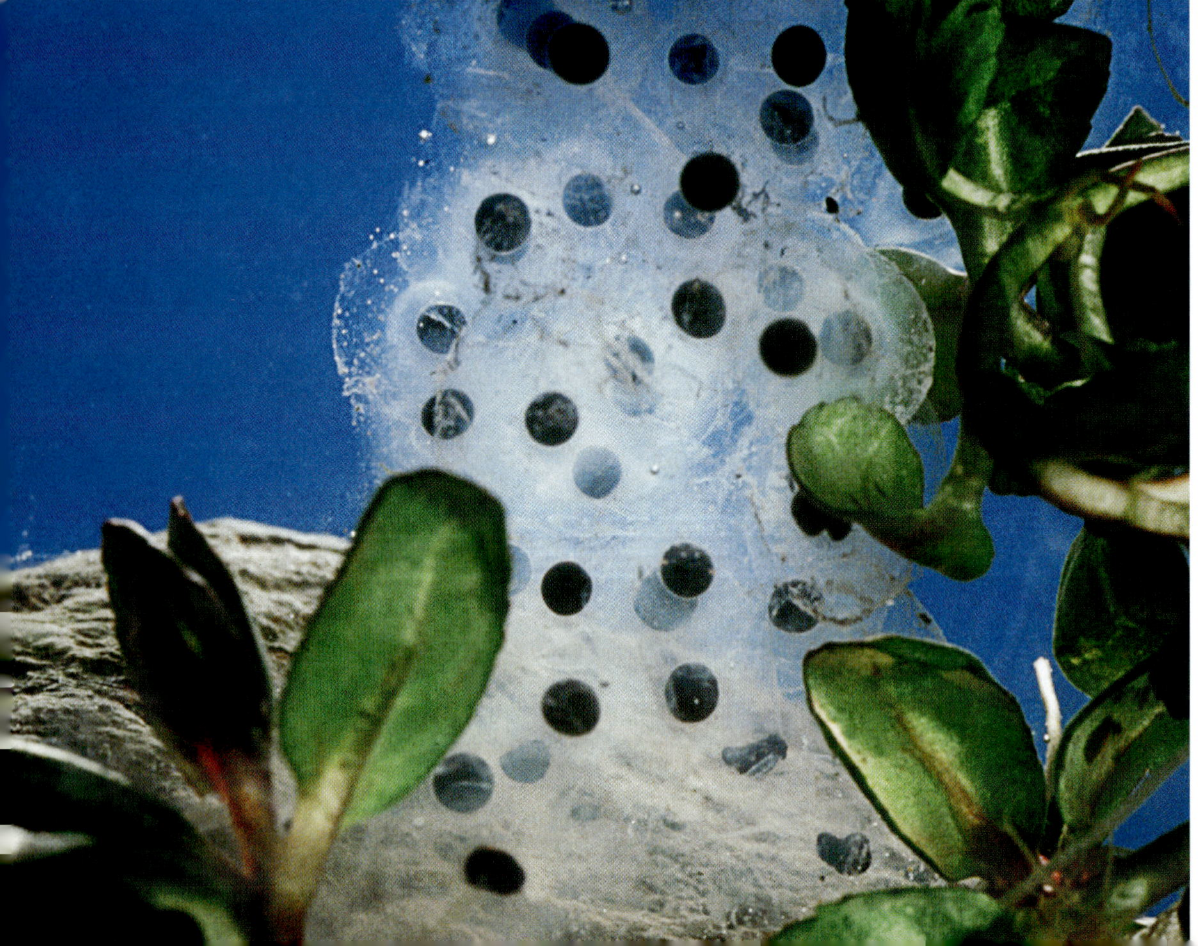

Larva of a Blackfly, *Simulium sp.* (above left)

Blackflies are small and stout-bodied, with more than 1,800 known species worldwide, belonging to the family Simuliidae. Adult flies of both sexes feed on nectar, but females usually require a blood meal to supply proteins necessary to mature their eggs. When blood feeding, a female anchors its proboscis to her victim with small hooks on the mouthparts. The mandibles then cut into the surface of the skin with rapid scissor movements, causing blood to flow freely. The flies tend to frequent areas near flowing water and the female scatters her eggs on the surface, after which they sink to settle on the bottom sediment.

Magnification: x20   *Macro-photograph*

Frogspawn (below left)

A familiar sight in a garden pond in early spring is the jelly-like mass of frogspawn, hundreds of eggs each coated with albumen that swells in the water to form a protective transparent capsule. Male and female frogs will have emerged from hibernation during March to seek out suitable ponds where they will mate. The most likely species of frog to colonise a garden pond is the common frog, *Rana temporaria*, but two other kinds of frog exist in Britain, the edible frog, *Rana esculenta* and the marsh frog, *Rana ridibunda*. All three frog species, along with newts and toads, have suffered a serious decline in numbers over the last fifty years due to pollution and habitat destruction, so garden ponds provide important oases that will help these animals to survive. Toad spawn can be distinguished from that of frogs because it consists of long, slender strands of eggs in transparent jelly that is wound around the stems of underwater plants. Newts lay their eggs singly, usually attached to plant leaves.

Magnification: x4   *Macro-photograph*

Sludge Worm, *Tubifex sp.* (right)

Tubifex inhabits the bottom sediment of ponds, lakes and rivers. These worms can survive even in quite polluted waters; they contain haemoglobin, giving them a red colour and enabling them to extract oxygen at low concentrations, and characteristically wave about to generate ventilating water currents.

Magnification: x20   *Macro-photograph*

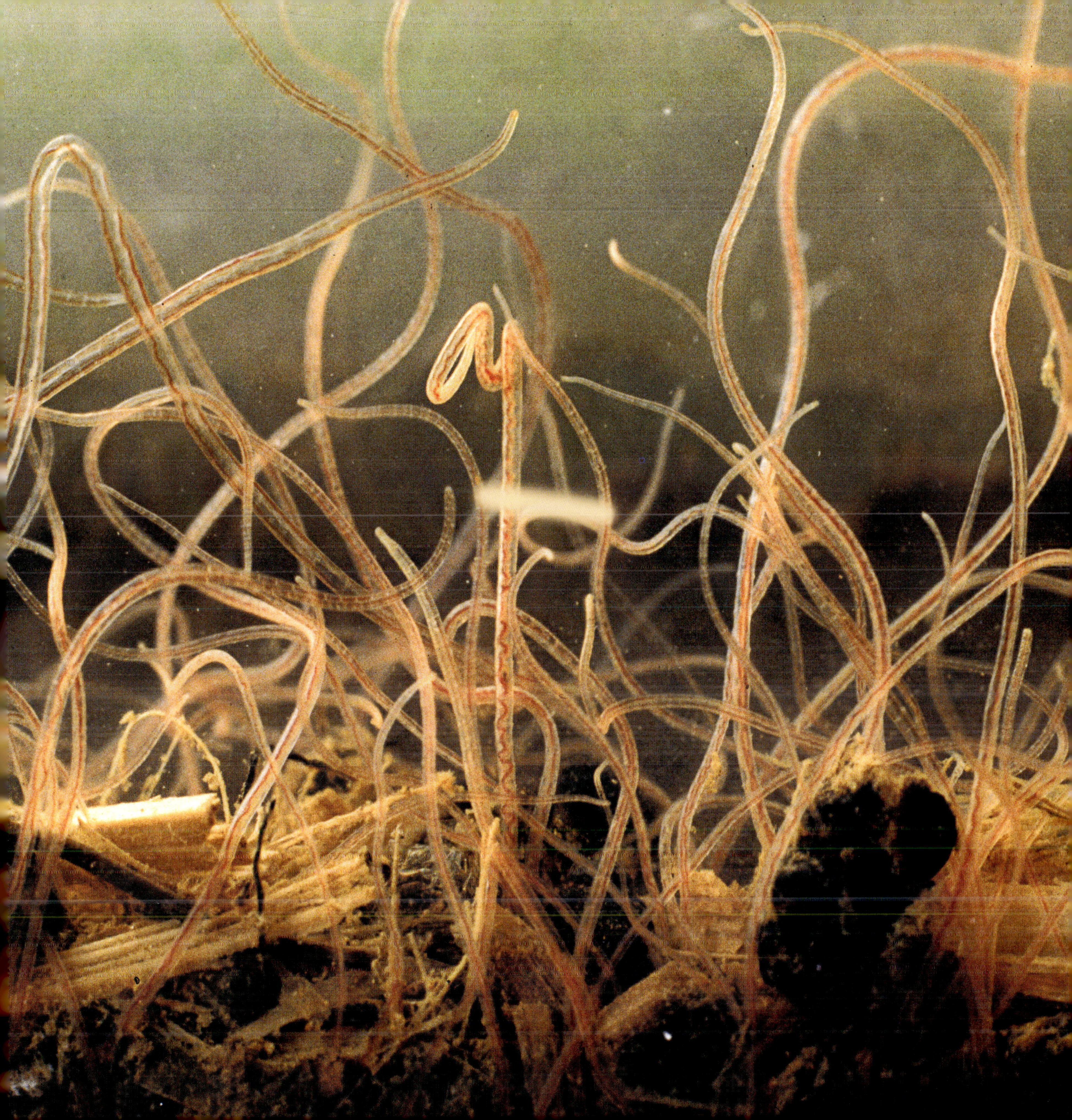

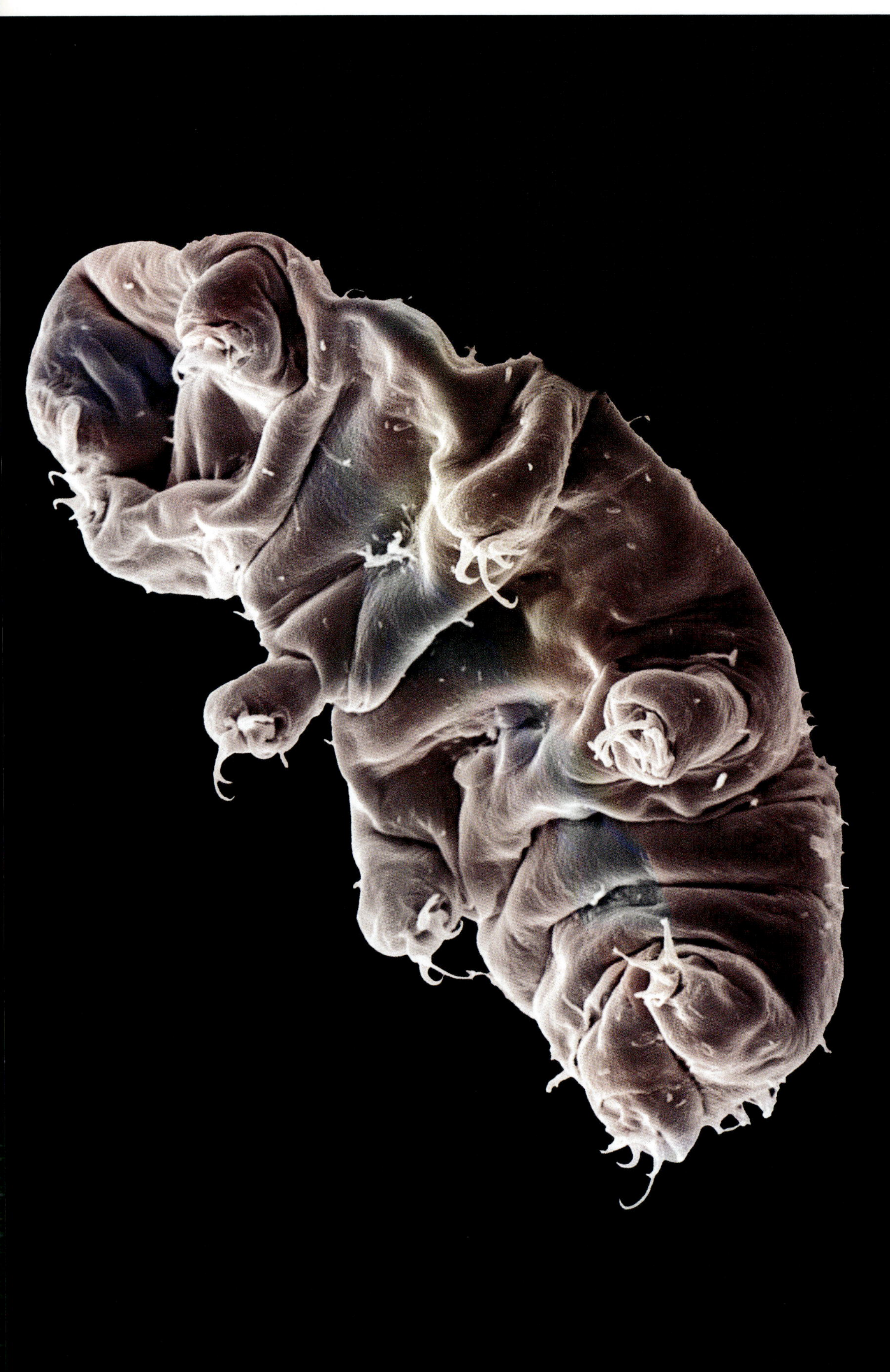

Tardigrades, also known as 'water bears' because of the way they look and walk, are unique animals. Normally less than 1 mm in length, their bodies are cylindrical in shape and almost transparent, supported by eight legs, which have tiny claws at their tips. Discovered in 1773 by German zoologist Johann August Ephraim Goeze, there are over 750 known species worldwide. They need a film of water around them in order to breathe, so as well as living in ponds they favour damp places, often in moist clumps of moss, lichen or liverworts. Having said that, they manage to survive almost everywhere, under polar ice, in hot springs, 4000 metres deep in the ocean, and up to 6000 metres high in the Himalayas. Under even more extreme conditions they can become cryptobiotic, lowering their water content to as little as 1% and shrinking down to a tiny barrel-shaped blob called a tun. With a reduced metabolic rate, they can survive in that state for years until brought back to life with a single drop of water. While in a cryptobiotic state, tardigrades have been shown, under experimental conditions, to be able to survive in a vacuum, under bombardment by radiation, in temperatures as high as 151°C and as low as minus 200°C. Female tardigrades deposit eggs that are so light they can be dispersed widely by animals or the wind. When the eggs hatch, there is no larval stage; the tardigrade starts its life as an adult. Most tardigrades feed on bacteria or plants, puncturing cells to devour the contents, but some, for example *Milnesium tardigradum*, are carnivorous, preying on tiny organisms such as rotifers, or even other tardigrades, by piercing cell membranes with their mouthparts to suck up the juices.

Tardigrade or 'Water Bear',
Ramazzotius oberhauseri (left)
Echiniscus testudo (right)

Tardigrades are unique, enigmatic creatures unlike anything else on Earth. They live in films of moisture on mosses and lichens in gutters and around pond edges, feeding on cell sap, which they reach using piercing mouthparts. Amazingly, they can survive long periods of hot, dry weather by shrinking down into a tiny barrel-shaped structure called a 'tun' (see right). When the rain returns, the tun swells, the legs pop out, and the tardigrade comes back to life.

Magnification: x1500   *Scanning Electron Micrograph* (left)
Magnification: x1225   *Scanning Electron Micrograph* (right)

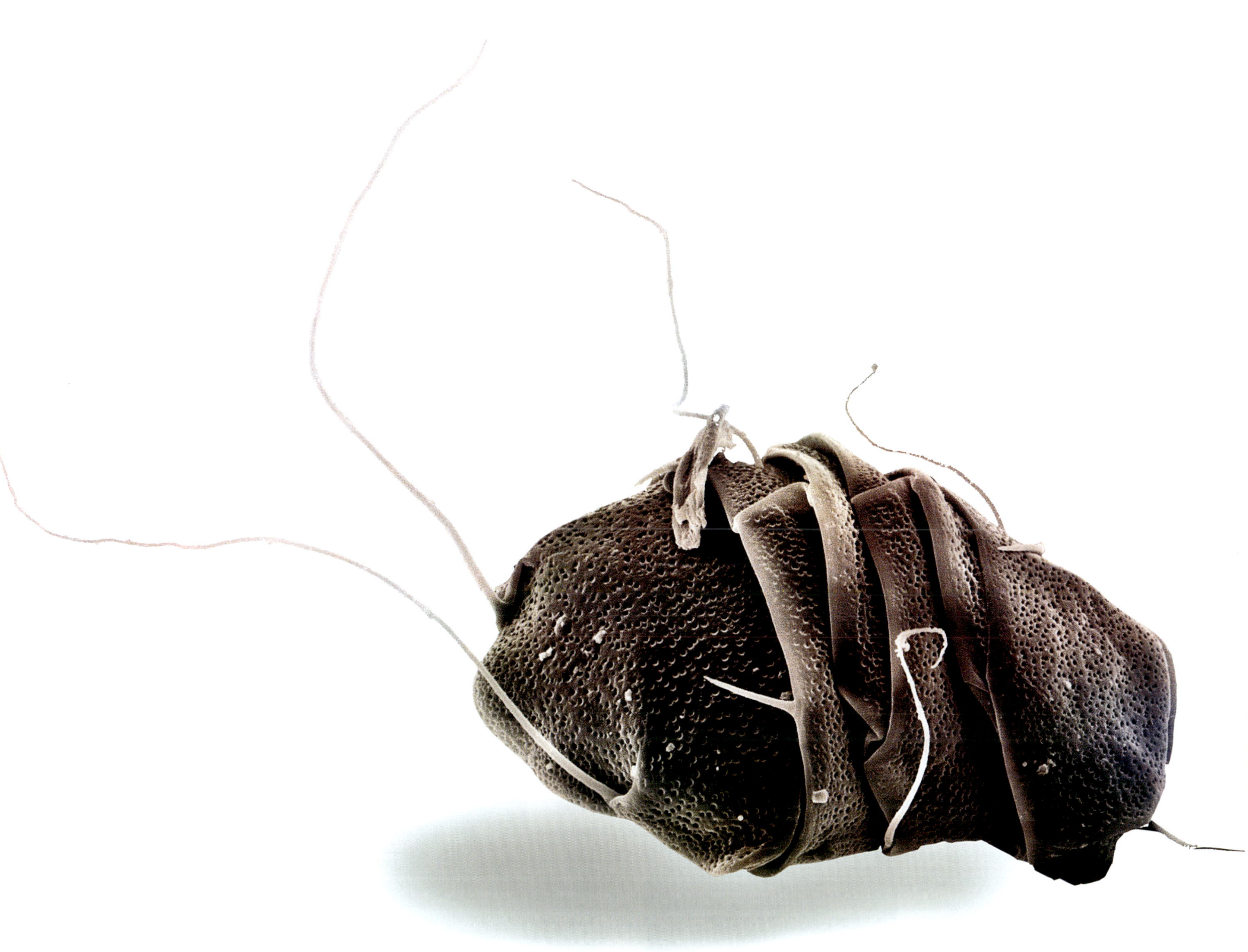

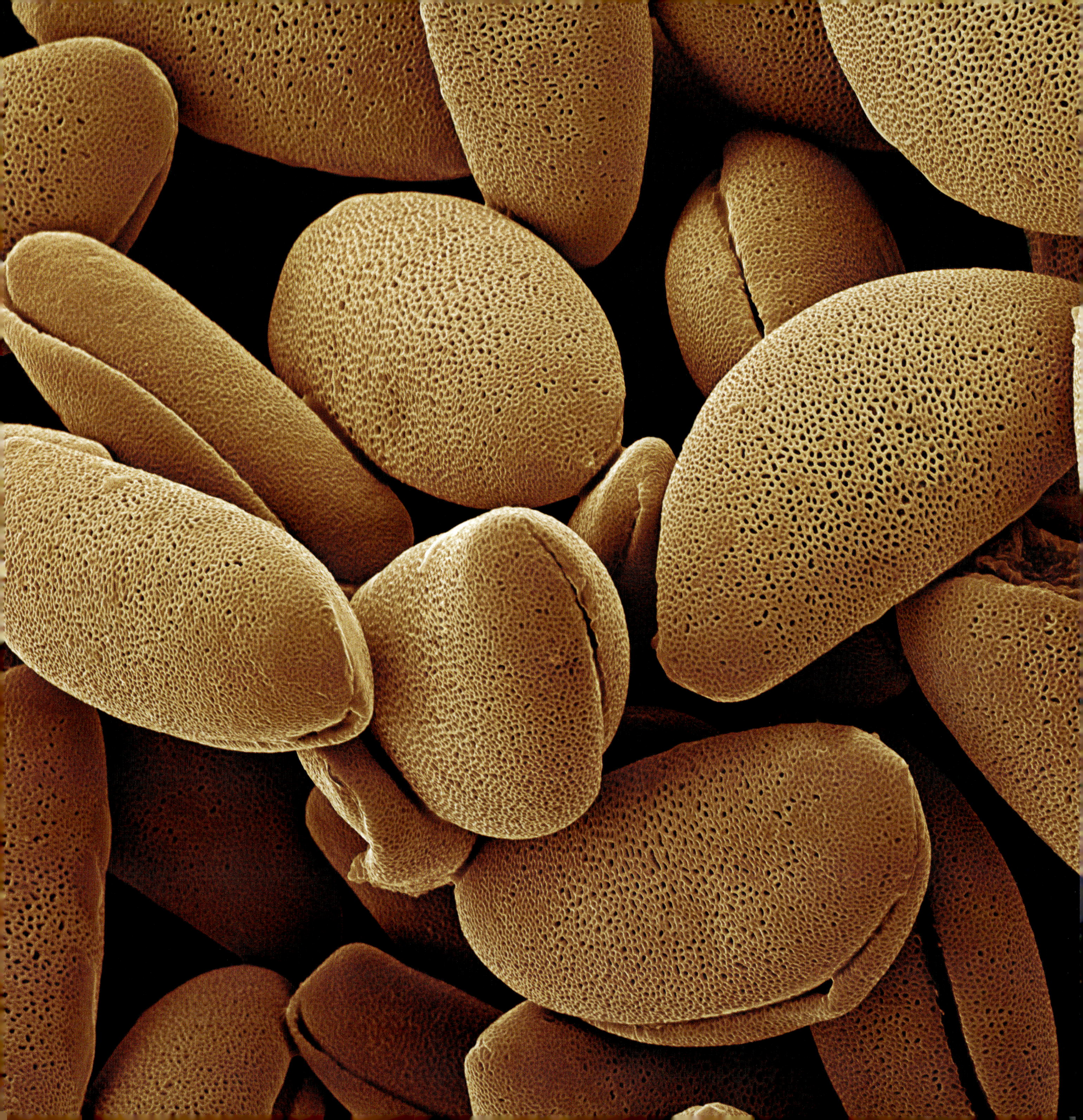

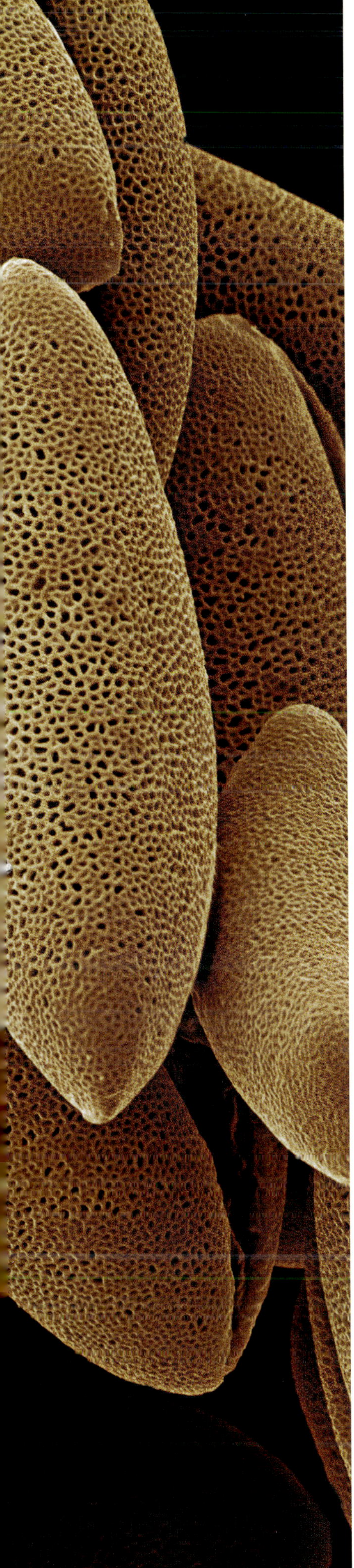

ACKNOWLEDGEMENTS

Had it not been for her extraordinary creative and computer skills, not to mention her singular dedication and tireless energy, "Unseen Companions" would not have been possible without Dae Sasitorn. Not only did she undertake the painstaking post-production work on the images, but she also designed the layout of the book. The authors also wish to thank Paul Cook for additional SEM photography of some of the specimens; Will Brett who assisted with some of the image colouring and designing; and Patrick O'Rourke who wrote drafts for chapter introductions when the book was at an early conceptual stage. Professor Mike Wren, Caroline Carder and Pam Nye of the Department of Microbiology at University College Hospital, London provided bacterial specimens; Kate Turner of Rentokil provided household pests; at the London School of Hygiene and Tropical Medicine, Cheryl Whitehorn supplied mites and fleas, while Nigel Hill and John Williams provided mosquitoes and endoparasites respectively. Dr. Gerald Legg, Dr. Jette Eibye-Jacobsen and Professor Wayne Maddison for helping with some of the identifications; and David McCarthy of the School of Pharmacy in London helped with microscopy facilities when our SEM machine was temporarily out of action. Dr Brian Ridout and David Halford both made very valuable suggestions and amendments to the text. The authors sincerely apologise for any inaccuracies that have inadvertently slipped through the net.

Lily Pollen, (left)

Magnification: x2000
Scanning Electron Micrograph

Jumping Spider, *Portia sp.* (overleaf)

Magnification: x110   *Scanning Electron Micrograph*

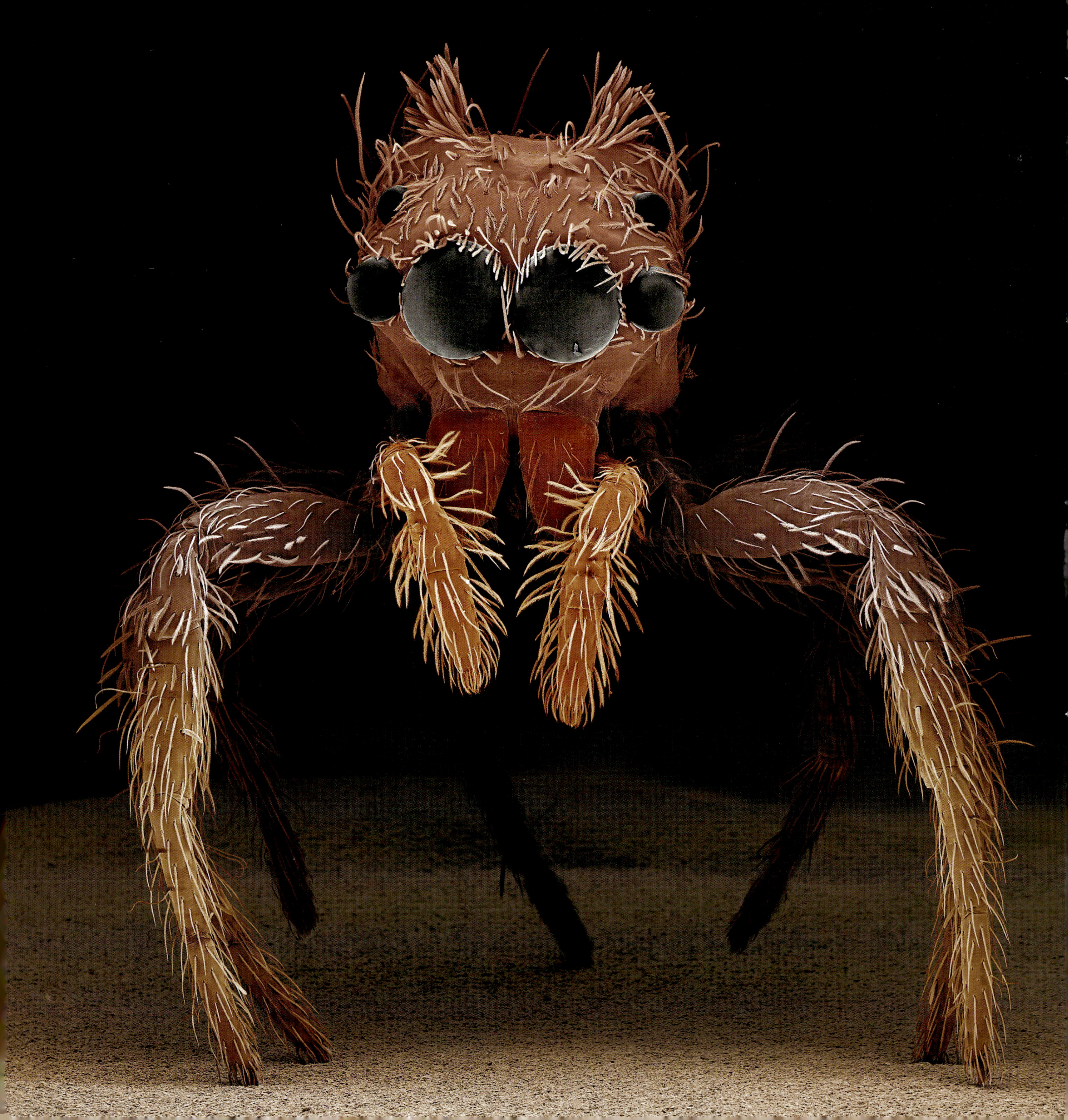